Hermann Schweickert

Voith Power Transmission – 100 Years of the Föttinger Principle

Hermann Schweickert
On behalf of
Voith Turbo GmbH & Co. KG
Alexanderstraße 2
89522 Heidenheim, Germany
Telephone +49 7321 37-0
Telefax   +49 7321 37-70 00
info.voithturbo@voith.com
www.voithturbo.com

Bibliographical information of Deutsche Bibliothek.
This publication is listed by Deutsche Bibliothek in Deutsche National-
bibliographie; detailed bibliographical data can be obtained from the
internet on http://dnb.ddb.de

ISBN 10 3-540-68784-X Springer Berlin Heidelberg New York
ISBN 13 978-3-540-68784-9 Springer Berlin Heidelberg New York

Springer is a company of Springer Science+Business Media
Springer.de
© of this edition with Springer-Verlag Berlin Heidelberg 2005.
Printed in Germany.

Illustrations: Baden-Württemberg Business Archives, Stuttgart,
Stock B 80 (Voith AG), as well as Voith Turbo GmbH & Co. KG, Heidenheim

Visual concept and overall graphic design:
Manfred Schindler, Aalen; Nicola Schindler, Neu-Ulm.

Layout, digital image processing and lithography:
MSW Manfred Schindler Werbeagentur OHG, Aalen.

Translated by Veronika Binoeder.

Printing and binding:
C. Maurer, Druckzentrum, Geislingen, Baden-Württemberg.

The paper of this book was produced on a Voith paper machine.

Voith Power Transmission

100 Years of the Föttinger Principle

Edited by Hermann Schweickert

Contents

Preface

It was in Stettin on 24 June 1905 when Dr. Hermann Föttinger was granted Patent No. 22 14 22 of the German Reich for a *"hydrodynamic transmission with one or several driving and one or several driven turbine wheels for power transmission between adjacent shafts"*. Originally intended for marine drives, this hydrodynamic power transmission principle was at first celebrated as an "epoch-making" invention, only to be driven out of its field of application within a few years by mechanical units with improved gear technology. The German shipbuilding industry subsequently put the Föttinger principle "ad acta". A quarter of a century went past, before the "Maschinenbauanstalt Johann Matthäus Voith" in Heidenheim, south Germany, signed a licensing agreement with the inventor on 17 April 1929, enabling Voith to utilize his idea. What were the reasons behind this?

As early as 1906, Voith engineers from the turbine department had supported Föttinger with their experiences. Through this contact, they became acquainted with the function and the possibilities of the Föttinger principle. Twenty years later, when Voith received an order for the pumps and the turbines of the "Herdecke" pumped storage plant, Voith resorted to the specific advantages of a hydrodynamic coupling to enable the transmission of 36 000 HP.

In 1930, "Herdecke" took up service as the most advanced pumped storage plant of its time. The Voith turbo coupling was hailed as a new form of efficient, reliable power transmission. Only three years later, this breakthrough was followed by the spectacularly successful application of a Voith turbo transmission in a railbus of Austrian Railways. This was the moment when committed engineers laid the foundation stone for the Voith Product Area "Power Transmission". The persistence and vision of these men of the first hour led to today's Group Division "Voith Turbo" of the Swabian, internationally active engineering company Voith that is still in family ownership.

100 years of the Föttinger principle and 75 years of Voith Power Transmission! This double-anniversary in 2005 offers a welcome opportunity to look into the interesting history of a superior technical idea and its exciting realization. It is unlikely that "Voith Turbo" would exist today without the "Föttinger Principle". Without "Voith Turbo", Föttinger's ingenious vision probably would have had no future – a future that is in no ways completed but will be continued on the basis of the great challenges of our time and the increasing demand for more economical uses of energy resources. A future that has a past! This fills us with pride and will be both an incentive and an obligation – today and also in the future.

Peter Edelmann, Member of the Board of Voith AG and Chairman of the Board of Voith Turbo.

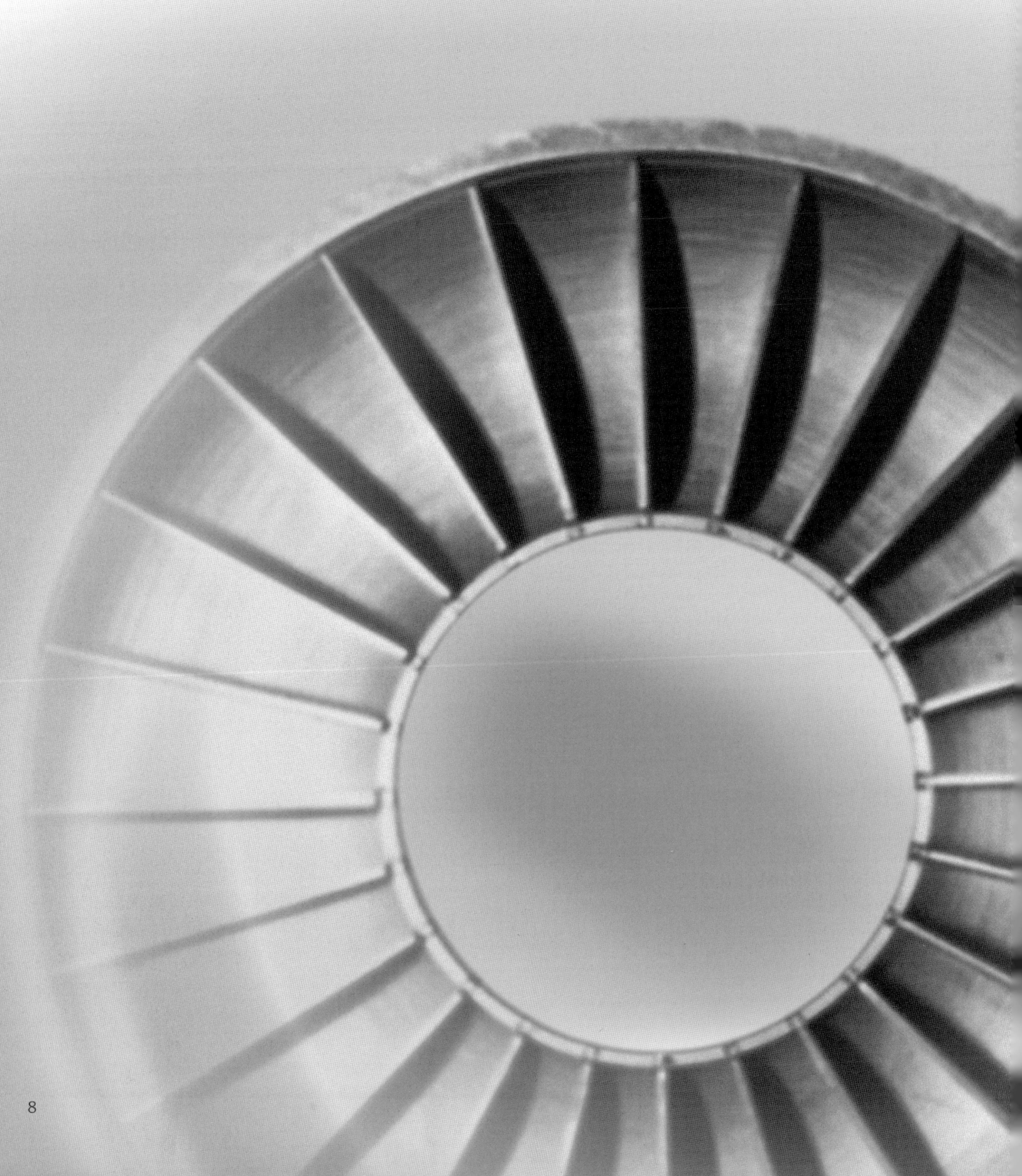

Hermann Schweickert

A New Way

The Development of Hydrodynamic Power Transmission

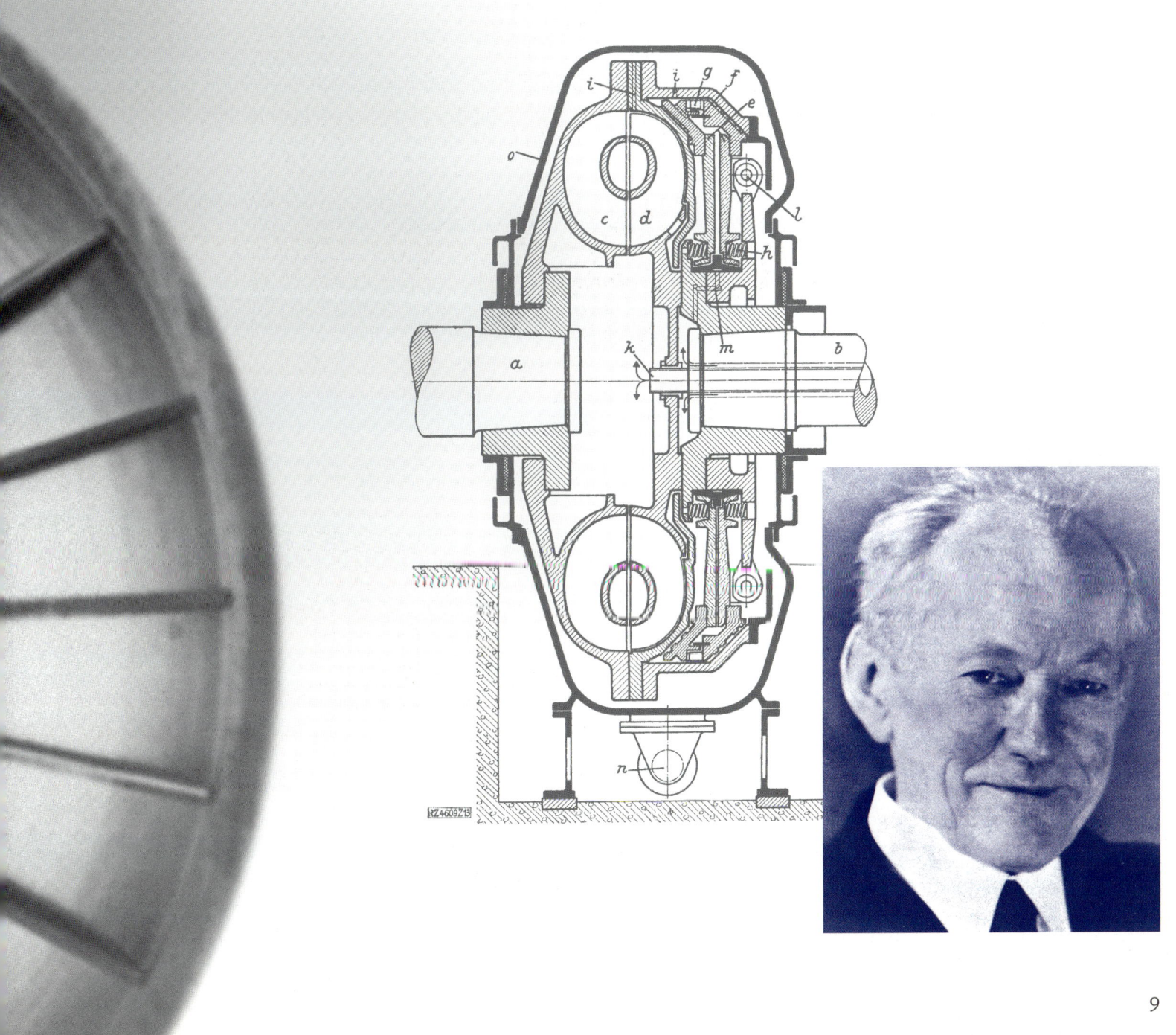

Who is Voith?

Voith is a household name all over the world for people who are involved in paper making technology, hydro power stations or hydrodynamic drive systems. Nearly 30 000 employees are working for this company.

More than 40 percent of the world's paper production is manufactured by Voith paper machines. Voith is the global leader for recycled paper making technology. Two thirds of all daily newspapers are already produced on recycled newsprint.

Thirty percent of the world's hydro turbines are manufactured by Voith. Globally, Voith is an active participant in the construction of high-performance hydro power stations, be it in Brazil, China, or other parts of the world.

Without reliable transmissions and safe brakes, today's busy road and rail traffic would be unthinkable. Millions of vehicles are driving with hydrodynamic transmissions, couplings and brakes. Countless hydrodynamic couplings and torque converters are installed in industrial applications. They are used for starting, controlling and securing sensitive technical processes.

These highly diverse company products were not developed independently from each other. A short review of the beginnings and important milestones of the company's history illustrates how they are related.

1881: Installation plan of the first Voith paper machine for the Raithelhuber paper mill in Gemmrigheim.

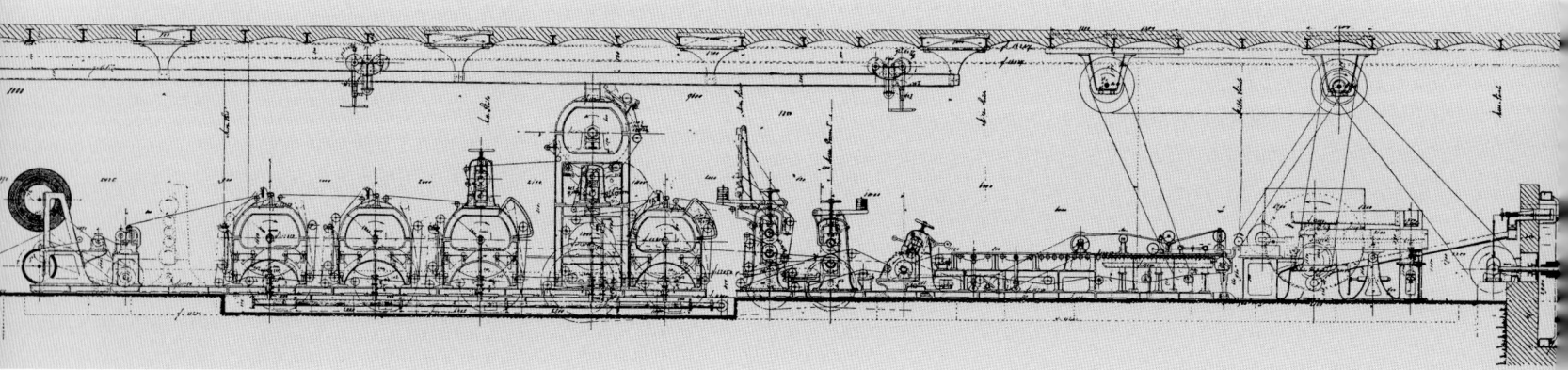

2004: Voith paper machine with a wire width of 8.9 meters for an annual production of 300 000 tons of high-quality printing paper from 100% recycled paper for Georg Leinfelder GmbH paper mill in Schwedt.

Voith in Heidenheim

At the beginning of the 19th century, France and England began constructing Fourdrinier ("long-wire") paper machines. In 1830, the first German machine manufactured by Johannes Widmann from Heilbronn "based on the English method" was commissioned at the paper mill of Rau and Voelter in Heidenheim. Among the installation team was also the "Locksmith and Mechanic" Johann Matthäus Voith with his four assistants. He repaired and improved imported textile and paper machines, built transmissions and was particularly interested in water-powered drive systems.

In 1867, he handed over his workshop, by then employing 35 people, to his son Friedrich who had just completed his engineering studies at the "Royal Polytechnikum" in Stuttgart. The year 1867 is regarded as the official founding of the company, after the son had entered the company into the register of the Royal Higher County Court of Heidenheim as "Johann Matthäus Voith, Maschinenfabrik" in honor of his father's pioneering work in 1870.

The first water turbine was built in 1870, the first Francis turbine in 1873 and in 1881, Voith built its first in-house designed paper machine. At the turn of the century, the company had 875 employees and was one of the largest engineering firms in Baden-Württemberg.

*Friedrich Voith, 1840 – 1913,
founder of the Maschinenfabrik
Johann Matthäus Voith.*

At the beginning of the new century, Friedrich Voith demonstrated his entrepreneurial skills and accepted an order for the delivery of 18 Francis turbines with dimensions that the company had never been built before for the two hydro power stations at Niagara Falls. But the extraordinarily high technical and financial risk paid off. The company turned into an internationally active global enterprise. During the same period, Friedrich Voith had established a plant in St. Pölten, Austria, and hence secured market leadership in the Danubian Monarchy beyond high customs barriers.

But the growth and the reputation of the company were also based on scientific foundations, which Friedrich Voith systematically pursued. His close contacts to the professors Robert Thomann and Carl von Bach of the Royal Polytech University in Stuttgart resulted in the recruitment of highly qualified junior engineers for Heidenheim. Professor Ernst Reichel of Berlin-Charlottenburg Technical University whose institute Voith had fitted with a test stand had also become a close technical advisor. The rank of outstanding engineers that had joined Voith during these years ensured excellent work within the company and beyond. Some of them were appointed as professors of their special fields at technical universities, others, such as Carlos Schmitthenner or later Hans Faic Canaan and Fritz Kugel, received honorary doctorates. In 1906, Berlin-Charlottenburg Technical University awarded Friedrich Voith with the academic title "Doctor of Engineering h. c." in recognition of his developments of hydrodynamic machines.

The large number of highly qualified engineers had earned the turbine division of Voith a reputation as a scientifically oriented department. This tradition continued and created the basis on which sophisticated hydrodynamic machines for power transmission and naval engineering were soon developed.

Friedrich Voith died in 1913. Shortly before his death he had converted J. M. Voith into a general partnership. Now his sons Walther, Hermann and Hanns managed the company jointly. At that time, the company had 2 280 employees.

The Development of Integrated Electrical Networks in Germany

By 1913, Voith had delivered as many as 6 500 water turbines. The invention of the electro-dynamic principle and the ensuing new generation of electric power generators resulted in a steep rise in demand for turbines. Other primary

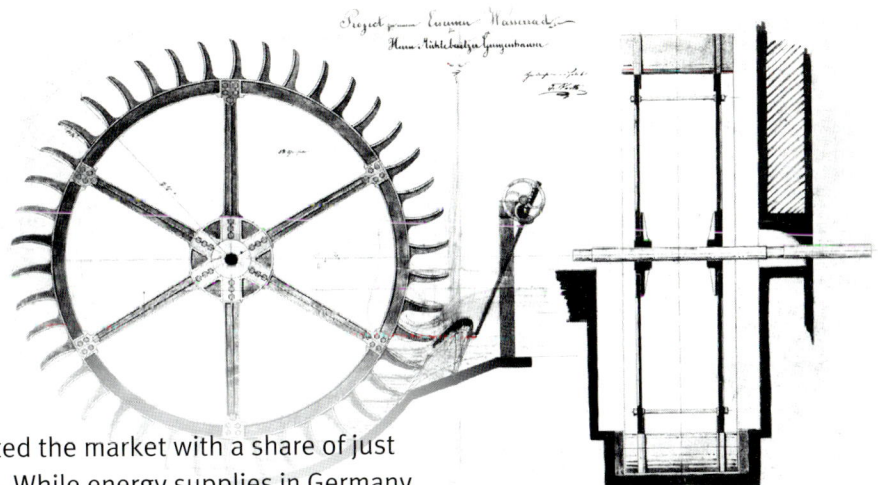

energy carriers, for example coal that dominated the market with a share of just under 80%, also benefited from this invention. While energy supplies in Germany had strongly increased as a result of these developments, they were still limited to municipal or urban power stations under local or regional control.

After the First World War, this situation radically changed. As a result of the war, Germany suffered from a severe and ongoing scarcity of coal. This fuel crisis forced people to accelerate the development of hydropower, utilizing the previously unused potential of streams and rivers in the flatlands. A newly developed turbine type, the Kaplan turbine, appeared to be the optimum solution in these regions.

Voith became very successful with this turbine type. As early as 1929, the Rheinkraftwerk Ryburg-Schwörstadt power station was fitted with four Voith units, each with a runner diameter of 7 400 mm and an output of 40 000 HP – at the time a world record for Kaplan turbines.

But turbines and power stations took on previously unparalleled dimensions. When coal was available again, thermal power stations also developed in this direction, and economic central super power stations were built.

Electricity transmission distances grew, leading to major combined systems that covered power stations of all primary energy types. With such combined

Top: Sketch of an iron water wheel by Friedrich Voith from 1865.

Left: The largest water turbines of their time with an output of 12 000 HP, built by Voith in 1904. Friedrich Voith is pictured at the center, in front of the machine.

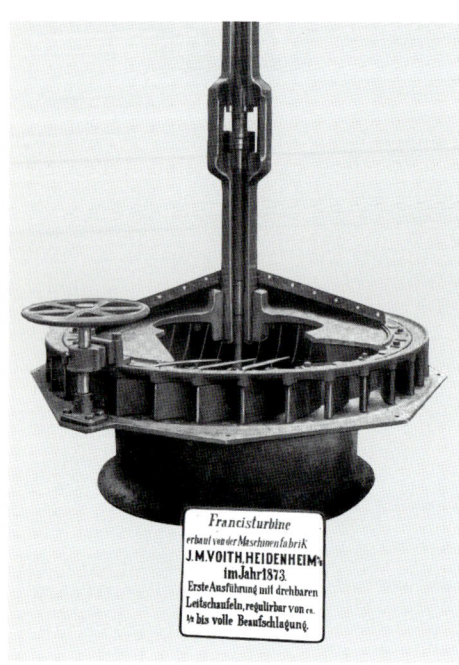

1873: First Voith Francis turbine, today at the German Museum in Munich.

systems, local hydropower stations were, for the first time, in a position to supply electricity to metropolitan areas much further away, and thus compensate for any water supply variations. Long-term or seasonal disparities were balanced out by supply lines between north and south Germany. With the help of pumped storage systems, thermal power stations could be continuously operated at optimum efficiency.

1978: Francis runner for Itaipu hydro power station in Brazil/Paraguay.
$P = 800$ MW, $H = 118.4$ m

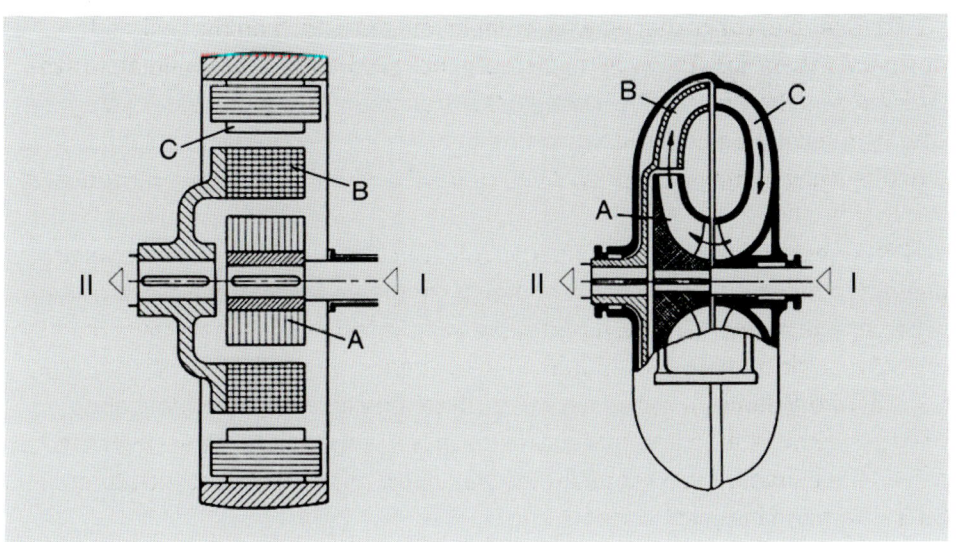

Basic schemes:

Differential dynamo
I   Primary Shaft
II Secondary shaft
A Dynamo anchor
B Engine anchor
C Stationary magnet stand

Turbo transformer
I   Primary shaft
II Secondary shaft
A Primary wheel
B Secondary wheel
C Stationary guide wheel,
    also functioning as housing

In 1903, these considerations resulted in the basic scheme of the "Turbo Transformer", as the invention was called later.

This instance of the analog conversion of an idea from electrical engineering into a mechanical application marked the birth of the Föttinger converter.

On page 18 of his essay, Föttinger described the "Operating Scheme" as the "Scheme of a primitive hydrodynamic power transmitting process". And he immediately explains, why he chose this phrase.

It was the efficiency of just under 70% that, except in special situations, did not permit the technical utilization of this fluid system: "Such arrangements that have been recently used even for ships and automobiles, can therefore only be considered for minor purposes, not to mention their space and weight requirements ...". It appears that Föttinger was less than enthralled by this concept and regarded it merely as a starting point for his further, decisive considerations.

He continues: "A totally different matter are the conditions of modern hydrodynamic power transmissions that will be the focal point of my explanations. It has been evident in all areas of technology that progress, a new effect beyond the ordinary, is never achieved by the thoughtless, mechanical arrangement of what we know, but by organic conversion, by mutual synergies, by adaptation and expansion of individual components. We want to see how this has been put into practice with the new power transmission system."

Although his statement may have been somewhat longwinded, Föttinger possessed the ability to explain his ideas in an understandable and demonstrative manner. He was once described as a "Man open to the world", as somebody who understood how to relate to young people. He had the great ability to write about complex subjects without being pretentious. This is the style he adopted to describe the path of the operating fluid between the pump wheel that he called "turbine wheel" and the turbine wheel that he described as "secondary turbine" and to point out the advantage of components located close to each other, requiring short connections. In his own words:

"The water flows only through the turbine wheels needed for immediate power transmission. These wheels are designed and arranged in such a way that a very close circuit, formed by the side wall of the individual wheels, is formed by the primary wheel after the secondary wheel and vice versa. This circuit has the appearance of a hollow vortex ring and can be compared with a smoke circle. The conventional shape of standard water turbine wheels has been completely abandoned. Instead, a new, free design is being used that is fully adapted to its actual purpose."

The "new power transmission", called "transformer" in Föttinger's essay, had already been registered by "Vulcan" in his name. The patent (number 221422) was filed on 24 June, 1905. This date is the reason for this book, first published in 2005, 100 years later.

By 1906, Föttinger had established most of the basic theory but nevertheless pointed out that "...this and everything else had to be critically established and selected over years of work". These basics indicated in theory efficiencies of 80 to 82% at four to five times the ratio of the converter, which later proved to be true. Föttinger was proud of this: "I mention this, because there is a widely held opinion that innovations in turbine construction are impossible without a model test laboratory. The only tool we had was a healthy, critically applied theory of two-dimensional flow that built upon the findings of Pfarr, Prasil, Lorenz-Bauersfeld and von Mises and modified what seemed to be suitable for the new forms." While still researching, he contacted "the largest German turbine companies" – Voith in Heidenheim and Briegleb & Hansen in Gotha – as a precaution, in order to be able to incorporate the efficiencies of modern water turbines into his calculations.

In recent years, gear technology had undergone rapid progress and now offered higher efficiencies at bigger ratios than the Föttinger converter had ever achieved: as early as 1909, a value of 98.5% had been measured in an English ship, the "Vespasian", compared to the maximum value of 91% achieved by the turbo converter in the "Wiesbaden". Föttinger resigned himself to this fact in 1930: "Since 1918, the first design of the turbo transmission in its original field had been superseded by gear wheels: not so much regarding its efficiency, which is at least equal to electrical transmission, but with a view to the limited ratio". In a series of sea trips, the "Vespasian" had additionally proven the "absolute reliability of the gear drive". Eventually, gear turbines proved themselves as simpler and less expensive. Föttinger tried once more to save his converter by suggesting from Gdansk to use gears for the forward gear and a turbo converter for reverse. But this proposal was never taken up.

As a result, the "demise" of the turbo converter – using the expression of Gustav Bauer – had finally and irreversibly arrived. The life cycle of the turbo converter in its role as a transmission between steam turbine and ship's propeller had ended after only nine years.

The decision-makers at "Vulcan", too, had recognized that the turbo converter had no future in marine propulsion. As ever in such cases, the company calculated how high the development cost had been. The commercial side of "Vulcan" had to state that more than three million Marks had been spent on it between 1907 and 1914. It has not been documented whether these findings influenced the termination of Föttinger's consultancy contract in 1915. Yet in those days, Föttinger, too, had complained about the assumed indifference of "Vulcan" as far as the further development and the marketing of the patents were concerned.

Follow-Up Development at "Hamburger Vulcan"

In the field of locomotive drives, "Vulcan" had carried out a short trial with the hydrodynamic converter in 1924, which turned out to be unsuccessful. The company therefore had to find a new direction. It was seeking different paths not only from a technological aspect, but also because there had been far-reaching internal changes. The company had moved to a shipyard in Hamburg that it had set up between 1905 and 1908 mainly for infrastructural reasons. But there was also a decisive political reason, as the Imperial Navy had requested to have a shipyard in Hamburg that was familiar with warship construction.

After abandoning the production of converters, Vulcan, now calling itself "Hamburger Vulcan" expressed a special interest in hydrodynamic couplings for which it had owned the patent rights since 1905. From approximately 1910, tests had been carried out to examine efficiencies and blade contours, but had been interrupted because of the First World War. The invention did not appear to be sufficiently attractive to be utilized commercially. To the management, it took on a secondary role in the light of the success with the converter. In his patent specification, Föttinger himself had mentioned it in just one sentence, when he pointed out the hydraulic advantages offered by an additional, stationary guide blade ring between both sides. It should be omitted only if the speed of the secondary side "was, by a certain degree, lower than that of the primary side. Yet such cases are of minor relevance." Although the unit had already been a coupling in those days, the registering parties did not use this word; patents rights for it were consequently not claimed.

As a precaution, the patent office of "Vulcan" had, however, applied for separate patent protection for the coupling in the company's name on 24 June 1905. As Föttinger had never regarded it as relevant, he is likely to have agreed that his name was be absent. In this patent specification, the term "coupling" is not used. The object of the invention is referred to as "Hydrodynamic transmission".

Now, after losing the converter business, the company could consider itself fortunate to own the coupling patent. After the war, from 1919 until the end of 1922, it carried out comprehensive research with the coupling. "The results, proving the superiority of radial blades, were summarized on a number of curve sheets, from which circuit diameters and number of blades for a turbine with a given performance, speed and transmission efficiency could be extracted at any time". Subsequently, the company management stipulated that the newly developed coupling type should be designated "Vulcan Kupplung". This name resulted in a long dispute between Föttinger and "Vulcan" that ended only in 1931 in the form of a financial settlement. The name was, however, retained. At the time, the contract partner was no longer "Vulcan" but Deutsche Schiff- und Maschinenbau AG (Deschimag) in Bremen with which Vulcan had been affiliated since 1926.

The tests at "Vulcan" had shown "that torque fluctuations and torsional vibrations of diesel engine were either eliminated or reduced to a safe level" – a fact that became very important in view of the increasing application of large, slow-speed marine diesel engines. The company, utilizing these findings, combined the gear drives with a Vulcan coupling and launched this combination onto the market under the name "Vulcan Transmission" in 1924.

With the onset of the global economic crisis of 1930, Vulcan in Hamburg and Stetting were closed down. The further production of Vulcan transmissions and Vulcan couplings was carried out by A. G. Weser, the main shipyard of Deschimag. "Vulcan" ceased to exist at this time.

After his consultancy role ended in 1915, Föttinger was no longer involved in these developments. He left Gdansk in 1924, following a call of the Technical University in Berlin-Charlottenburg to the new chair for hydro physics and turbo machinery that had been created specifically for him. At his new workplace, hydrodynamic converters were no longer the focus of his activities. In 1942 he received the Commemorative Golden Coin of the Naval Engineering Society.

Hermann Föttinger died on 28 April 1945 after being hit by a shell splinter on the premises of his university in the last days of the war. A year later, the institute was renamed "Hermann Föttinger Institut für Strömungstechnik" (Institute for Fluid Technology).

The Herdecke Project –
Voith and Föttinger Get Together

When Hermann Föttinger contacted Voith in Heidenheim for the first time in order to inform himself about the state of technology in water turbine hydraulics, the young man had been helped in the "most amicable manner". He had been provided with ready access to efficiency data established in "exemplary test station". Without the application of experimental results, Föttinger might otherwise not have dared to put his own theories into practice, especially as it had been as recently as 1897 that Georg Adolf Pfarr, whom he quoted on several occasions, had carried out the famous "Braking test of Königsbronn" – a purely experimental examination of the hydraulic values of a water turbine.

The test stand of Voith at "Alte Bleiche" in Heidenheim complied with the latest state of technology, but cannot not be compared with the test laboratories in Hermaringen and Brunnenmühle built in 1907 and 1908. Yet it had been used for tests that enabled Voith to build the 18 turbines for the two super power stations Hamilton and Ontario at the Niagara Falls. Föttinger had thus been furnished with reliable information. His writings do not state whether he received this assistance in writing or whether he had come personally to Heidenheim. There are two reasons that suggest that he had undertaken the long journey to the Swabian Jura. On the one hand, the subject was rather complex and Föttinger had neither sufficient knowledge of water turbines nor did anybody at Voith know much about transformers. A series of letters discussing the complex issue is unlikely to have resulted in an early and satisfactory reply to Föttinger's questions. On the other hand, Heidenheim is in southern Germany and, viewed from Stettin, very close to Föttinger's hometown Nuremberg. It is therefore assumed that he did travel to Heidenheim, not least because his other destination, Gotha, could be reached from there quite easily.

The scientific approach of the company cultivated by Friedrich Voith lends ample credibility to Föttinger's impression of being welcomed by Voith. Although it was probably not the company proprietor himself who discussed the subject with him, the engineer who had recently gained his doctorate was most likely to have had Dr. Ing. Fritz Oesterlen at his side. Approximately the same age as Föttinger, Oesterlen was in charge of the development of the runners of Francis turbines and hence the perfect contact for Föttinger. Additionally, both had the same scientific background. Their professional careers after these probable consultations were virtually identical.

Both became university professors in mechanical engineering: Föttinger in Gdansk for marine propulsions, Oesterlen in Hanover for water turbines. It can be assumed that the two men communicated well with each other.

While the talks proved satisfactory for Föttinger, neither of the two engineers could know that his invention would be of immense significance for Voith some 25 years later and open up a completely new business field. Even their next contact was more than 20 years away.

The Constructive Development of the Start-Up Coupling

In 1927 Voith had received an order for the delivery of the entire hydraulic equipment for the Herdecke pumped storage power station. Voith also celebrated a little anniversary on the side – one of the four turbines carried the (continuous) production number 10 000.

The delivery also included three special start-up clutches between engine/generator and pump, intended to bring the pump from standstill to synchronization output at full engine speed. They were also meant to separate the relevant pump from the engine/generator at the transition of pump to turbine or phase modifier operation, and then bring this pump to a halt. They had to be able to be disengaged during operation. The power to be transmitted was high – 36 000 HP at full load. The starting capacity of the filled pumps alone was 10 000 HP. This was unavoidable in those days. Starting processes with an empty ("blown out") pump and leakage water cooler were not yet available.

Installation plan of the hydraulic equipment of Herdecke pumped storage power station.

1 Turbine
2 Motor/Generator
3 Coupling
4 Pump

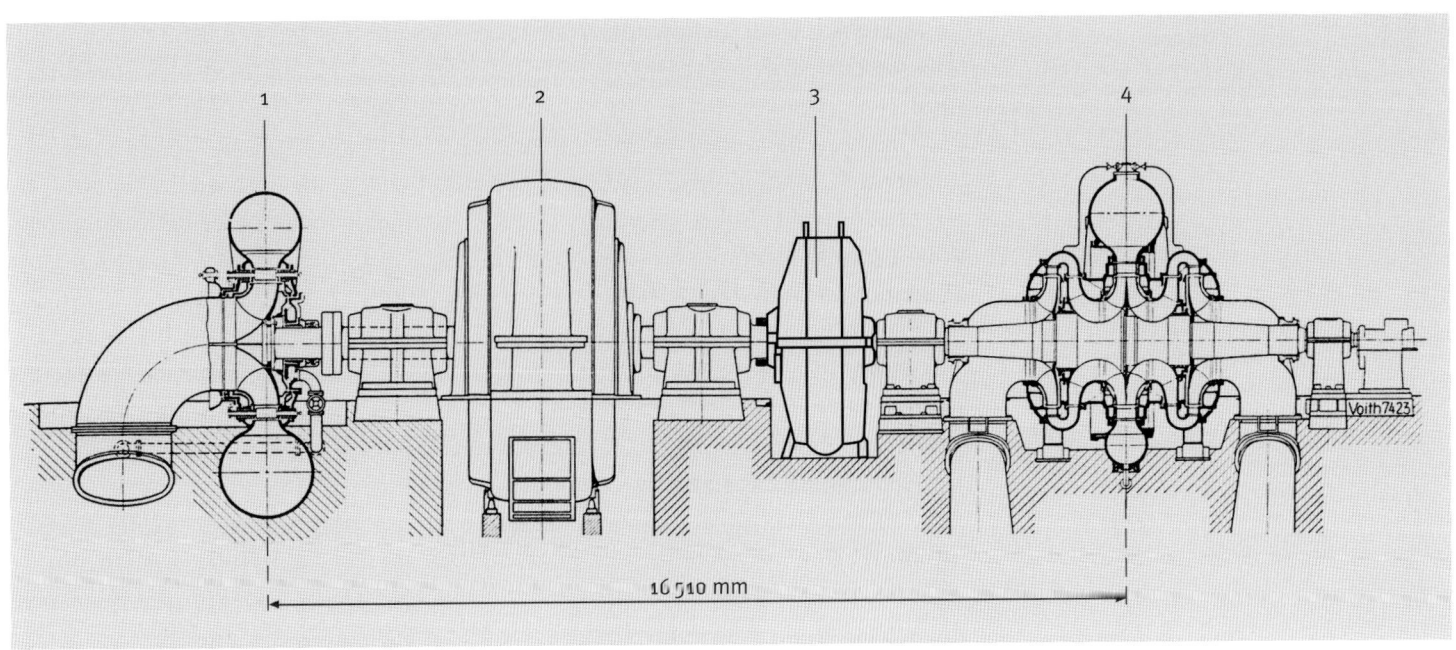

16 510 mm

When the order was placed (!), nobody knew what the machines should actually look like. In an "Explanatory report on the hydraulic equipment of the pumped storage power station Hengstey", Rheinisch-Westfälische Elektrizitäts-werke merely listed a few possible solutions and stated their advantages and disadvantages. After the acceleration phase, an added magnetic clutch of Mag-netwerke Eisenach was intended to ensure the transmission of lock-up torque. It was also anticipated that this clutch did not have to provide "any further accel-erating power" and could therefore be kept small. But a prerequisite for satisfac-tory operation was "that the current exciting the magnets during pump opera-tion is not interrupted, as this would cause immediate decoupling and disen-gage the pressurized pump from the centrifugal torque of the engine generator". This would indeed have resulted in a highly undesirable operating condition.

Voith had already considered this issue: "In the knowledge that the coupling is one of the most important components of pumped storage operation, Voith has undertaken to build a coupling through which the acceleration period is overcome by hydrodynamic friction and which is switched over to a mechanical friction coupling after this acceleration period."

The first draft by Voith included an oil-filled rotary housing in which the rotary motion of the driving shaft was transmitted to the secondary shaft via co-rotating discs and the viscosity of the oil. From a certain maximum speed of the secondary shaft, the further acceleration of the pump impeller was intended to take place via a friction connection that had built up between the co-rotating discs and the driving shaft.

This first draft had little in common with a Föttinger coupling. At the same time, the hydrodynamic friction process was highly unlikely to run the water-filled pump up to synchronization speed. Yet the absence of an electro-magnetic friction coupling subject to power cuts is interesting. The vertical force was meant to be created automatically by oil pressure.

This draft was the starting point for all further developments. These were largely determined by Wilhelm Hahn and his coworker Ernst Seibold.

At the time, the ambitious engineer Dr. Wilhelm Hahn was in charge of the turbine construction and development department and, a few years later, was promoted as Director of Turbine Construction. For the young turbine engineer Ernst Seibold, the task was a completely new challenge. As the product was new to Hahn also, they immersed themselves intensively into this project.

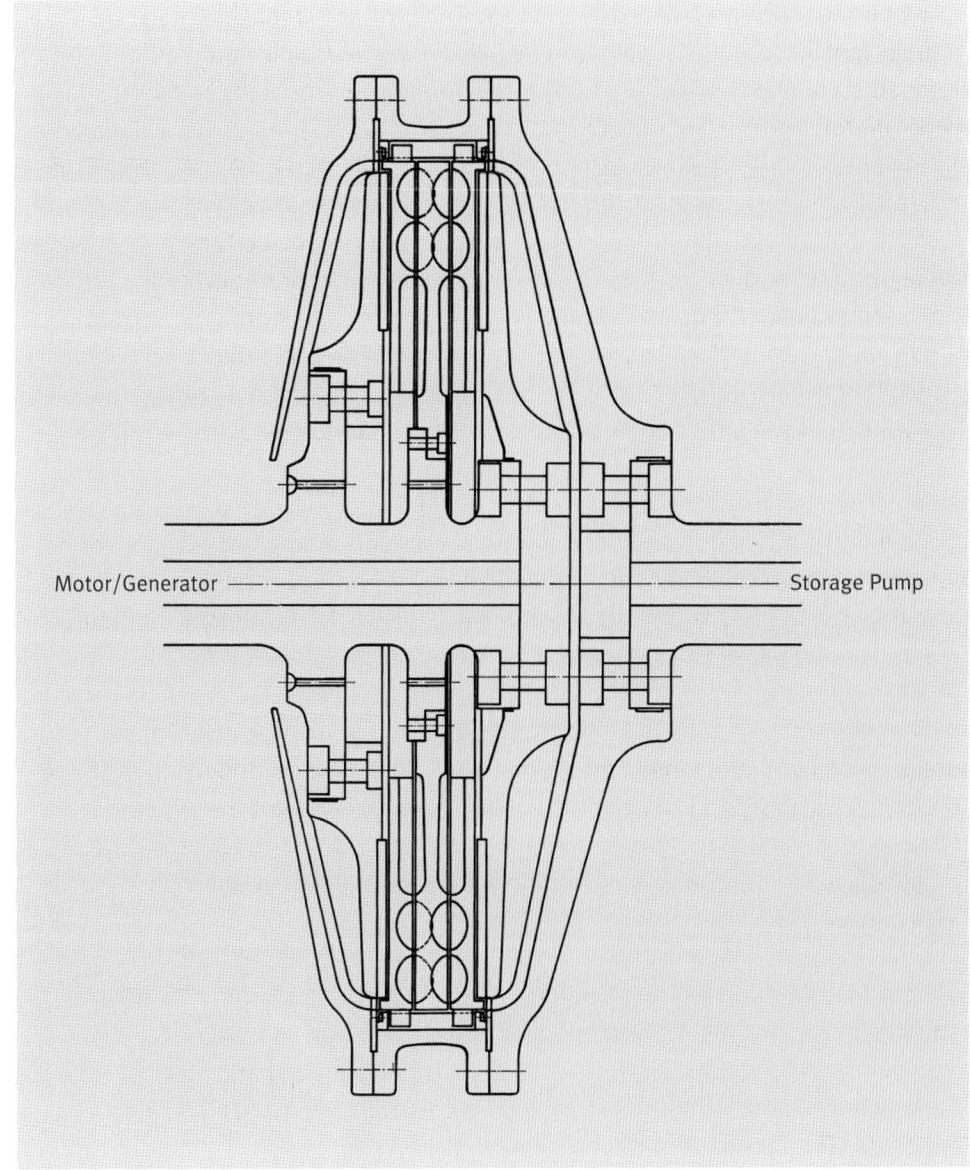

Motor/Generator ———————————————————— Storage Pump

"Hydraulic coupling"
Voith draft design dated
9 May 1927.

Wilhelm Hahn, 1882 – 1939

Ernst Seibold, 1900 – 1977

He was a passionate and gifted design engineer who recognized his career opportunities at Voith and grasped them accordingly. The successful conclusion of such an enormous task under his auspices was another good opportunity to do so. His commitment is documented in an essay that had been published after the commissioning of the pumped storage plant in Herdecke. It describes in great detail what the task of the coupling was and how it progressed in terms of construction and testing. Ernst Seibold and his colleague Hugo Schlebach were thus not on their own.

The project was indeed given special attention. With a view to the increasing importance of pumped storage plants in Germany's growing integrated networks it was anticipated that the requirements for this type of coupling would increase.

1929 – Föttinger's Second Visit to Heidenheim

Voith had not yet built a coupling to this specification. "The friction couplings that had been used so far appeared to be unsuitable for dealing with such high outputs; therefore a completely new coupling had to be developed", Hahn wrote. By then, no contact had been made with Föttinger. It was only in the course of the developments that someone had the idea of getting in touch with him and utilizing those of his inventions that were essential for running up the storage pumps. Since 9 May 1919, he had owned a patent for a "Combined asynchronous coupling and synchronous coupling" (patent number 374 259), since 2 July 1927 patents for a "Device for the operation of an integrated coupling" (patent number 479 119), as well as a "Device for switching combined asynchronous and synchronous couplings" (patent number 559 972) in which Voith was interested. This interest was even greater, as his name had already been mentioned several times in connection with the coupling in the "Explanations", and time was scarce.

Relatively late, on 17 April 1929, Voith eventually signed a licensing agreement with the inventor after further improvements of the draft concept. For this purpose, Föttinger made the journey from Berlin to Heidenheim. The subject of the licensing contract were the two inventions mentioned earlier, but not "exclusively".

On 17 April and on the following day, intensive and detailed technical talks took place: "On the basis of the experiences with the hydraulic slip couplings invented by Professor Föttinger, it was decided to design a combination with a hydraulically activated friction coupling, in order to avoid the losses caused by the continuous slip in this coupling type", wrote Hahn.

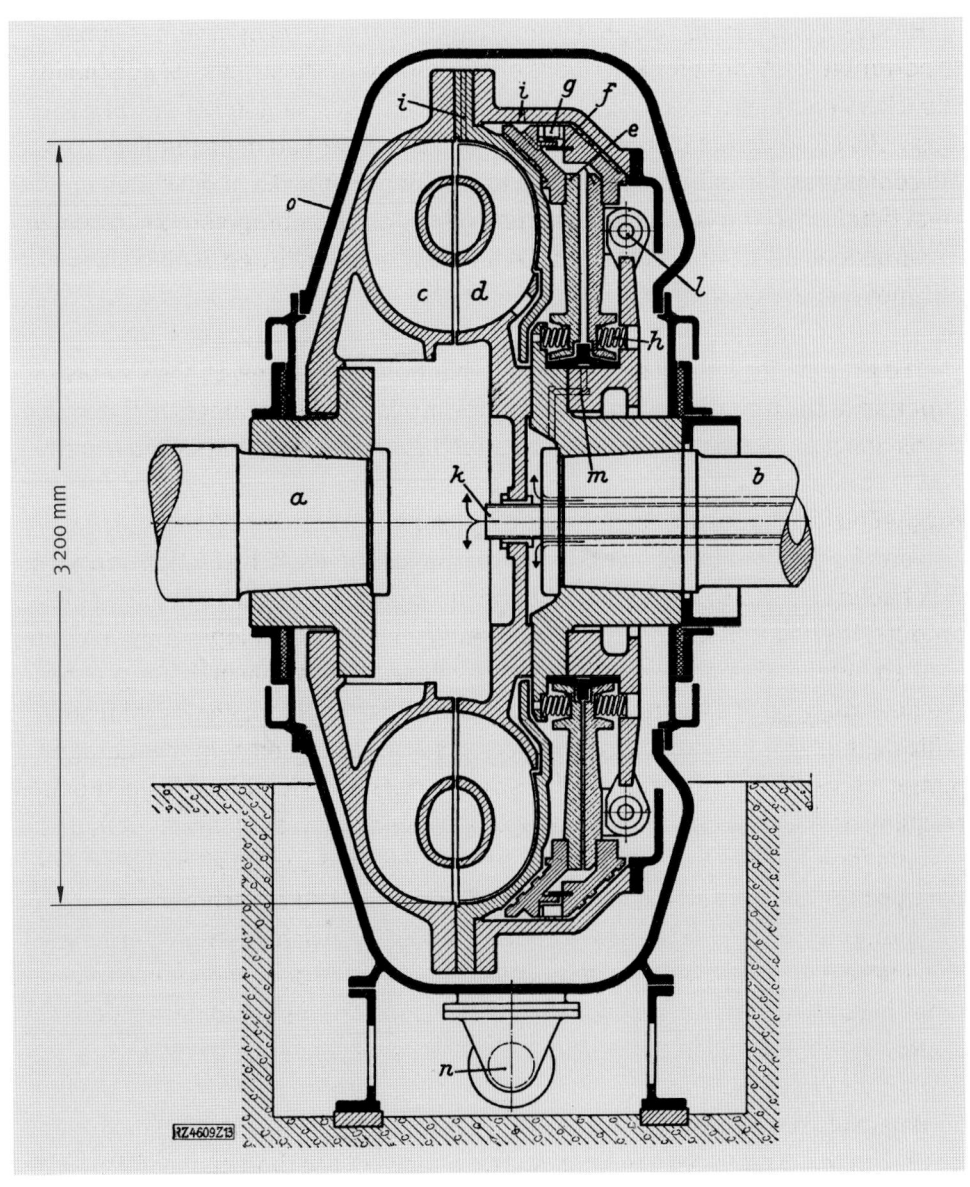

3 200 mm

Sectional drawing of the
Voith Föttinger coupling

a	Motor/Generator shaft
b	Pump shaft
c/d	Turbo coupling
e-h	Friction coupling
l and n	Water outlets
k/m	Water inlets
l	Shock-absorbing spring coupling
o	Air-tight housing

The final dimensions of the hydraulic part of the coupling were specified jointly. After discussing further individual hydraulic issues, the coupling was now ready for construction. It barely resembled the first draft from 9 May 1927.

In the end, the friction-mechanical part, too, looked totally different from the version patented by Föttinger. As a further element, an elastic intermediate link had been added, in order to ensure jolt-free transmission of torque. These new ideas were described as "Hydraulic Coupling" in patent specification number 567 156, and protected for Voith with effect from 19 April 1929, one day after Föttinger's visit.

Model Tests and Production of the Coupling

Yet Wilhelm Hahn was still not prepared to release the new design for production. There needed to be further observations regarding its operating safety, and it was intended to identify and eliminate possible weakness through model tests. The Brunnenmühle turbine test laboratory was the ideal place for this. The measuring results were encouraging. "Due to model tests during the construction of the pump, this Voith-Föttinger coupling had been developed to such perfection that it operated trouble-free from its very first commissioning in the power station", Hahn reported. This was the first time that the name "Voith Föttinger Coupling" had been mentioned in a publication. It was certainly justified, as this unit combines inventions by both Voith and Föttinger. Collectively it will, nevertheless, always be remembered as the "Herdecke Coupling".

A comprehensive program had been completed. After the construction phase, the hydraulic and the mechanical part ("Friction part") of the coupling had been tested separately on a specifically constructed test stand. After individual optimizations, a complete coupling model was tested, "i. e., some 2000 shifting movements for engaging and disengaging the coupling were carried out in uninterrupted sequence. The operation of both the hydraulic and the mechanical part was completely satisfactory. A particularly remarkable feature was that the material selected for the friction surfaces showed hardly any wear." This result was particularly rewarding for Hahn and Seibold.

Now nothing stood in the way of the actual construction. "After studying all currently known coupling types in detail we have decided in favor of the newly constructed Voith Föttinger system ...".

And it was high time, too! The date on which the sectional drawing was made, 5 September 1929, illustrates the tight schedule. Voith was under enormous time pressure. While work on other components had begun immediately after receiving the contract in late 1929, the development of the coupling had only just begun. Its main dimensions were specified when other major parts of the complete order had already entered production. Construction departments and workshops must have collaborated at high speed: "Dispatch immediately after production!". There are no documents stating that the commissioning of the pumped storage power station was delayed because of the coupling.

Operating Experience and Consequences

The operation of the new coupling that had been developed specifically for Herdecke was indeed excellent. Unfortunately, the service life of the friction parts was not quite as long as Voith had hoped and expected. By March 1940, all friction parts had been exchanged and/or overhauled for already the second time. In January 1942, the fifth spare parts shipment was required, and in February 1948 the seventh. Ensuring the operating safety of the pumped storage power station was so labor-intensive that the synchronization process could not be continued in this form

The power station operator, RWE, therefore proposed a different solution, when storage pump number four, planned right from the beginning, was to be installed after the war. RWE abandoned the synchronization method by installing a conventional coupling, although it describes this unit in retrospect as "sensational for the designs available at the time". Yet the high wear in the friction part of the coupling eventually made way for a different method that Voith had developed by then. The coupling was now run up by a small Pelton turbine in combination with a subsequently disengaged gear coupling. This had become possible, after the process of "blowing out" the pump allowed accelerating the impeller in air, reducing the required starting power to a fraction of its previous level. Between 1955 and 1958, the three existing couplings were gradually replaced in the same way. The Voith Föttinger coupling as a starting device for pump synchronization remained a unique application. It was never again used for this task in power station operation.

The Hydrodynamic Starting Converter in a Pumped Storage Power Station – Generation "Lünersee" 1954

Although the hydrodynamic coupling had soon reached its technical limits, Föttinger's invention remained an indispensable element for storage pumps. An increasing number of pumped storage power stations with high heads and delivery heights used and still use hydrodynamic converters exclusively as starting devices for storage pumps. In these applications, the pumps have several stages and require a high counter pressure ("Supply Pressure") on the suction side. A starting turbine is no longer sufficient here. It requires a water-free pump for its operation. The safe, uniform "blow-out" of a multi-stage high-pressure storage pump with its narrow flow ducts and numerous split ring seals is quite complex in itself. Additionally, the high counter-pressure would require an air chamber that would not be able to guarantee the required short switching times, even if it was designed with large dimensions.

This first and largest hydrodynamic coupling based on the Föttinger principle leaves the Voith plant in Heidenheim.

Machine set Lünersee

1 *Motor/Generator*
2 *Shut-off device*
3 *Free-jet turbine*
4 *Starting converter*
5 *Storage pump*
6 *Thrust bearing*

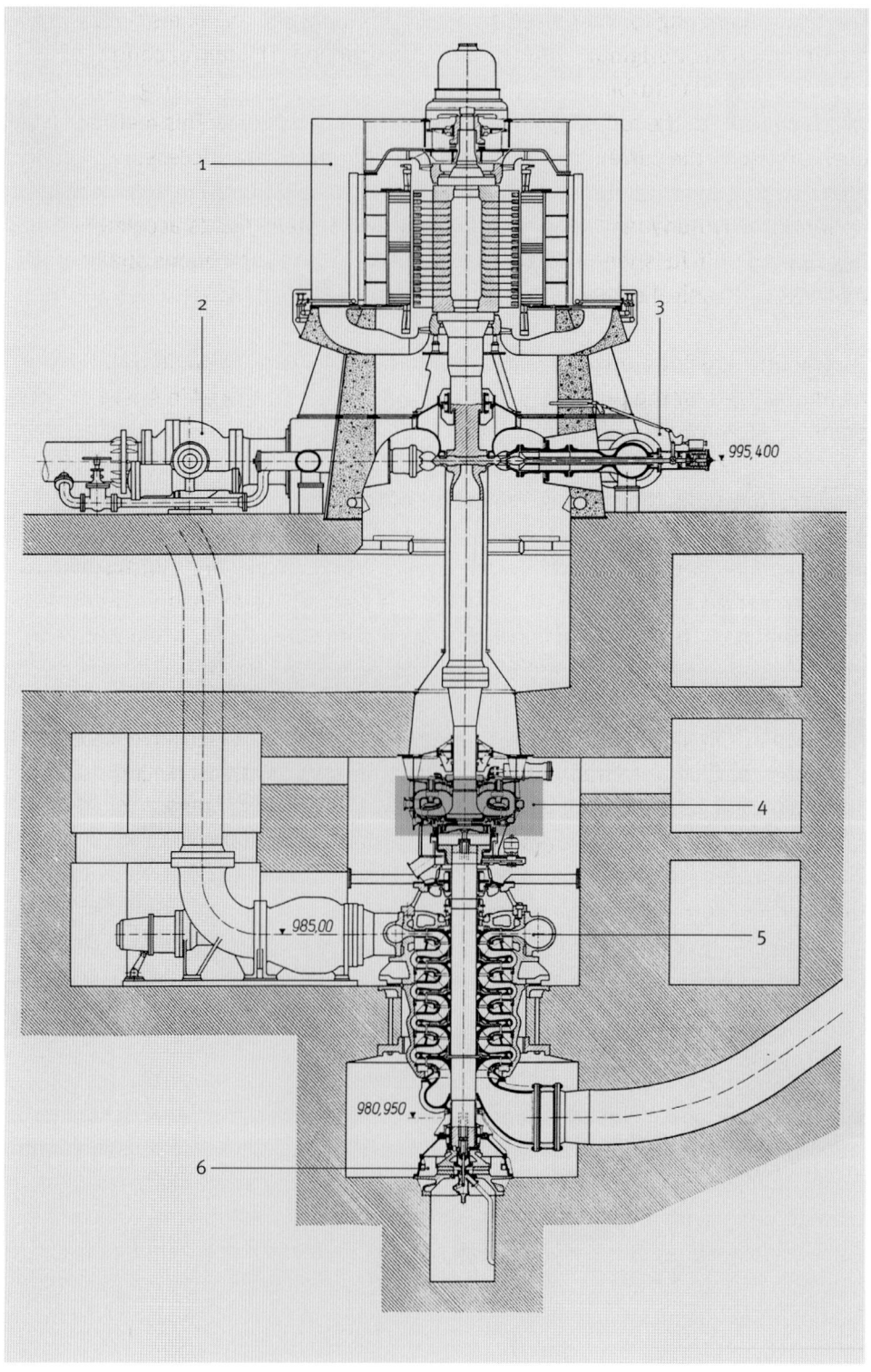

Whilst searching for better solutions, the Voith design engineers finally remembered their original craft – turbine construction. The power control of Francis and Kaplan turbines occurs via an adjustable guide vane ring, and Voith was familiar with the function and the technology of this unit. This method was now applied to the current problem and brought immediate success.

Modern, hydrodynamic converters in pumped storage plants accelerate the filled pump until its speed exactly matches that of the synchronous speed of the motor. A mechanical device ensures rigid coupling on both sides.

The first starting converter with adjustable guide vane ring built by Voith had been designed for the pumped storage power plant "Lünersee" in Austria. The plant was commissioned in 1957.

Since this time, the starting converters of high-pressure pumped storage plants have been equipped with adjustable guide vane rings. Among them is also the plant "Wehr", or, as it is referred to at Voith, "Hornberg". With a diameter of 3650 mm and an output of 140 MW, this converter remains the largest ever designed by Voith. Most recently, a converter designated "Kops II" has been added, transmitting 85 MW at the same diameter of 3650 mm.

For heads of up to approximately 800 meters, pump turbines have become a serious competitor to traditional storage pump sets. With this machine type, starting in the direction of pump rotation is carried out either directly via the motor generator, by a starting motor or a starting turbine.

Hydrodynamics in Drive Technology –
A Short Introduction

This chapter provides a very concise introduction to the significance of hydro-dynamic components in drive technology. It is meant to familiarize the reader with this fascinating technology.

As soon as the first simple machines emerged in ancient history, there was a need for adaptation or compensation in the power transmission process between the driving power source and the driven machine or a certain operating procedure. The irrigation systems in ancient Egypt and medieval grist mills used water wheels or windmills, while devices for lifting heavy loads or scooping water were moved by draft animals.

The operating process involved required that the speed of a driven machine was either increased, decreased or diverted to the rotating axle to match the output of the driving machine. Simple mechanical transmissions appeared on the scene. The demand for further intermediate driving components started to grow.

The necessity for installing an adapting component into the driveline always remained, even when the modern driven machines of our time were introduced, for example steam engines, electric motors, combustion engines and steam or gas turbines. The questions raised in this context today are basically identical to those asked in prehistoric times:

- What is the operating process for which the drive is needed and how important is the process?
- Which process quantities are to be influenced by the driven machine in which way?
- Which type of machine is to drive the process?
- And finally, important for Voith: which characteristics of the driving and driven machine are unsuitable for the process and have to be compensated or adapted by a hydrodynamic element in the driveline?

Typical Processes and Their Drives

The following three operating processes and three common driving machines are used as an example of the wide range of possible applications in drive technology:

Conventional driving machines

Electric motor

Combustion engine

Steam or gas turbine

Power range $P_1 = M_1 \cdot \omega_1$

Drive components

Torque converter

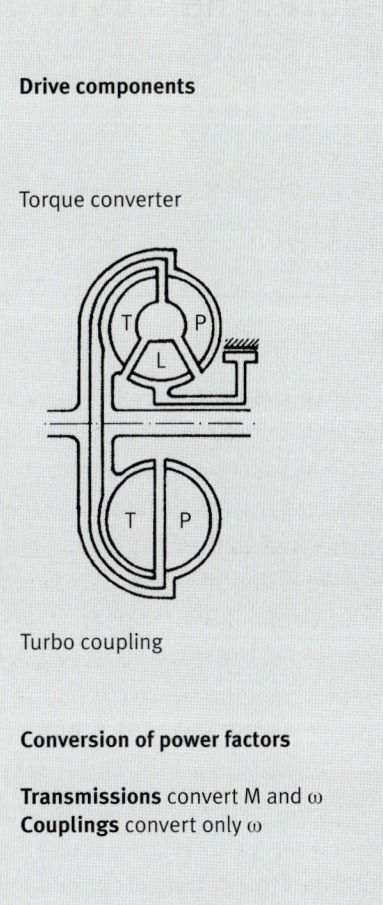

Turbo coupling

Conversion of power factors

Transmissions convert M and ω
Couplings convert only ω

Typical Driven Machines

Constant torque

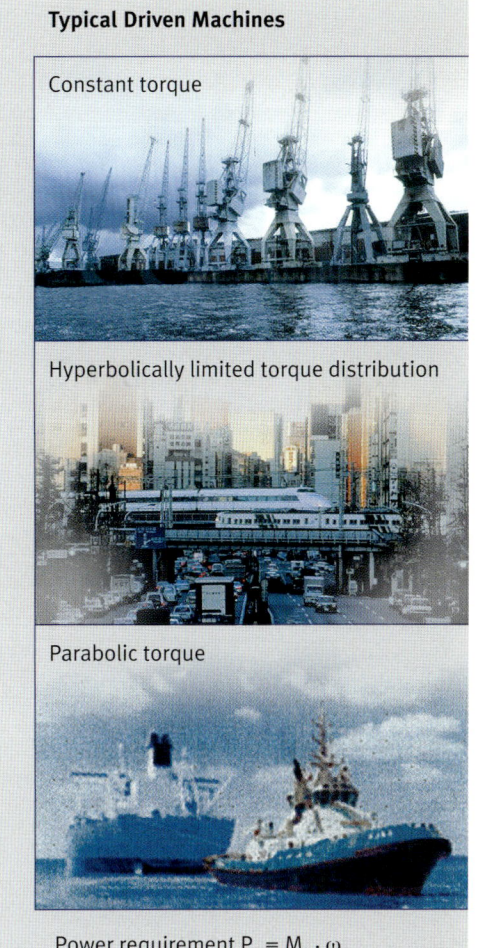

Hyperbolically limited torque distribution

Parabolic torque

Power requirement $P_2 = M_2 \cdot \omega_2$

Föttinger converters or couplings, shown in the middle of the illustration above, transmit the output of the driving machine to the driven machine while adapting themselves to the process characteristics of this driven machine. Brakes (later also called "Retarders") function in the opposite way. They are rotated by the mass of the driven machine to be braked, resulting in energy being converted into heat.

The central requirement in this context is that engineers know the characteristics and the characteristic curves of the two machine types. A straightforward selection of machines on the basis of their nominal output can only serve as a guide value, but never as a decisive factor. Using this basic knowledge, engineers develop especially customized hydrodynamic power transmission components, or they resort to proven technologies and adapt or upgrade them.

Consequently, power transmission components such as hydraulic couplings, converters and retarders, are able to demonstrate the full range of their features and benefits in standard operation and under unusual conditions: they steplessly transmit their nominal outputs accurately and without wear, they control output speeds – or, in the case of converters, also torque – dissipate heat, reduce torsional vibrations in the driveline, start machines smoothly, brake continuously without signs of fatigue and protect sensitive drive systems against shocks.

Right: Operating processes with constant, hyperbolically limited and parabolic torque distribution.

Left: Squirrel-cage motors are the most commonly used electric motors in stationary drive technology, diesel engines are the most popular combustion units in rail and road vehicles, gas and steam turbines are frequently found in stationary and mobile applications.

Depending on their design, they operate in very different areas. As **torque converters** they are used mainly in rail and road applications, as **turbo couplings** they are essential elements both for rail and road vehicles and for stationary industrial drives. As **hydrodynamic brakes** or **retarders** they ensure safety in trains, buses and trucks.

Despite their rather different tasks and application areas, these three hydrodynamic machines look quite similar at first glance. This is not surprising, as they all originate from the same source, the Föttinger converter from 1905. The only difference in the design of the Föttinger converter and the Föttinger coupling that was registered as a patent by Stettiner Vulcan at the same time is a stationary guide apparatus in the converter. The retarder (brake) is set up like the coupling, but during operation only the bladed wheel to be retarded rotates, while the other one is stationary. The high number of Föttinger units used in industrial applications hence always relate to these three basic forms.

The differences in hydrodynamic behavior are nevertheless enormous. Their characteristics in terms of heat balance, control and design and also the structure of their end users were increasingly diverse. Voith therefore decided to establish several product groups, and eventually, two decades ago, three independent market divisions within the Voith Group Division Voith Turbo: **Rail, Industry** and **Road**.

Basic Hydrodynamic Characteristics of Transmitting Components

Hydrodynamic transmission – upon recommendation of the VDI Association of German Engineers also called Föttinger transmissions – convert the power factors "torque" and "speed" within the power flow from a driving machine to a driven machine based on their own, indirect principle. The mechanical power introduced by a circular pump is transmitted to a fluid that transports it as hydraulic energy and forwards it to the turbine. There, it is reconverted into its mechanical form.

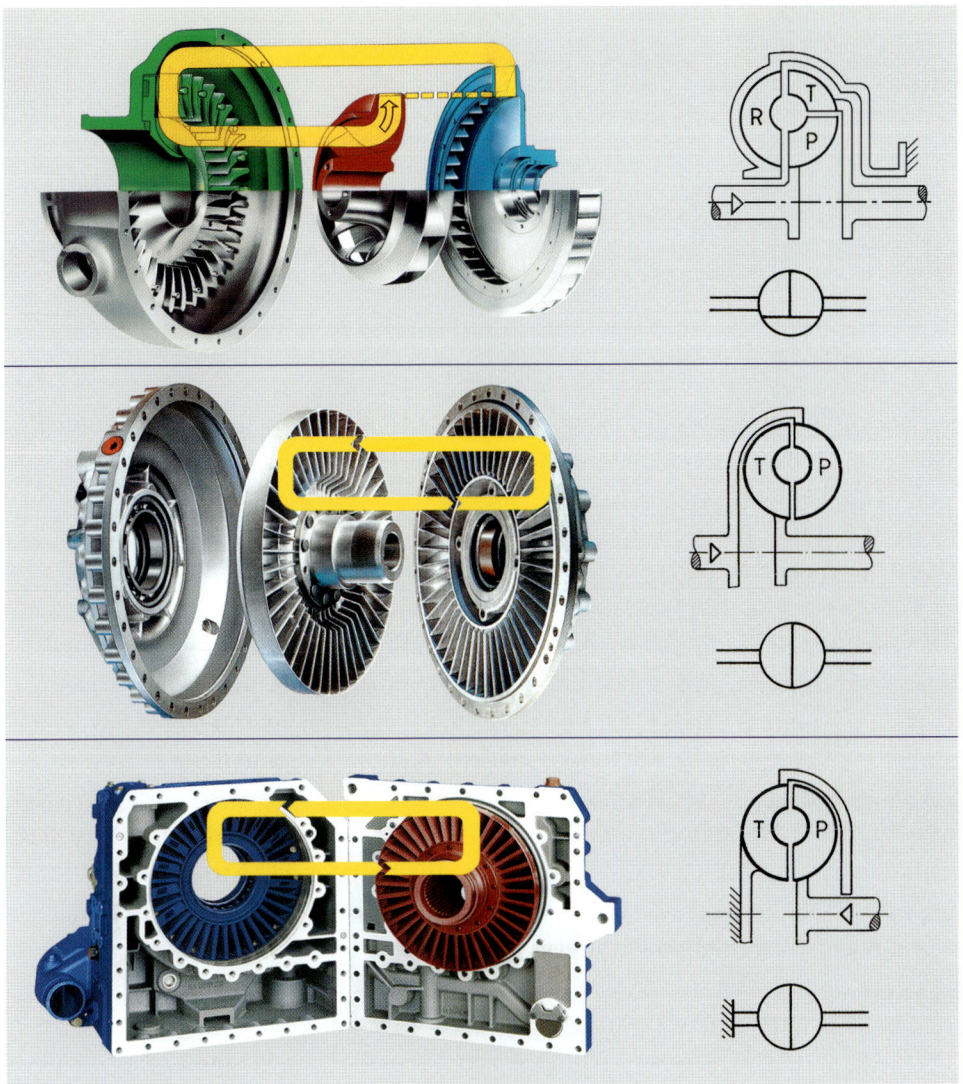

Pumps and turbines in hydrodynamic transmissions are fluid machines.
Their hydraulic output P_h can generally be described on the basis of a formula
developed by the Swiss mathematician Leonhard Euler:

$$P_h = \dot{m} \cdot \Delta\,(r \cdot c_u) \cdot \omega$$

Which means:
$\dot{m}$ = Mass flow
$\Delta\,(r \cdot c_u)$ = Change of momentum between inlet and outlet of the flow grid
ω = Angular velocity of bladed wheel

Using the laws of fluid dynamics, a similar formula for the pump output P_P
can be developed for practical use on the basis of the classic Euler formula:

$$P_P = \lambda \cdot \rho \cdot D_P^5 \cdot \omega_P^3 \quad \text{and} \quad M_P = \lambda \cdot \rho \cdot D_P^5 \cdot \omega_P^2$$

This means:
D_P = Outer diameter of pump profile
ρ = Density of operating fluid
ω_P = Angular velocity of pump wheel
M_P = Pump torque
λ = Characteristic power figure (proportional figure)

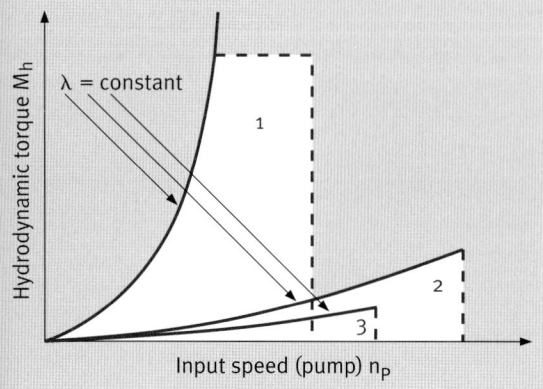

Hydrodynamic torque M_h

λ = constant

1

2

3

Input speed (pump) n_P

Primary performance graphs
of three Föttinger units with
similar size

1 Brake (retarder)
2 Coupling
3 Converter

with typical limit values of the
coefficient of performance λ
and maximum pump speeds,
related to the same wheel
diameter. Depending on the
process parameters, the
primary operating torque lies
on parabolas below the design
limit. The required coefficient
of performance λ is determined
by the secondary field.

The coefficient of performance λ is a dimensionless quantity determining the design and the type of the unit (converter, coupling or brake). It is used for describing the hydrodynamic characteristics in the primary and the secondary field.

For constant performance coefficients λ, the **primary performance curve** of the three Föttinger units describes the capability of the pump to provide parabolically increasing, hydraulic torque at rising input speed.

Föttinger units are characterized by the collaboration of at least two fluid machines (pump and turbine). As the interconnection of the two rotating bladed wheels occurs via hydrodynamic forces only, the relevant operating point of a hydrodynamic transmission (converter, coupling) depends on the load and is self-adjusting. Therefore the operating behavior is determined by the characteristic power figure Lambda vs. speed ratio in conjunction with load vs. speed. This also applies to controlled quantities such as guide blade adjustment, filling volume or scoop tube position.

For a constant input speed, the **secondary performance curve** describes the dependence of the coefficient of performance λ, on the speed ratio v. This interrelation is also referred to as the "first characteristic relation" or as the characteristic or performance curve of Föttinger units.

The development of the secondary curve is largely influenced by the turbine. If the flow is centripetal (from the outside to the inside) – for example with turbo couplings or Trilok converters – the pressure between pump and turbine drops while the turbine speed is rising, which also results in a reduction of the circulating mass flow required for the transmission of energy. The curves are characterized by a zero divergence at $v = 1$. The situation is different with the classic Föttinger converter and its centrifugal turbine flow. Here, the turbine shows hardly any reverse effect regarding the power uptake of the pump. Axially arranged converter turbines increase the uptake of the pump with rising turbine speeds, the pump is actually "fed". The type of flow on the turbine wheels of couplings is always centripetal as a result of their function.

Another typical characteristic of torque converters is a static bladed wheel (guide wheel) that is arranged opposite the flow circulation between two rotating bladed wheels. The change in momentum caused by entering or exiting

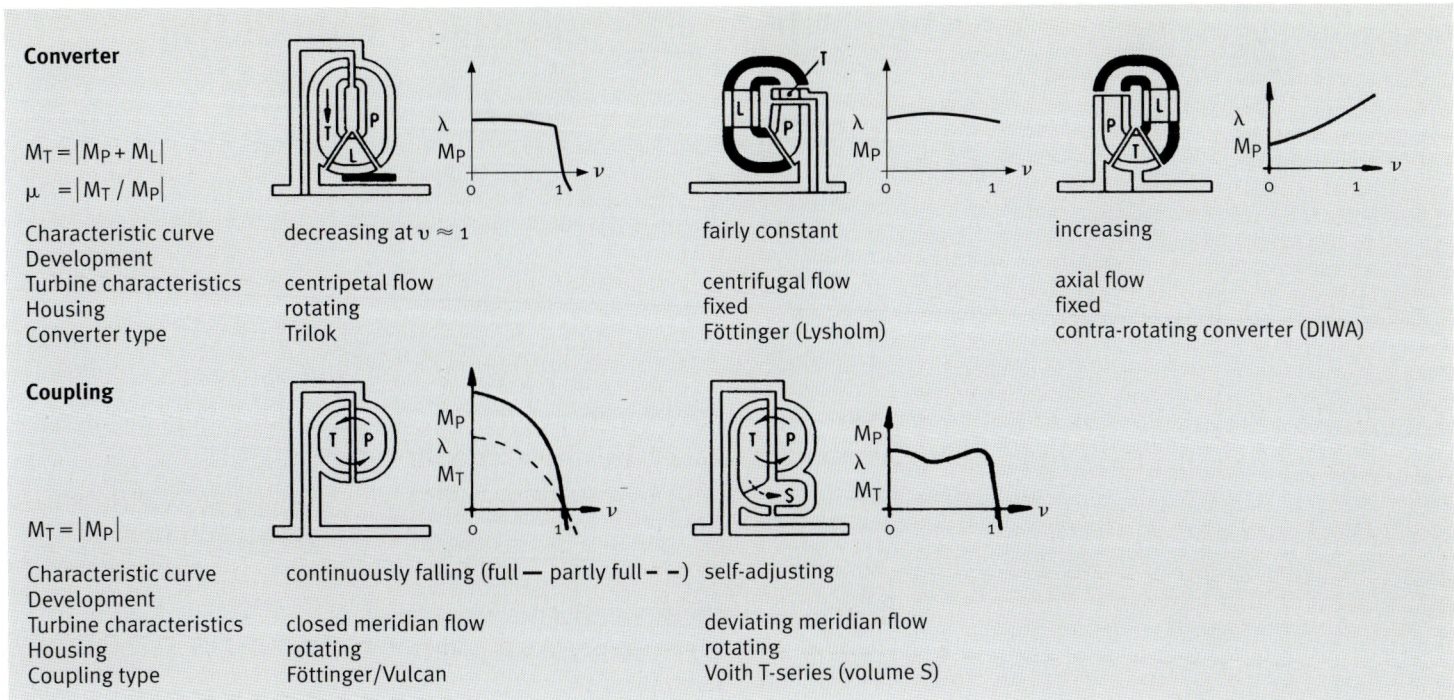

Converter

$M_T = |M_P + M_L|$

$\mu = |M_T / M_P|$

Characteristic curve Development	decreasing at $v \approx 1$	fairly constant	increasing
Turbine characteristics	centripetal flow	centrifugal flow	axial flow
Housing	rotating	fixed	fixed
Converter type	Trilok	Föttinger (Lysholm)	contra-rotating converter (DIWA)

Coupling

$M_T = |M_P|$

Characteristic curve Development	continuously falling (full — partly full - -)	self-adjusting
Turbine characteristics	closed meridian flow	deviating meridian flow
Housing	rotating	rotating
Coupling type	Föttinger/Vulcan	Voith T-series (volume S)

the grid results in a guide wheel torque M_L, which is introduced into the transmission housing. The amount of this supporting torque corresponds to the torque difference between pump and turbine (torque set of mechanics $M_P + M_T + M_L = 0$). Like the coefficient of performance λ, this characteristic of the hydrodynamic torque converter referred to as conversion μ = absolute value of turbine torque/pump torque, depends on the speed ratio v and the position of the guide blades.

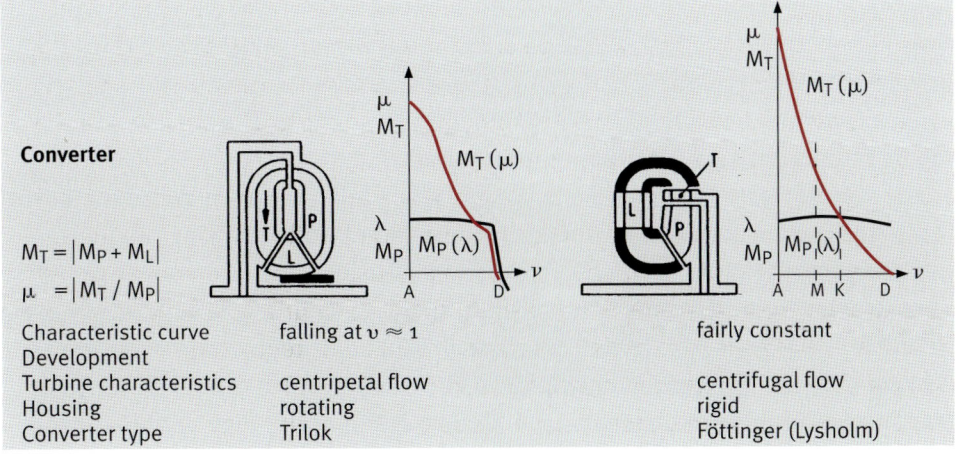

Converter

$M_T = |M_P + M_L|$

$\mu = |M_T / M_P|$

Characteristic curve Development	falling at $v \approx 1$	fairly constant
Turbine characteristics	centripetal flow	centrifugal flow
Housing	rotating	rigid
Converter type	Trilok	Föttinger (Lysholm)

For torque converters according to the Föttinger design, the load-dependent hydrodynamic power transmission characteristic is reflected by the development of the turbine torque. The torque development corresponds to that of a classic water turbine whose blade geometry has been optimized a certain speed ratio M (operating speed with best efficiency). The highest turbine torque is created at start-up point A and with the turbine at standstill, due to the high redirection and/or the change in momentum of the volume flow. At speed D, the volume flow is not diverted when it passes through the fast rotating turbine grid, and hence does not generate torque.

The Föttinger Principle Gains Popularity

With the development of his fluid transmissions and the ensuing fluid couplings, Föttinger's outstanding innovative achievement resides in having created two steplessly adaptable drive elements with a theoretically infinite performance. His patent from 1905 already lists some of the essential characteristics of future converter developments:

- Freely selectable ratio
- Adaptable number of stages (several turbines or pumps)
- Controllability (vanes, sleeve valves)
- Reversing the direction of rotation (contra-rotating converter due to reversing grid).

The rapid further development in the field of marine drives focused primarily on higher input powers, maximum efficiency and higher transmission ratios. However, until the end of the First World War the hydrodynamic converters lost the competition against conventional gear drives in marine applications.

Only in the 1920s, as a result of increasing traffic on the roads, the idea of hydrodynamic torque conversion was revisited by several people at the same time.

The following account of some remarkable developments from this time is meant to demonstrate that, after a hesitant start, the market gradually warmed to the idea of hydrodynamic power transmission, and eventually made the new fluid machine attractive for Voith as an independent and promising product.

The Rieseler Hydrodynamic Transmission 1925

While it was true that converters were no longer attractive for marine drives, this application represented only a small niche in the wide field of the technical possibilities of this product. Hermann Rieseler, a former colleague of Föttinger, explored new avenues that enabled him to utilize the original design of the "Vulcan". On this basis he developed a converter with two turbines intended to operate steplessly in a passenger car. He was granted patent no. 441549 for this invention.

In 1927, Rieseler's hydrodynamic transmission was tested in a Mercedes. In combination with the combustion engine with its limited speed control capability and the specific operating parameters of a car, a hydrodynamic converter demonstrated for the first time the advantages of stepless operation.

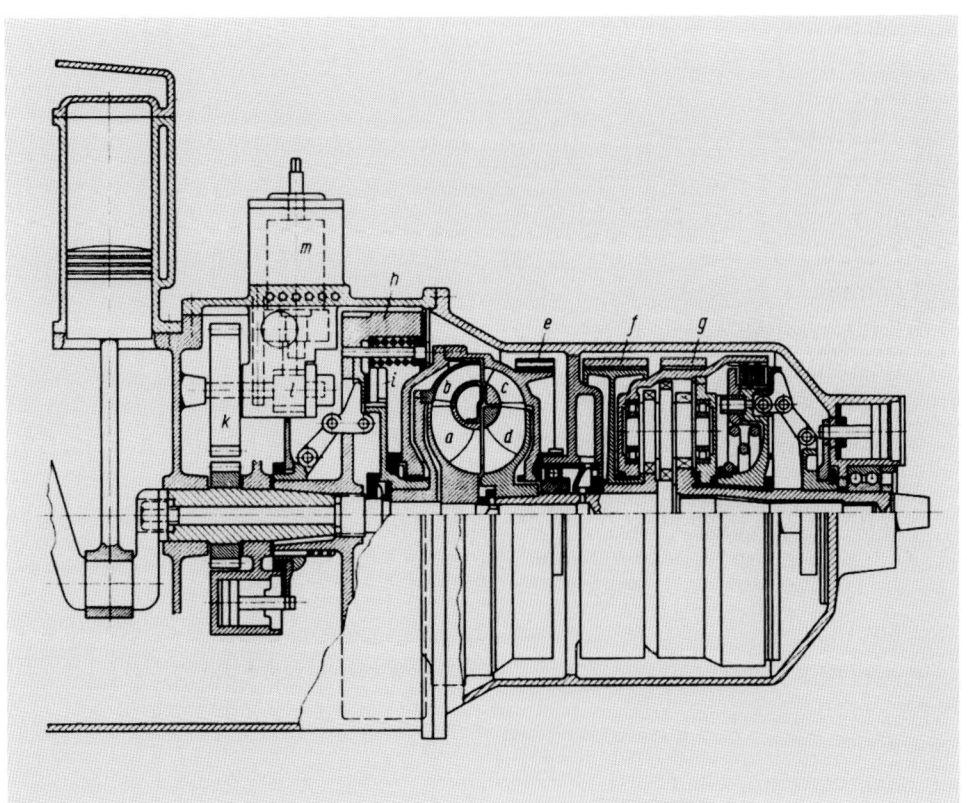

Rieseler transmission

In 1925 it already comprised many components of modern automatic transmissions:

– Hydrodynamic converter
– Planetary gear
– Multi-disc clutch
– Band brake
– Lock-up clutch

But the time was not ready yet for these modern automatic transmissions. The problem was that torque converters in road vehicles could only cover the required wide speed range at acceptable fuel rates if they were combined with a gear unit. Although Rieseler tried to combine the converter with a planetary gear set, the technical possibilities for practical driving had not been developed sufficiently.

The Lysholm Hydrodynamic Transmission 1926
The Swede Alf James Rudolf Lysholm went considerably further with his research efforts than Rieseler. In 1927 he was granted a patent (no. 628545) for a hydrodynamic changing transmission based on Föttingers idea for "agricultural and passenger vehicles". His converter had two pump wheels (2 and 3), four turbine wheels (4, 5, 6 and 7) and two guide wheels (8 and 9).

The idea was to extend the total transmission ratio with a large number of vanes while retaining efficiency, so that the vehicle would be able to operate exclusively via the converter. The secondary gear stage was merely used to change the driving direction.

Lysholm converter

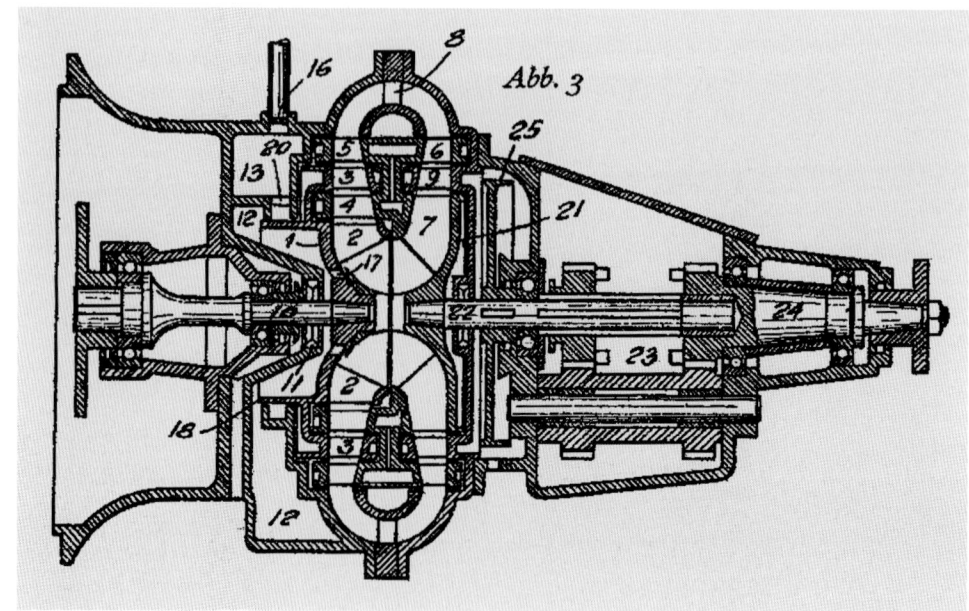

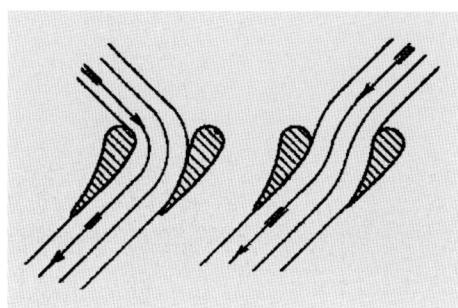

Single curved profiled blades.

A special feature of this converter were the short, single curved and profiled blades used by Lysholm who worked for a steam turbine company. With such a grid, shock losses during alternating flow angles are considerably lower than with long and unprofiled blades. The converter therefore offered the inventor a much higher ratio than the Föttinger converter. This basic concept was later used by British Leyland where it was developed further into the Lysholm-Smith transmission. Krupp used the concept for rail vehicles.

The Trilok Converter 1928

With the professors Wilhelm Spannhake, once Föttinger's closest colleague, Hans Kluge, a former employee of Vulcan shipyards in Hamburg and the mathematician Kurt von Sanden, three Föttinger enthusiasts got together at the winter semester at Karlsruhe University in 1924/25. Like Rieseler and Lysholm they focused on automobiles – apparently an attractive field in those years. They wanted to improve the still quite unsatisfactory collaboration between combustion engines and unsynchronized shift gear units by interposing a Föttinger transformer. This trio later called themselves "TRILOK" – an acronym of "Three" and "Locomotive" which symbolized the team's goal to develop innovative direct diesel drives.

Soon, an operating principle with bladed wheels was registered and patented under number 558445. Its special feature was that individual or all of the bladed wheels can swap roles during operation or at standstill.

At Voith, this resulted in "lengthy debate and meetings, not least with Heidenheim, because a good test laboratory was imperative for the development of such a machine", as Walther Voith, the boss, described it. He had good reason to be hesitant. Due to its small dimensions, the new product was unlikely to make a significant profit to cover the costs of the huge processing machines. Things had been completely different with the Kaplan turbine and the Voith Schneider Propeller, both of which received his spontaneous backing when these products were proposed for integration into the product portfolio.

Yet the hydrodynamic transmissions definitely fulfilled the requirements for adding a new product to the sales program. While the individual production in the turbine and paper machine business had always been characterized by periodic and sometimes drastically fluctuating order intakes, the production of the new transmissions promised a more steady business. Therefore, as Walther Voith remembers, "we finally decided to get into this field". This decision was the logical consequence of considerations that Walther Voith had entertained as early as 1918. A significant item on the agenda at the "new orientation" envisaged at the time was the appeal to the employees to look for a "high-quality mass product".

The Economic Situation
The extremely difficult situation did not allow for any further discussions. Like the rest of the country, Voith suffered strongly under the worldwide economic crisis. In both locations, sales had dropped to less than half of the previous year, and the order intake stagnated at a greatly reduced level. Taking up a new product was absolutely paramount.

Walther Voith, head of both Voith plants, took to this task with enormous energy. He focused on the Heidenheim factory with its modern turbine test laboratory, the "Brunnenmühle", which he regarded as absolutely essential for the development of the new product.

Additionally, the financial situation in Heidenheim was much better. The hydro department achieved considerably higher sales figures, in 1932 its share in the total Voith result was visibly on the rise. At the same time, the company was considering relocating the future development of the Voith Schneider Propeller to Heidenheim – which was eventually put into practice. Similar deliberations led to the technical responsibility for the Föttinger units being delegated from the very beginning to Heidenheim rather than St. Pölten.

Orders received in tons of total output (measuring unit at the time).

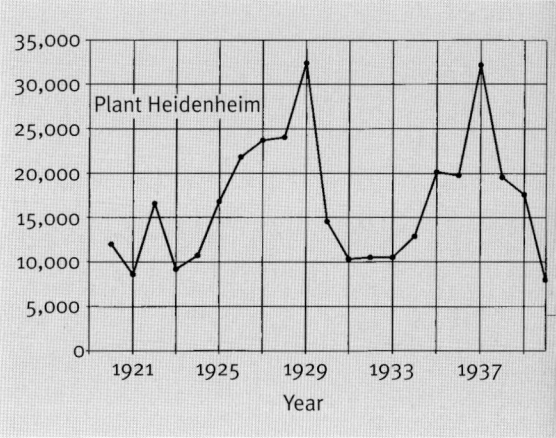

Product Integration at Heidenheim

An important technical reason for the entrepreneurial move of introducing this new product was also the fact that it was a fluid machine, the hydraulic principle of which was quite familiar to the turbine department. At the Heidenheim turbine test laboratory, it could now be further developed on the basis of in-house experience.

Reassured that Voith was going to build a hydrodynamic transmission for his railbus drive, Hacker succeeded in winning the previously somewhat hesitant Austrian Federal Railways' confidence in the project. The railway company was, however, still not sure whether the hydrodynamic transmission presented a better solution compared to the proven mechanical unit. Austrian Federal Railways were not prepared to accept an interruption to their operating routines should the hydrodynamic transmission fail. Austro-Daimler therefore had to commit themselves to building an identical second railbus with a mechanical transmission that would be ready for operation if required. The order with the code "Austrotrieb" (Austro Drive) was placed in 1932. In early 1933, the vehicles entered service. The hydrodynamic version immediately proved itself. In his speech on the 25th anniversary of Voith turbo transmissions and turbo couplings in 1959, Hacker was able to "report with great satisfaction that recourse to the mechanical version had never been necessary". Also in 1933, Voith received its first series order under the code "Längstrieb" (longitudinal drive) for 24 turbo transmissions each rated at 80 HP for Austrian Federal Railways and other railway companies.

The first hydrodynamic rail transmission from Voith was therefore an instant success. Its principle of filling and draining in order to switch from converter to coupling would become the key characteristic of the Voith turbo transmission until today. Also in 1932, in anticipation of the significance of hydrodynamic couplings and transmissions in the Heidenheim factory, the hydro turbine division established a separate department "TG" that was to focus exclusively on these products.

Licensing Agreement with Harold Sinclair

In 1934, the TG department also took up the development and production of hydrodynamic couplings. For this purpose, it had signed a licensing agreement with the Englishman Harold Sinclair. This agreement was followed by decades of friendly cooperation.

Sinclair, at the time still working for the British company Vickers, had already shown an interest in the torsional vibration-damping Vulcan coupling and therefore visited the Hamburg shipyard in 1926. The visit led to a licensing agreement between Vickers and Vulcan. Sinclair recognized that the coupling offered numerous other application possibilities. He became a pioneer in this field, after being granted the rights in the Vickers licenses for hydrodynamic couplings and establishing the company "Hydraulic Coupling Patents Ltd." (H.C.P) near London in 1928. When business with the "Vulcan Sinclair Couplings" expanded, Sinclair founded the production company "Hydraulic Coupling & Engineering Company Ltd." which took the development of the product further and operated a large test stand. Later, the company changed its name to Fluidrive Engineering Co. Ltd. (FE).

Harold Sinclair

On January 26, 1934, Voith began the production of Vulcan-Sinclair couplings following a licensing agreement with H.C.P. They had been marketed under the brand "Voith Sinclair couplings". Despite its earlier involvement with the product, Voith regards this date as the starting point of its strong and successful development of hydrodynamic couplings. Voith, like other licensees, initially built the couplings with minor design modifications, but from 1947 gradually replaced them by their own models incorporating new functions. This led to a new licensing agreement in 1956, according to which Voith now also granted licenses to Sinclair. The storage chamber principle that had become very popular within the German mining industry was a prominent feature of this agreement.

The agreement expired in 1964, which did not, however, affect the friendly relations between the two companies. In 1982, Harold Sinclair passed away aged eighty-four. In 2000, Voith eventually took over Fluidrive, including the licensing products of the French partner SIME.

Overview of Development

Hydrodynamic couplings, converters and the various transmission types deriving from them were not taken up by Voith all at the same time. They were the consequence of market demands that influenced the company at differing periods. This step-by-step expansion of the new product area of hydrodynamic power transmission spanned a period of more than twenty-five years. Hydrodynamics at Voith emerged from six different starting points. The overview shows which elementary Föttinger units were developed at Voith in which year, their origins and their continuity, to each other.

The earliest starting point (1) is the Föttinger patent itself. In the wake of the order for Herdecke power station, its application led to the development of the hydrodynamic Föttinger coupling for starting and synchronizing the storage pumps. Thirty years later, similar reasons led to the development of the synchronous converter. Unlike the coupling, the converter is able to control the synchronous speed accurately using hydrodynamics and couples the storage pump mechanically without wear. This converter is a relatively rare, heavy engineering product. It was part of the delivery program of the hydro turbine department from the beginning and is still used today in pumped storage plants with high heads.

A subsequent starting point (2) was the decision of the three proprietors Walther, Hermann and Hanns Voith to add "turbo transmissions" to the company's product portfolio. Small in size and suitable for large-scale production, it was meant to alleviate the negative effects of the cyclically fluctuating plant engineering business. It gave way to a number of products for a wide range of applications, including rail transmissions, industrial converters and retarders.

Two years later, in 1934, Harold Sinclair acquired a license for developing and manufacturing constant-fill start-up couplings (T) and indirectly adjustable variable-speed couplings (S). This resulted in a third starting point (3). In 1942, the cooperation with Junkers led to the fourth starting point (4), the directly adjustable variable-speed coupling (SV). Both starting points formed the basis for industrial applications.

The DIWAbus transmission with a synchronous converter developed in the early post-war years, became the important fifth starting point (5) for the bus transmissions of the following decades. Finally, a sixth initial point (6) was the acquisition of the product area ILOmatic of ILO in Pinneberg creating the foundation of the new DIWAmatic series.

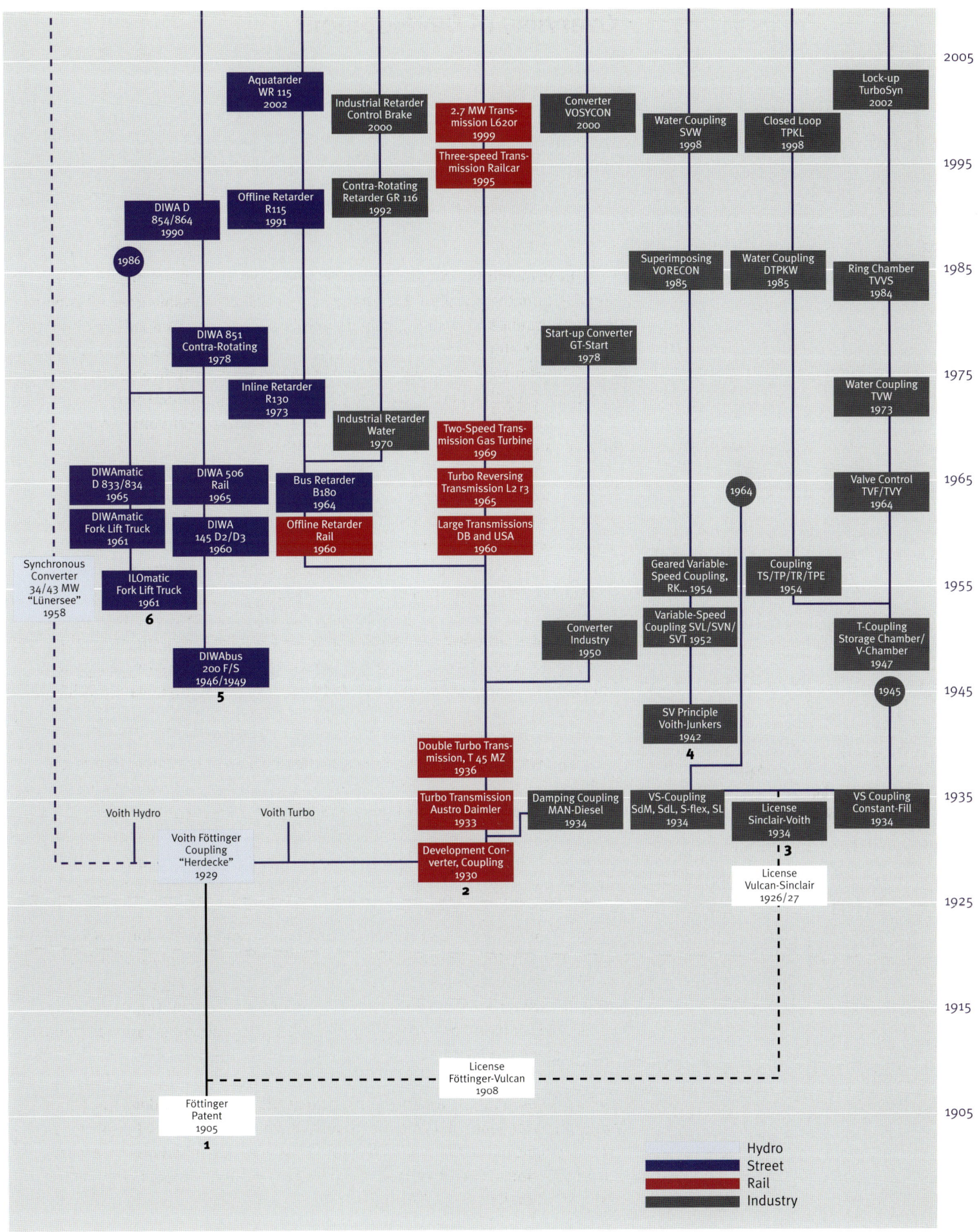

2005

Aquatarder
WR 115
2002

Industrial Retarder
Control Brake
2000

2.7 MW Trans-
mission L62or
1999

Converter
VOSYCON
2000

Water Coupling
SVW
1998

Closed Loop
TPKL
1998

Lock-up
TurboSyn
2002

1995

Three-speed Trans-
mission Railcar
1995

DIWA D
854/864
1990

Offline Retarder
R115
1991

Contra-Rotating
Retarder GR 116
1992

1985

1986

Superimposing
VORECON
1985

Water Coupling
DTPKW
1985

Ring Chamber
TVVS
1984

DIWA 851
Contra-Rotating
1978

Start-up Converter
GT-Start
1978

1975

Inline Retarder
R130
1973

Water Coupling
TVW
1973

Industrial Retarder
Water
1970

Two-Speed Trans-
mission Gas Turbine
1969

1965

DIWAmatic
D 833/834
1965

DIWA 506
Rail
1965

Bus Retarder
B180
1964

Turbo Reversing
Transmission L2 r3
1965

1964

Valve Control
TVF/TVY
1964

DIWAmatic
Fork Lift Truck
1961

DIWA
145 D2/D3
1960

Offline Retarder
Rail
1960

Large Transmissions
DB and USA
1960

Synchronous
Converter
34/43 MW
"Lünersee"
1958

Geared Variable-
Speed Coupling,
RK... 1954

Coupling
TS/TP/TR/TPE
1954

1955

ILOmatic
Fork Lift Truck
1961

6

Variable-Speed
Coupling SVL/SVN/
SVT 1952

Converter
Industry
1950

T-Coupling
Storage Chamber/
V-Chamber
1947

DIWAbus
200 F/S
1946/1949

5

1945

1945

SV Principle
Voith-Junkers
1942

4

Double Turbo Trans-
mission, T 45 MZ
1936

Voith Hydro

Voith Turbo

Turbo Transmission
Austro Daimler
1933

Damping Coupling
MAN-Diesel
1934

VS-Coupling
SdM, SdL, S-flex, SL
1934

License
Sinclair-Voith
1934

VS Coupling
Constant-Fill
1934

1935

Voith Föttinger
Coupling
"Herdecke"
1929

Development Con-
verter, Coupling
1930

2

3

License
Vulcan-Sinclair
1926/27

1925

1915

License
Föttinger-Vulcan
1908

1905

Föttinger
Patent
1905

1

Hydro
Street
Rail
Industry

Wolfgang Paetzold

On the Rails of the World

Voith Turbo Transmissions in Railway Technology

The Pioneering Years 1932 – 1944

As the three large Herdecke couplings from 1929 were predominantly based on the design principles of Hermann Föttinger's marine drives, Voith pursued a totally different path regarding its vehicle drives; after all, the company had decades of experience in the hydraulic development of water turbines and centrifugal pumps. The office developing the turbine wheels calculated and designed the blade profiles, while at the Brunnenmühle test laboratory the hydrodynamic characteristics were examined. From the beginning of the thirties, smaller torque converters and turbo couplings were developed that promised to be successful as transmitting elements for vehicle drives, especially in rail vehicles. In the twenties, the newly established Deutsche Reichsbahn-Gesellschaft

The ABL 2.8e1 turbo transmission of the first series with its main components. Code word "Austrotrieb". March 1933

had, for economical reasons, began to convert its steam trains to diesel engine railcars on routes with low passengers levels. Initially, these vehicles had only mechanical gearboxes combined with reversing final drive gearboxes – for rail traffic this solution was only suitable to a limited degree. This was an incentive for Voith to develop a superior transmission system.

The impetus for the construction of an 80 HP turbo transmission came from Oscar H. Hacker in the middle of 1932. At the time, Hacker managed the vehicle construction department at Austro-Daimler in Vienna. He did not want pure mechanical power transmission for his new light railcars but something completely new. He contacted Voith in St. Pölten, where Walther Voith decided to hand over this interesting task to the water turbine department in Heidenheim. Under the auspices of Wilhelm Hahn and Ernst Seibold, a turbo transmission with a converter-coupling design was built. The first test run took place at the Brunnenmühle on 7 December 1932.

The new turbo transmission had several remarkable characteristics:
- Simple design
- Low weight, suitable for flanging to the engine
- Smooth, wear-free acceleration of the vehicle
- Changing from converter to the coupling using Föttinger's filling and draining principle
- Coasting effect while circuits are drained
- Unlike Föttinger, mineral oil replaced water as the operating medium.

The first ten ABL2.8e1 test transmissions were delivered to Austro-Daimler free of charge in early 1933. Six transmissions were installed in three, two-axle, railbuses with two power units each of 80 HP. The four remaining transmissions went into all-terrain armored reconnaissance vehicles of the Austrian Federal Army. By the end of 1933, Austro-Daimler had taken delivery of another 37 transmissions.

The Voith transmissions met the high expectations of Austro-Daimler and Austrian Federal Railways, due to the fact that they offered numerous advantages when compared with the mechanical gearboxes. Moreover, the innovative high-speed railcars received much attention at home and abroad, and resulted in a series of new orders for Voith. As installation conditions and power requirements varied significantly from order to order, Voith was consequently forced to develop ever more transmission types at considerable development cost.

Demonstration run of the first Austro-Daimler railbus with Voith turbo transmission in Innsbruck. March 1933

This photograph was sent to Ernst Seibold by Director Hacker with the inscription: "In memory of one of the definitive turning points in transport technology".

During the period up to the Second World War, a defining period for the newly arrived technology of diesel-hydraulics, Voith with its Turbo Transmissions had to assert itself against competitive transmission systems. It managed to do so, because turbo transmissions were more reliable than mechanical drives; and they were lighter and less costly than diesel electrics.

Transmission efficiency, heat dissipation, starting torque and adaptation to differing engine speeds were gradually improved, and the manual speed-change system was replaced by an automatic one.

The technical imperatives for the success of the turbo transmissions were primarily created by Wilhelm Hahn the manager of the water turbine department, his expanding team, as well as his successor Hans Faic Canaan and the divisional managers Ernst Seibold and Fritz Kugel. In the early forties – during the war – the young Wilhelm Gsching succeeded with the development of a single-stage torque converter with higher efficiencies than earlier models with cast blades.

At the helm, chief executive Walther Voith ensured that the required development funds were available even in critical phases, as he had an unassailable faith in the future of the new department "Turbo transmissions/Turbo couplings TG/VSK" established in 1934.

From the many pre-war constructions, this chapter will take a closer look at three particular developments, for they opened up new areas of application.

By 1935, the era of the single-shaft transmission, converter-coupling design without step-up gear had come to an end. The rated speeds of diesel engines being only half that of petrol engines definitely required a step-up gear. Without the step-up gear, the circuit diameter would have had to be uneconomically large and filling times too long. The engine outputs that had risen to 400 – 600 HP required two wheel sets to be driven instead of the previous single set. In co-operation with Maybach-Motoren-Werke in Friedrichshafen, a power bogie incorporating engine, transmission and final drives was developed for railcars. The cooling system remained within the railcar frame.

For the Voith-Maybach system, Voith developed a three-speed converter transmission with two rotors, each one driving one wheel set within the traction bogie. Thus avoiding coupling of the wheel sets and the inevitable locked-in torque and additional stresses that occurs with different wheel diameters.

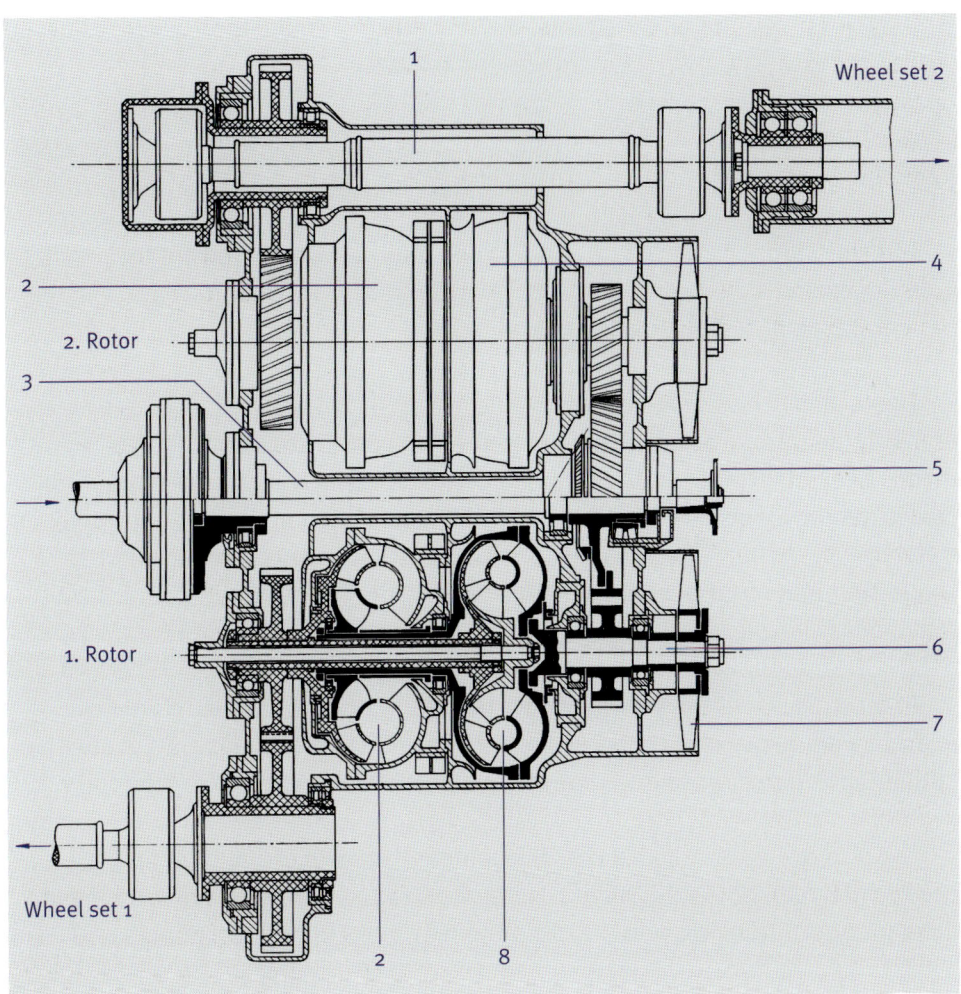

Wheel set 2

1

2. Rotor

3

1. Rotor

2   8

Wheel set 1

4

5

6

7

*T 45 MZ twin turbo trans-
mission with four torque converters for hydrodynamically
decoupled driving of two wheel
sets. Built in 1936*

*From 1937 the cooling of the
housing by two axial fans was
replaced by direct oil cooling
via heat exchangers.*

1   *Output shaft*
2   *Starting converter, speed 1*
3   *Main drive shaft*
4   *Cruising converter, speed 3*
5   *Fan drive for water cooler*
6   *Primary shaft*
7   *Fan for transmission cooling*
8   *Cruising converter, speed 2*

*Tractive effort diagram of
Deutsche Reichsbahn's high-
speed three-car unit fitted with
two T 45 MZ transmissions for
express traffic in the Ruhr
region. In speed 1, four wheel
sets are driven. Wheel set load
15 t each.
W = Train resistance.*

In speed range I, both starting converters were filled where they were subjected to only half the engine output and drove both wheel-sets. In speed range
II, the cruising converter of the first rotor drove wheel set 1, while in speed range
III the cruising converter of the second rotor drove wheel-set number 2, in each
case at full power. For a three-car DMU rated at 2 x 410 HP and a maximum
speed of 120 km/h, this resulted in the tractive effort diagram shown.

In total, 173 twin turbo transmissions were delivered to the State Railways of
Germany, Austria and Belgium.

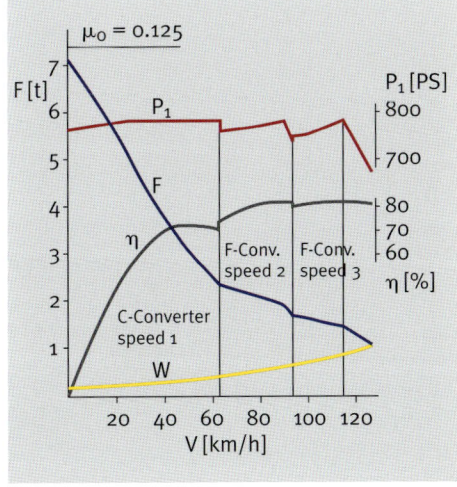

Deutsche Reichsbahn's three-car DMU VT 137 283, built in 1939 with two T 45 MZ twin turbo transmissions. Code word: "Ruhrwagen".

Of all the turbo transmissions of the pre-war years the technical highlight was the JJG 251 turbo transmission from 1935 with its converter-coupling-coupling configuration. It was intended for a 1 400 HP output, 1C1 mainline locomotive of Deutsche Reichsbahn-Gesellschaft and delivered by Krauss-Maffei within an incredible eight months. Delivery times for Voith were equally short – just six months from receipt of order.

As a result of the large torque converter diameter of 750 mm, the transmission with its mechanical reversing section had to be bolted rigidly to the two longitudinal frames and hence constituted part of the locomotive frame.

The V 16 01 (later V 140 01) diesel locomotive, shown for the first time at the German transport exhibition in Nuremberg in 1935, proved that Voith hydrodynamic power transmission was equally well suited for higher outputs.

The large JJG 251 transmission remained a one-off. In contrast, the smaller locomotive transmissions for 100 to 500 HP were both a great technical and commercial success. These incorporated a starting converter and two turbo

The world's first successful high-power locomotive after completion by Krauss-Maffei in Munich-Allach. July 1935

Assembly of the new high-power transmission at the Heidenheim workshop. Code word: "Krausslok". June 1935

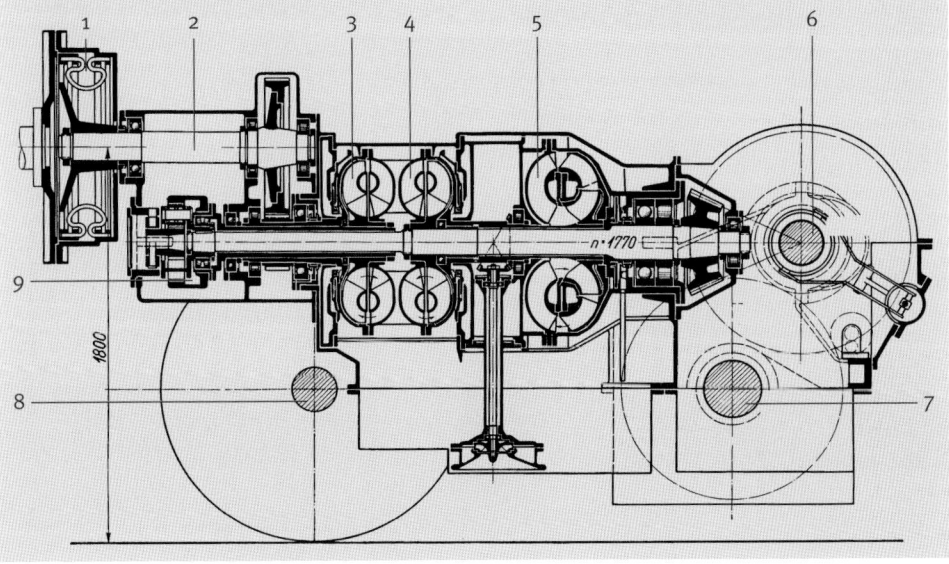

JJG 251 with step-up gear and jack shaft reversing gearbox.

1 Voith-Maurer coupling
2 Input shaft $n = 700 \ min^{-1}$
3 Coupling I
4 Coupling II
5 Converter
6 Reversing shaft
7 Jack shaft $n = 380 \ min^{-1}$
8 Driving axle $V_x = 100 \ km/h$
9 BHS-Stoeckicht gearbox
  1.41:1

couplings with mechanically staggered ratios. Step-up gear, high-speed rotor group and output gear reduction resulted in the classical three-shaft transmission with a flanged mechanical jackshaft gearbox containing the reversing section.

This concept proved to be extraordinarily successful. Thanks to their simple, robust construction, and reliability, these locomotive turbo transmissions paved the way for a successful post war renaissance.

Longitudinal section of the
L37U built for the WR 360 C 14
army locomotive in high
numbers.
With torsionally flexible
coupling 1 (Voith-Maurer),
rigid drive shaft 2 and
Voith pin coupling 3.
Built in 1939

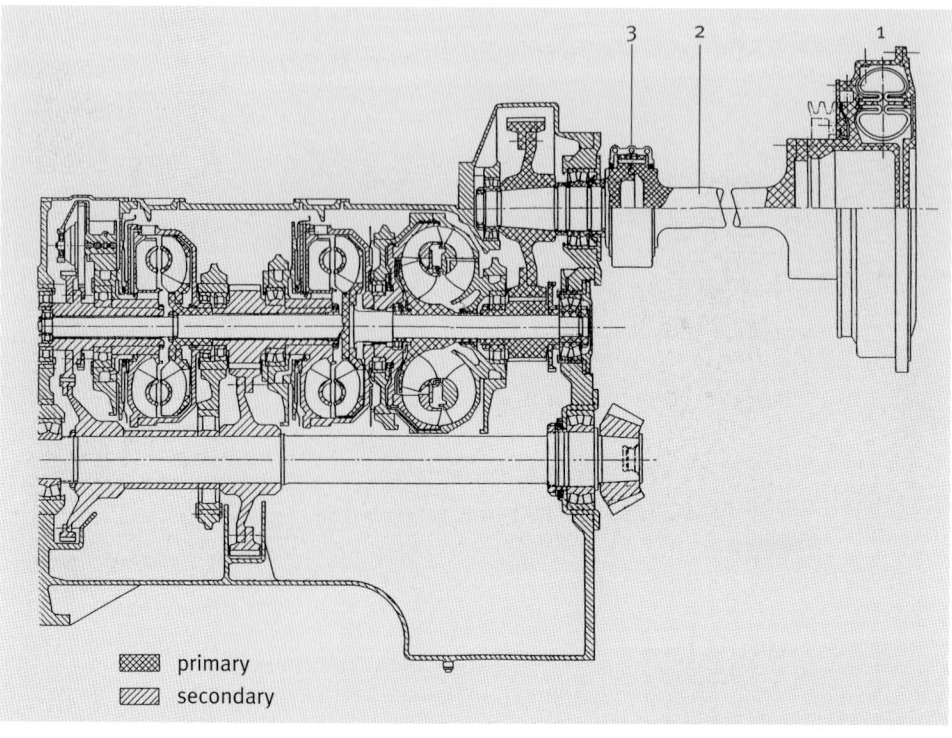

primary

secondary

Successful locomotive transmissions from the pioneering
years. From 1937 until well
into the war these were built
in large numbers.
L 37 (500 HP)
L 33y (200 HP)
L 33 (110 HP)
L 22 (130 HP)

HR 360 C 12 prototype loco-
motive built by Orenstein &
Koppel, Babelsberg.
May 1937

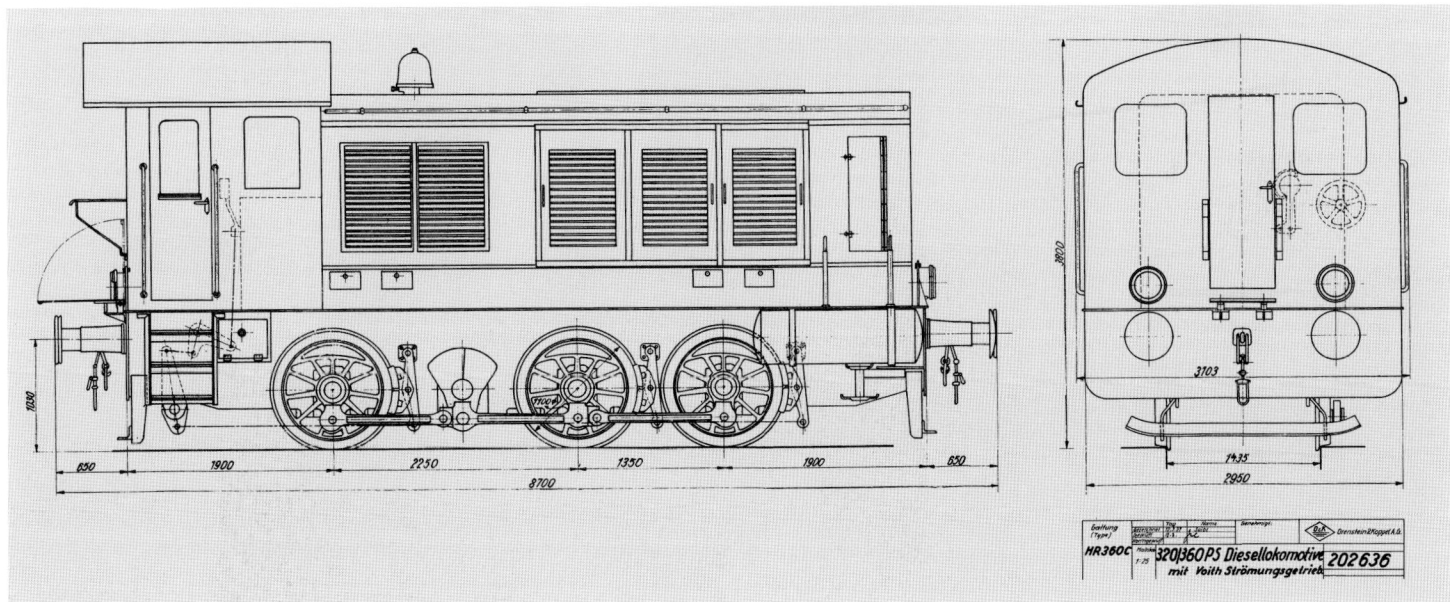

Hydrodynamic Developments for Voith Turbo Transmissions

From their early beginnings, the extraordinary success of Voith turbo transmissions for rail vehicles is due to a considerable degree to the ongoing evolu-
tion of hydrodynamic circuit design at Voith. The individual developments of
appropriate torque converters and turbo couplings should therefore be given a
closer look.

Turbo Couplings

Initially, the small turbo couplings were largely influenced by their "big sisters". The latter had a core ring that was regarded as indispensable for guiding the flow. Fritz Kugel's first coupling for the first turbo transmission, the type "A", was therefore a Herdecke coupling with a reduced symmetrical profile and a profile diameter of 292 mm.

The A-coupling and the similarly designed E-coupling were filled via the hub and were drained via a series of outlet bores at their circumference. This resulted in a sluggish behavior during change-over. For some transmission types, the subsequent I-coupling was therefore fitted with quick-emptying valves. These were improved in the O-coupling that had a more rounded cross-section profile due to an enlarged inner diameter. After the war, type "U" was launched with the

Turbo coupling type "A". The secondary wheel 2 is enclosed by a housing 1 that rotates with the primary wheel 3.

new Chrysler profile and no core ring and for the first time installed into a mining locomotive in 1952. This profile is still used in current turbo transmissions.

As turbo couplings for rail vehicles are operated only in a narrow slip range of 2 to 8%, their torque characteristic plays only a secondary role. From a maximum value at 100% slip it decreases to its nominal value of 3% slip.

Torque Converters

Turbo couplings with their radial blades have a simple design. For applications in turbo transmissions it was sufficient to focus on improving the filling and draining process. Hydrodynamic torque converters allow an almost endless variety of blade designs to meet individual requirements. It is therefore no

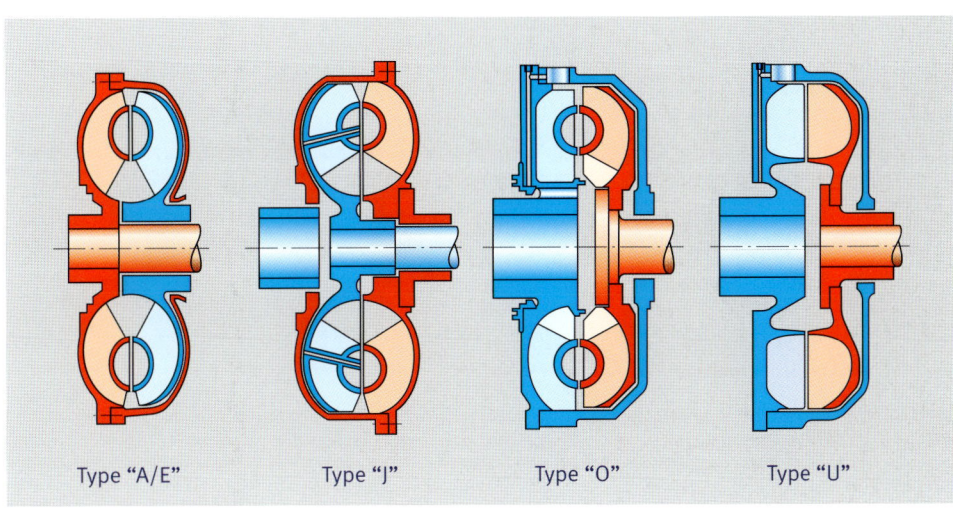

Type "A/E"   Type "J"   Type "O"   Type "U"

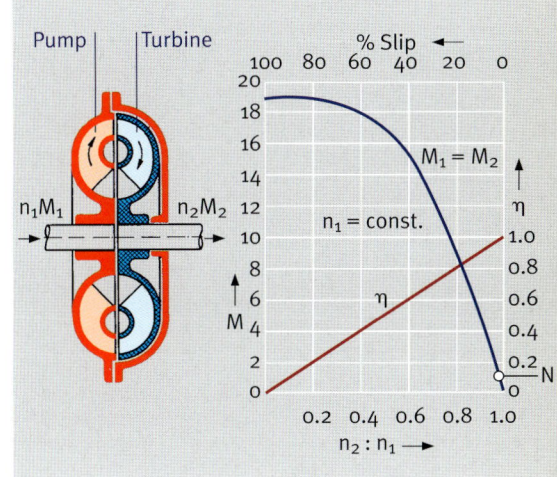

surprise that, next to transmission-specific research, torque converters took up the major share of the hydraulic developments at Voith.

As the flow processes in a torque converter are highly complex and require high calculatory efforts to be understood, the new torque converters required for turbo transmissions had to be designed empirically using theoretical considerations and experiences from water turbine construction. The designs were proven on the test bed.

Since the development of the two-stage B-converter and the single-stage C-converter in 1932, Voith had accumulated a wealth of experience over the years. The company's comprehensive development efforts resulted in the best and most efficient single-stage converters with the highest power density available for installation in turbo transmissions.

In the thirties, Voith torque converters were still largely based on the designs used in water turbine construction. They had cast bladed wheels with spatially curved blades that had to undergo a time-consuming manual finishing process.

While the first turbo transmissions were two-speed units with a converter and a coupling, soon three-speed transmissions with a type "D" starting converter and two turbo couplings were used for locomotive transmissions. For railcars the twin turbine transmissions were also available as genuine converter transmissions with a starting converter and type "F" cruising converters. The latter were derived from a turbo coupling. Fast-running and with high efficiency,

The development of the hydrodynamic coupling. From the original Föttinger design (A) to the Chrysler profile (U).

Turbo coupling with symmetrical profile and its torque/slip curve. N = Rating point with 3% slip.

Voith converter test bed in the thirties.

its engine speed lug-down capacities were "inherited" – albeit only to a certain degree – from the coupling.

The pre-war development was thus completed. In the early forties, Wilhelm Gsching developed new torque converters. The starting point was based on studies by Lysholm who had developed a new type of hydrodynamic torque converter at the Swedish steam turbine manufacturer Ljungström. He used wheels normally used in steam turbines that had essentially cylindrical blades. His three-stage converter with six blade rings had a high specific power absorption and achieved a flat efficiency curve owing to its ability to lug down the engine.

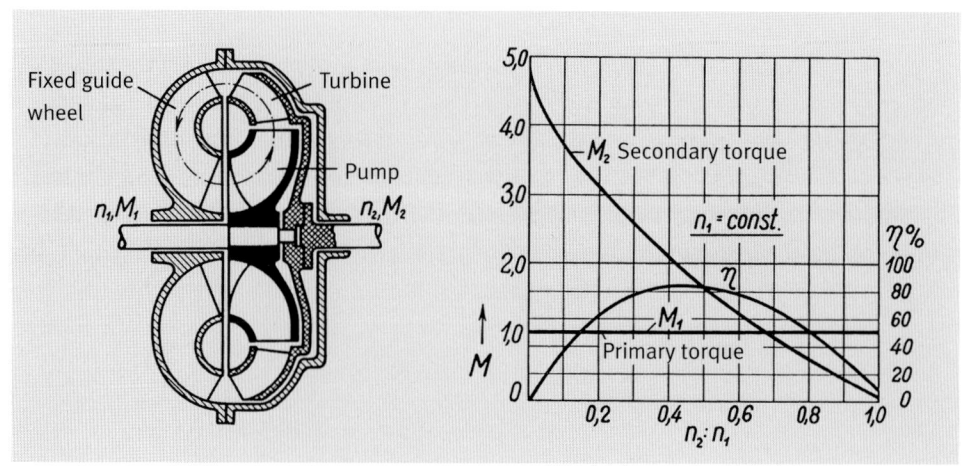

66

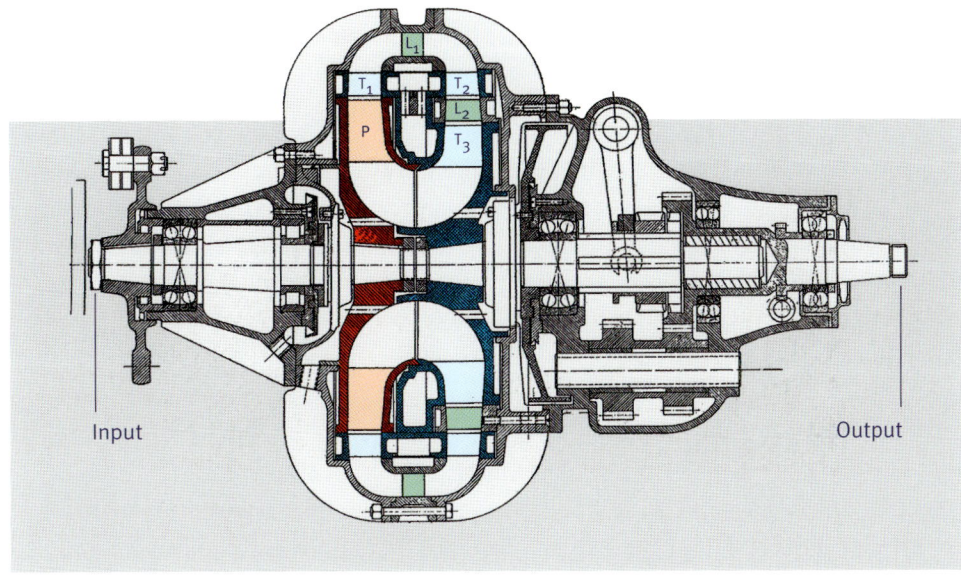

Input

Output

L   P   T

It was noted that fast-running cruising converters offered higher efficiencies. In 1940/41, Wilhelm Gsching developed two converter types that had cylindrical blades for the turbine wheel and the stator extruded from profile bars. These blades were riveted into cover discs. Compared to previous converters with cast blades, the significant increase in the number of blades and the enhanced surface quality greatly improved efficiency. The permissible circumferential speed of the turbine wheels could be increased, making the converter smaller and lighter. This was the basis for converter transmissions with a wide speed range

In 1942 the new cruising converters were used, for the first time, in a T 24 turbo transmission. This transmission had an integrated reversing section. In view of the transmission rating of 630 HP, their profile diameter of 384 mm was rather small, resulting in a pump speed of 4 130 min^{-1}. In view of the then prevailing standards, this resulted in a very compact transmission, weighing only 1 150 kg including two lateral outputs and a Silumin housing.

Additionally, the "D" type starting converter was developed during the war, with the intention of replacing the cast bladed wheels. In 1946/47, Rolf Keller, at the time head of hydraulic development, introduced an improved type that became the standard starting converter for all turbo transmissions of the post-war generation. It marked the beginning of a long development of continually improving starting converters.

The Realization of an Idea –
Post-War Developments 1945 – 1962

The Voith plant in Heidenheim survived the war unscathed, but the Russians expropriated Walther Voith's works in St. Pölten. Germany with its infrastructure in ruins was shattered, and it was impossible to build locomotives and railcars – consequently there was no demand for turbo transmissions or turbo couplings. Voith in Heidenheim had to concentrate on the elimination of war damage. In 1945, the company restored blown up bridges, and in 1946 it repaired damaged steam locomotives of the Deutsche Reichsbahn.

It was only after the arrival of the currency reform in July 1948 that the situation for the transmission department improved. For locomotive transmissions, the company had a solid foundation particularly with its pre-war transmissions for smaller diesel locomotives. These were the first vehicles to be in demand after the war. Industrial railways, too, gradually abandoned steam and converted to diesel traction.

For railcar transmissions, the situation was different. Here, Deutsche Bundesbahn (DB) – newly established in October 1949 – not only needed turbo transmissions and final drives for the rehabilitation of damaged pre-war vehicles, but also for the construction of new more powerful diesel multiple units. This required the development of new diesel engines and new transmissions in the 800/1 000 HP power range.

Repair of damaged locomotives in the Great Turbine Hall of Voith in 1946.

Due to its pre-war experiences, DB had decided in favor of diesel-hydraulics with high-speed engines rather than diesel-electrics with medium-speed engines. The new drive system offered advantages in terms of weight, volume and price and excelled in its superior acceleration behavior.

This was a far-reaching decision, providing an enormous incentive for the future development of transmission design at Voith under the management of Fritz Kugel. As a result, the manufacture of rail vehicles for Germany and the export market continued to be oriented to diesel-hydraulics for decades.

Above all the wishes of DB had priority. For the new 1 000 HP drive systems intended for both locomotives and railcars, Voith had to develop a completely new three-converter transmission. As an alternative to the hydro-mechanical Mekydro transmission of Maybach it was intended for installation into the power bogies of diesel railcars.

DB insisted on two independent transmission suppliers. This started a controversy about which was the better transmission concept – full hydraulic or hydro-mechanical – only to end many years later in favor of the Voith principle.

The design of the new T 36 r turned out to be a difficult task, as the installation space was prescribed the Mekydro transmissions. Four-speed, hydro-mechanical transmissions can be designed much more compactly than hydro-dynamic units with three circuits and a large oil sump. Additionally, the high transmission power of 1 000 HP and moderately sized converter diameters dictated very high rotor speeds. Finally the reversing section, two primary outputs for auxiliaries and an optional lateral, or lower, output had to be accommodated.

In view of the request by DB for a low-weight unit based on the confines of the Mekydro transmission, Voith retained the light "Silumin" housing that had proven itself in pre-war railcars. This led to a light transmission weighing only 2 300 kg including heat exchangers fitted on top.

Operating experience soon revealed that the service life of the T 36 r was insufficient. Although it was improved in two rebuild campaigns, it remained low. The anticipation for 500 000 km operation without the need for removal could not be achieved.

The two halves of the turbine wheel of the starting converter were milled out of two steel rings and riveted together to form a single-piece. May 1951

Scheme of T 36r turbo transmission for DB with starting converter I and two cruising converters II/III.

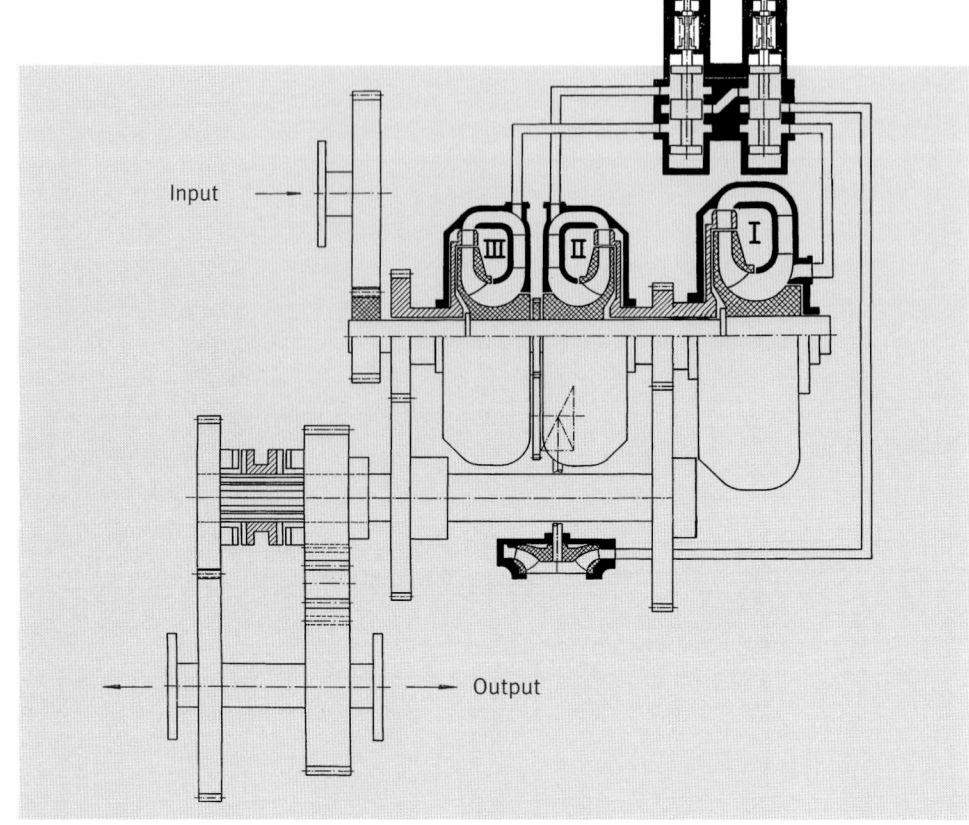

Input

Output

LT 306 r transmission for DB in the version for the V200 with two lower output flanges. Built in 1955

DB continued to increase their pressure on Voith. Eventually, the management urged Fritz Kugel to get Ernst Seibold to return to Voith. In the middle of 1953, Seibold rejoined Voith as chief engineer and head of the turbo transmission design department, having left the company in 1937. His first and foremost task was the development and design of a reliable successor transmission, in order to restore Voith's somewhat flawed reputation within DB.

In 1954, the new LT 306 r was developed under his auspices. Offering a higher starting tractive effort by means of an improved starting converter, it had a housing made from gray-cast iron, a reversing section with coupling gears mounted in the housing and a new control system with a centrifugal governor instead of measuring pumps. Even if its weight of 3 300 kg meant that it had become heavier, it represented an enormous step forward and proved itself from the first day. Its service life now complied with the expectations of DB; in VT 08 diesel multiple units, distances of 500 000 kms were achieved without the need for removal.

But this was not the end of the 1 000 HP transmission development. At the end of the fifties, the V 200 diesel locomotives no longer had to haul light high-

speed trains with a maximum trailing load of 200 t, but interzonal trains with trailing loads of up to 600 t. This resulted in a very different load-cycle with a high proportion of full-load operation and intense full load starts. Ernst Seibold redesigned the LT 306 r and developed the L 306 rb employing only one lower output in view of its exclusive use in locomotives. It had fewer, yet wider gears, improved converters and a reinforced filler pump.

The East German V 180 takes over the interzonal Cologne-Berlin service from the West German V 200 at Helmstedt railway station. Both locomotives are fitted with Voith L 306 rb turbo transmissions.

The L 306 rb marked the final development stage of the classic three-converter transmissions with secondary intermediate shafts. In the DB's V 200, with its lower output, the L306 rb gradually replaced the LT 306 r and was produced in high quantities for the East German V 180 at the Traismauer works of Voith St. Pölten. A total of 480 L 306 rb transmissions were built in Heidenheim and Traismauer.

L 37z Ub for DB Shunting Locomotives

In 1963 with the L 306 rb we are actually ahead of our time. The bread-and-butter transmissions of the post-war years were the L 37 and L 37 y locomotive transmissions developed by Dieter Gößler in the thirties. These were systematically adapted to the prevailing state of technology, without changing the proven

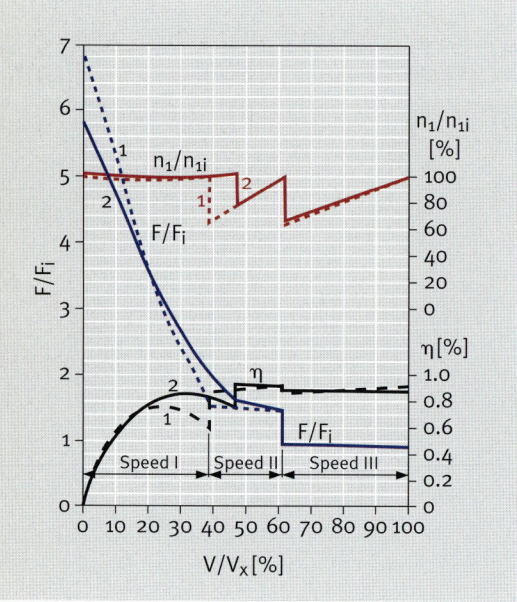

Comparison of tractive effort and transmission efficiency for the improved L 37 transmission compared to the pre-war design

1   *Original transmission from 1937*
2   *Modernized transmission from 1953*
F/F_i *Tractive effort*
n_1/n_{1i} *Engine speed*
η *Transmission efficiency*
V/V_x *Vehicle speed*

All values dimensionless.

The standard V 60 shunting locomotives of DB were series-fitted with the Voith L 37z Ub.

basic concept. Focal points for improvement were the shaft bearings and, most of all, the introduction of oil press fits for the shaft-hub connections. The converter blades were adapted to the latest state of technology, and the two couplings acquired Chrysler profiles without core rings.

In the course of its dieselization program, DB German Railways aimed for more economical shunting operations. This resulted in the purchase of numerous small locomotives, the "Kleinlokomotiven" Köf II, fitted with Voith turbo transmissions. What DB urgently needed was, however, a medium-heavy shunting locomotive for its hump yards.

As there had been good experiences with the V 36 locomotives, DB stayed with three-axle, rigid-frame locomotives with jackshaft and connecting rod drives. Whilst retaining the vehicle speed of 30/60 km/h, the power output was almost doubled by the installation of a high-speed diesel engine with a rated speed of $1\,400$ min^{-1}.

For the new V 60, the L 37 z, introduced in 1951, had to undergo further improvements. The L 37 zUb acquired a reinforced filler pump to shorten shifting times, a secondary lube pump for towing operation and an added variable-speed turbo coupling to drive the brake compressor at constant speed, independent of the engine speed.

When it became apparent in 1951 that there would be major orders from DB for V 60 transmissions, the board of Voith, headed by Hanns Voith and Hugo Rupf, decided to build a new transmission plant. The limited capacity in the old workshops would have been unsuitable for series production. Additionally, the early fifties saw a dramatic rise in demand by the German mining industry for constant fill turbo couplings of the Tv 1 type.

The future of hydrodynamic power transmission looked promising and justified the required investment. The opening of the transmission plant in May 1953, a factory incorporating the latest state of production technology and infrastructures, marked a new era for Voith power transmission. It had finally detached itself from the water turbine division. The new plant accommodated production lines for turbo transmissions, final drives, turbo couplings, torque converters and the new DIWAbus transmissions.

The L 37 zUb was flanged to a jackshaft two-speed reversing gearbox, produced according to the instructions of the V 60-project team by the locomotive manufacturers building the V 60. Combined, the two transmissions resulted in a massive unit weighing 7.5 t.

Between 1954 and 1964, the German locomotive industry supplied some 1 000, V 60 locomotives. The final 30 locomotives built by Krupp in Essen, were delivered to the Greek State Railways but with the L 27 zUb two-converter transmission fitted, an alternative to the L 37 zUb without turbo couplings.

Transmission unit of the V 60.

L 216 rs, the First Transmission of the New Generation for the V 100 of DB

The classic three-shaft transmission for locomotives was supplied to DB until the mid-sixties. As early as 1958, Fritz Kugel and Ernst Seibold had launched a new generation of high-performance transmissions for the new medium-heavy and heavy mainline locomotives of DB. They were fundamentally different from the LT 306 r, the largest transmission so far.

Fritz Kugel aimed at a radical simplification of the new transmissions. His idea was to keep the hydraulic section with the three hydrodynamic circuits and the oil sump including filler pump free from interfering gears, a concept that required the development of new torque converters. They had to be designed with different hydraulic ratios in order to drive one common secondary shaft. As a result, the intermediate shaft and four gears could be omitted. Additionally, high speeds of the empty starting converter's turbine was avoided in the high speed range of the transmission, allowing the rotor speed to be significantly increased without exceeding permissible circumferential speeds.

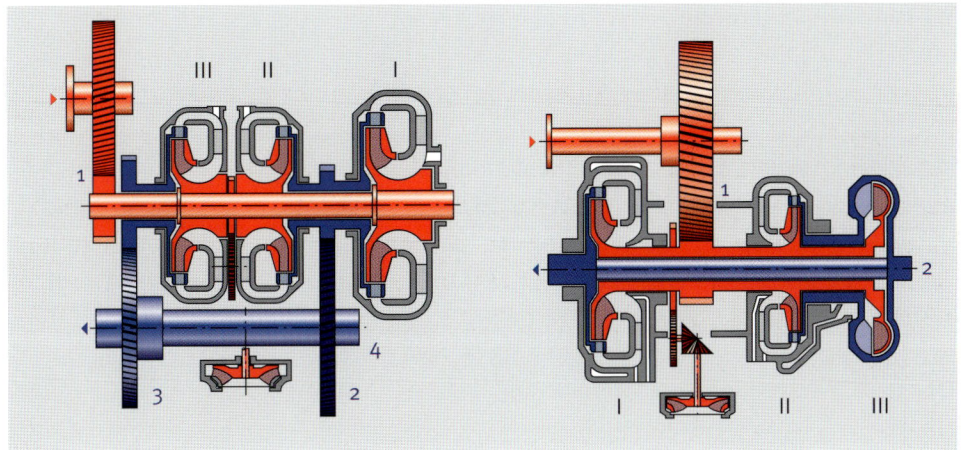

Different transmission concepts.
Left: LT 306 r – 1954
<u>With</u> secondary gears in the hydraulic section.
1 Step-up gear with common primary shaft
2 Slow speed range with converters I and II
3 High-speed range with converter III
4 Intermediate shaft
Right: L 216 rs – 1958
<u>Without</u> secondary gears in the hydraulic section.
1 Step-up gear with primary quill shaft
2 Common secondary shaft

With an output of 1 300 HP, the new L 216 rs was tailor-made for the V 100, DB's new locomotive class. To replace steam traction on branch lines DB needed a lighter locomotive with an output in excess of 1 000 HP. This requirement was initially to be covered by the V 80 from 1953, which did, however, not comply with DB's ideas of a simple and reliable diesel locomotive.

MaK in Kiel in co-operation with the "Central Office" of DB in Munich developed the new V 100; it was a milestone on the way to modern, diesel-hydraulic locomotives. It was the first locomotive in which the four wheel-sets were directly coupled together by cardan shafts. The range of high-speed 1 100 HP engines that had already proven themselves in the VT 08 and the V 200 was supplemented by a new 1 350 HP Daimler-Benz engine providing sufficient power even for medium-heavy goods trains.

Voith developed a new transmission concept for the V 100 branch line locomotive of Deutsche Bundesbahn rated at 1 100/1 350 HP and maximum speeds of 65/100 km/h.

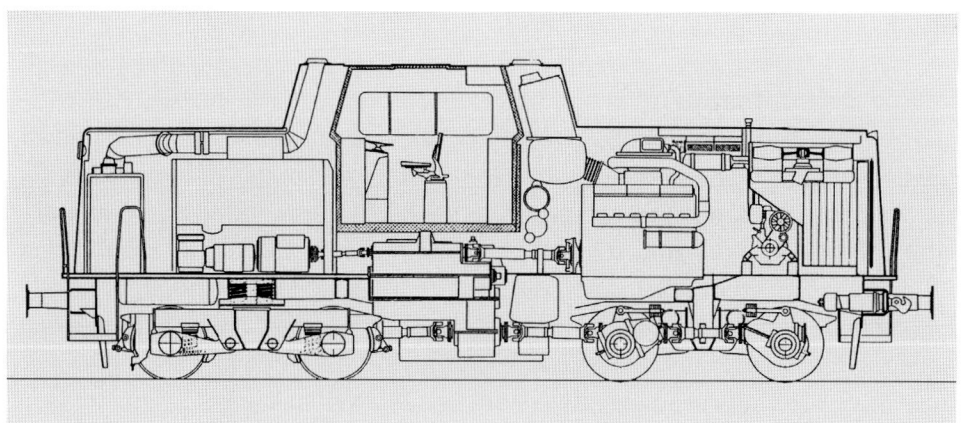

The L 216 rs three-speed transmission had a mechanical range-change gearing for two speed ranges. The third converter for the upper speed range was replaced by a turbo coupling having a higher efficiency during frequent part-load operation. Construction began in early 1956. Apart from the design of the transmission, an entirely new converter with new blading had to be developed as a hydrodynamic necessity for the simplified design of the new transmission generation.

The L 216 rs underwent a 100-hour test run with an input power of 1 500 HP on the Voith test bed, and was then tested at the DB Munich test laboratory in early 1959. After testing six prototype locomotives, series production of the 1 100 HP, V 100[10] and the 1 350 HP, V 100[20] began in 1961. By 1965 DB had purchased a total of 745, V 100 locomotives, from MaK, KHD and Jung.

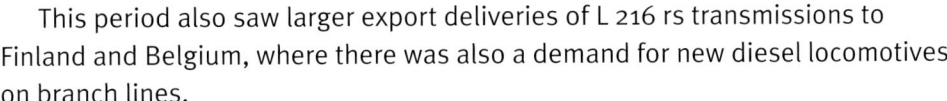

This period also saw larger export deliveries of L 216 rs transmissions to Finland and Belgium, where there was also a demand for new diesel locomotives on branch lines.

L 218 rs for the V 160 of DB

However DB required an even more powerful transmission for the new single-engine mainline locomotive designated V 160. This locomotive fitted with an engine rated at 1 900 HP output was required to have significantly reduced procurement and maintenance costs when compared with the double-engine V 200 first built in 1953. Krupp in Essen was put in charge of the development of this locomotive.

Voith had to develop an 1800 HP transmission based on the concept of the L 216 rs. Work on the L 218 rs transmission began in early 1958 at the "New Design" department. While the basic configuration was similar to that of the smaller L 216 rs, the profile diameters of the converters were increased from 504/460 mm to 573/527 mm. The weight rose from 4 500 kg to 6 300 kg including the transmission oil heat exchanger mounted on the top of the transmission.

Unlike the L 216 rs, which did not undergo any substantial modifications while it was being produced, the L 218 rs marked the beginning of a new series of transmission types. The six prototype locomotives built by Krupp were the start of a family of similar locomotives that found its powerful 2 800 HP conclusion in the Class 218 locomotive. These were fitted with the L 820 rs, two-converter transmission.

The new starting converter of the L 216 rs. October 1958
1 *Converter housing with milled guide blades*
2 *Pump*
3 *Turbine wheel, consisting of two rings riveted to each other.*
The first turbo transmission of the new generation was delivered at the beginning of August 1958.
Weight: 4 300 kg
Oil filling: 200 kg

Final assembly of the 1 800 HP transmission L 218 rs, early 1960.
1 *Starting converter*
2 *Cruising converter*
3 *Turbo coupling*
4 *Gear wheels of the reversing section*

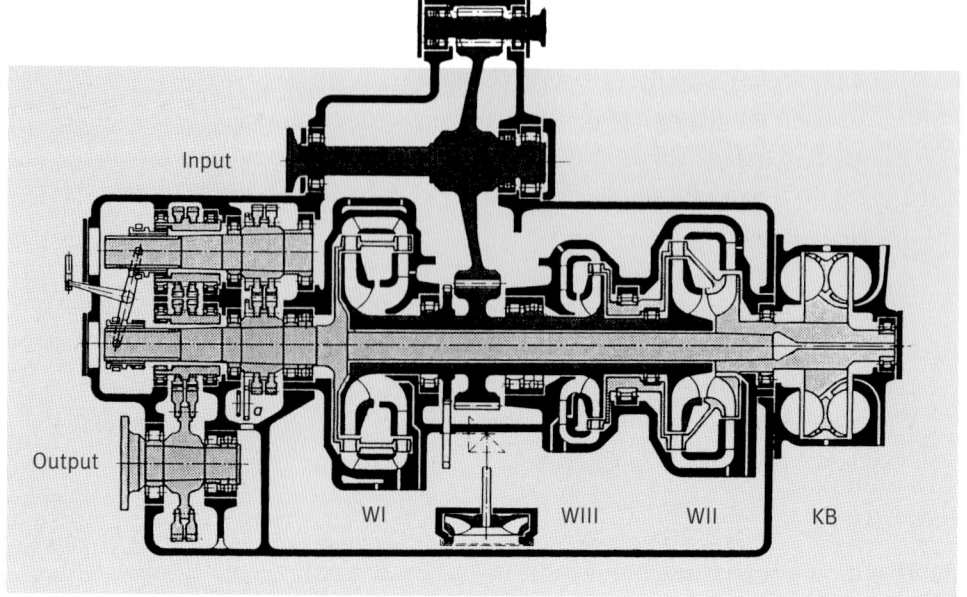

Photograph and transmission scheme of the large L 830 rU three-converter transmission with hydrodynamic brake. January 1961. Weight: 5 900 kg Oil filling: 360 kg

L 830 rU – the large transmission for the USA

The success of Deutsche Bundesbahn with its new diesel-hydraulic loco-motives did not go unnoticed among the large US railway companies. When Krauss-Maffei in Munich offered to double the output of a heavy six-axle diesel-electric North-American locomotive, Denver & Rio Grande Western Railroad and Southern Pacific Railroad (SP) took the opportunity. Each of them ordered three six-axle 4 000 HP locomotives with Voith supplying the hydrodynamic power transmission. This order from the USA in the early sixties was a sensation, because until then, North America had only diesel-electric locomotives with maximum outputs of 2 000 HP.

In line with the wishes of DB, the L 216 rs and L 218 rs transmissions with their two converters, a coupling and a range change gearbox had been chosen to meet the requirements for goods and passenger traffic. But when Fritz Kugel returned to Heidenheim after a trip to the USA in the middle of 1959, he had become aware that the large DB transmissions were not suitable for hauling heavy goods trains in North America. There, large trailing loads of up to 6 000 t had to be accelerated from zero to maximum speed. At the same time, the loco-motives had to be able to "crawl up" the Rocky Mountains at very low speeds without suffering thermal problems. Additionally, the locomotives had to con-tribute to braking the large trainloads down to their permitted speed on long descents. This could only be achieved using dynamic brakes.

For Voith, these operating conditions presented a huge challenge. They required a completely new transmission design. The elimination of the range-change gearing meant a return to the three-converter design. For this, new converters that hydraulically complement each other, had to be developed as quickly as possible:

- A super-starting converter with a particularly low speed ratio and a two-stage turbine wheel for starting. Due to its high stall torque ratio of $\mu_0 = 9$ it is still used as starting device for large single-shaft gas turbines.
- A through-drive converter, also employing two turbines, for speed range II.
- The extant cruising converter for speed range III.

Alongside the torque converters, a new hydrodynamic brake had to be developed. Fritz Kugel passed on this difficult task to one of the most creative heads in his team, Helmut Müller. He designed the KB 510, a double-flow brake with a profile diameter of 510 mm that was flanged to the rear of the transmission housing and driven at high speed by the secondary shaft of the transmission. In order to achieve the desired braking effort across the whole speed range, and to ensure stable control behavior during part-load operation, extensive tests were required to regulate the operation of the braking valve, the output valve and 16-position controller.

The new hydrodynamic brake had two shutters that were inserted between the rotor and the two stators, in order to reduce the ventilation losses during traction, when the brake was drained. In filled condition during braking, the shutters were withdrawn. The hydrodynamic brakes proved themselves excellently in demanding conditions proving superior to the electrical resistance brakes used in diesel-electric locomotives.

The operation of the six prototype locomotives was not as trouble-free as expected. At full load operation in the long tunnels on the routes of Denver Rio Grande, the Maybach engines had problems caused by the lack of induction air. The workshop criticized the maintenance routines that deviated from those of the DE locomotives, as well as the frequent need for repairs. Despite all these reservations, Southern Pacific was quite impressed by the enormous outputs of the German locomotives. It acquired the three test locomotives from Denver Rio Grande and ordered another 15 locomotives from Krauss-Maffei in 1963, albeit in a heavily Americanized format.

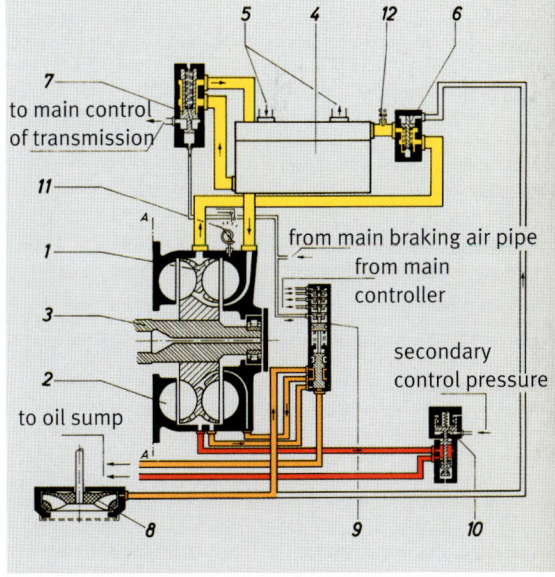

Hydrodynamic KB 510 brake with closed cooling circuit.
1   Rotor
2   Stator
3   Secondary shaft of turbo transmission (= drive shaft of brake)
4   Transmission oil heat exchanger
5   Connections on water side
6   Double return valve
7   Braking valve
8   Filler pump
9   16-position controller
10 Regulating valve
11 Contact thermometer
12 Oil temperature sensor

Second batch of ML 4000 CC locos with American bogies at Krauss-Maffei in Munich-Allach. March 1964

At the time the most powerful diesel locomotive in the world; built by Krauss-Maffei for the USA. Built in 1961 (pre-series):

1 *Maybach MD 870 engine rated at 2 000 HP*
2 *Voith L 830 rU transmission with KB 510 brake*
3 *Krauss-Maffei Intermediate gearbox*
4 *Maybach final drive*

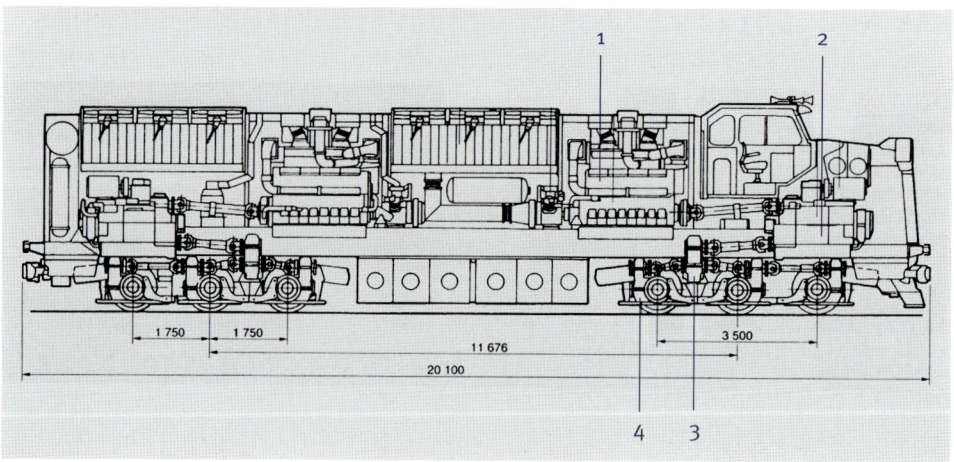

The 15 locomotives built in 1964 were in service at Southern Pacific for four years. Although they had to endure tough operating conditions, they did rather well with an availability of 80% during their first year. Unfortunately, the availability sank to almost 70% in the following three years, not least because the Maybach engines were not as robust as American medium-speed engines. For this reason, all 6 + 15 ML 4000 CC locomotives were taken out of service and scrapped in 1970.

The Voith transmissions, on the other hand, performed extraordinary well and their influence on the availability of the locomotives was not negative. An official statement by Southern Pacific in 1968 underlined the fact that the diesel-hydraulic transmission principle had been found to be healthy, but that all 6 + 15 locomotives were to cease operation prior to reaching their major overhaul dates because of problems with the high-speed engines.

As a pioneer among the numerous large US railway companies, SP nevertheless reached its most important goal, i. e. forcing the domestic locomotive industry (General Motors, General Electric and Alco) to develop more powerful diesel-electric locomotives.

In early 1963, Southern Pacific ordered three two-engine diesel-hydraulic locomotives from Alco, Schenectady, USA. These locomotives were typical examples of heavy American design. They had Alco medium-speed engines and bogies complete with final drives supplied by MaK, Kiel. Apart from L 830 rU transmissions with KB 510 brakes, Voith also supplied the roof cooling units and the transmission oil heat exchangers.

After a short operating time, the output of the three locomotives was increased from 2 x 2 000 HP to 2 x 2 150 HP. This meant that the Voith transmissions had to transmit input powers that had risen from 1 850 to 2 000 HP. For almost seven years of service, the transmissions coped perfectly with this power increase.

The three "Alcoholics" were strong as lions and reliable; in their first year of service they achieved an availability of 90%. They possessed the best technology that was available during the sixties. Due to their low number, they nevertheless remained outsiders, causing high maintenance cost. For this reason they were taken out of service in 1970. In the same year, Alco went into liquidation, and the planned 6 000 HP CC locomotive with L 920 rU transmissions never materialized. Sadly the commitment of Voith to gain a foothold for diesel-hydraulics in the North-American market never succeeded.

One of the two powertrains of the Southern Pacific 4 300 HP Alco locomotives with Voith L 830 rU + KB 510 turbo transmissions, as well as Voith cooling units.

Three Alco DH 643 diesel locomotives in front of a Southern Pacific 6 000 t goods train near Idio, California, in 1964.

Transmission Production is Booming
1963 – 1977

This period, probably the most successful for turbo transmissions, was characterized by Rolf Keller, who took over the rail transmission department from Ernst Seibold on 1 September 1965 and headed it until his retirement at the end of March 1978. It was a time dominated by the rapidly progressing dieselization of Deutsche Bundesbahn, imposing a huge responsibility on the transmission plant and its team. The conversion to electrical and diesel traction could only function if the newly developed hydrodynamic transmissions for diesel locomotives and railcars were robust and reliable.

Voith was inundated with new tasks and new orders from all directions. The transmission plant was bursting its seams. In 1963, the DIWA production had to be relocated to the new works in Garching near Munich. During the same year, the production of geared variable-speed couplings was moved to Crailsheim. The transmission factory had to concentrate fully on the series production of turbo transmissions and final drives.

L 206 rs for V 90 of Deutsche Bundesbahn

DB's heavy V 90 shunting locomotive was initially intended to have the same transmission as the V 100. Yet DB eventually decided in favor of a genuine two-converter transmission because shunting operations usually take place in the lower speed range, and goods train operating at high speed rarely drive in part-load mode. A turbo coupling in speed range III offered no advantages and was therefore not required.

The L 216 rs three-speed transmission was consequently modified into the L 206 rs two-speed version. On the outside hardly different from its predecessor, it was only slightly modified internally. The space for the turbo coupling was

The Deutsche Bundesbahn V 90 with L 206 rs turbo transmission deployed for heavy shunting service and medium-heavy good trains.

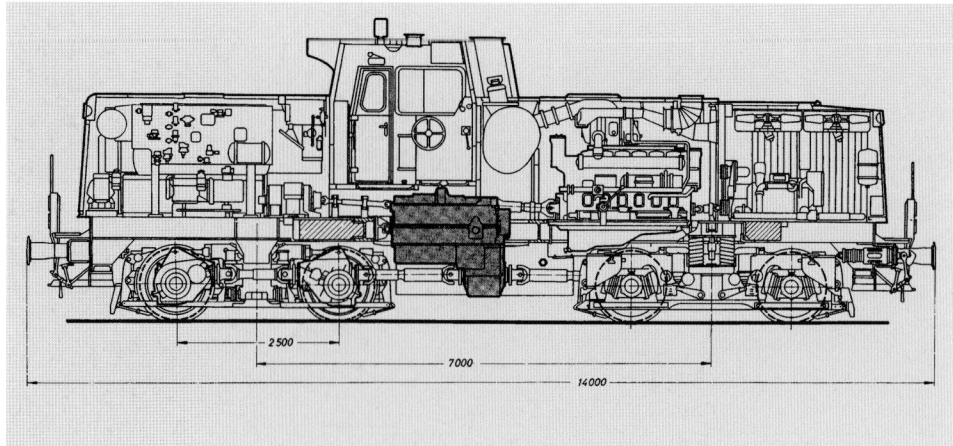

bridged by an intermediate part, the range-change gear ratio was increased from 1.464 to 1.735, and the main control system was simplified. In 1964 the L 206 rs was modernized. The bolted connections of the herringbone (double helix) gearing in the mechanical section, prone to torsional damage, were replaced by press fits. The selector rings of the range-change gear were redesigned, and the riveted brass cages of the cylindrical roller bearings made way for one-piece, broached cages made from forged bronze.

In 1970, a new converter combination was introduced offering better efficiencies and a greater high efficiency range. It had been developed after extensive tests by the converter design department under Elmar Rohne, and for decades served as the sound hydraulic basis for the many two-converter transmissions developed over the course of time.

The pre-series V 90 002, diesel locomotive, built by MaK, Kiel. July 1964

L 820 rs for Class 218 of DB

The transition from three-speed transmissions to simpler two-converter transmissions at DB also affected the V 160. The latter was initially fitted with 2 000 HP engines and a boiler for the steam heating system of the train. With the increasing electrification of main routes and the abandonment of steam traction in the sixties, passenger carriages were gradually retrofitted with electrical heating, while new carriages came ready-fitted with such systems. As the output of a 2 000 HP diesel engine was insufficient for both traction and the heating alternator, more powerful engines had to be developed. In 1967, MAN in Augsburg launched the new 2 500 HP V6 V23/23TL engine with a displacement volume of 115 liters. It was capable of providing 1 900 HP for traction, as well as generating 400 kVA for the heating alternator. In 1968, Krupp in Essen initially produced 12 pre-production Class 218 locomotives, which were tested thoroughly by DB.

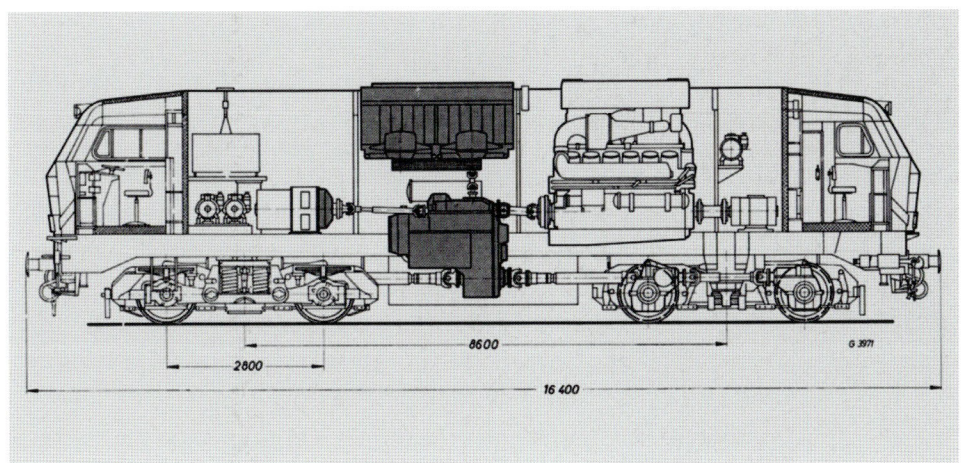

2 500 HP, Class 218 DB diesel locomotive with Voith L 820 rs transmission and KB 510/4 brake, Voith cooling unit and 510 TP turbo coupling for driving the heating alternator.

L 820 rs turbo transmission, with a welded steel/cast composite housing. May 1968

Series production started in 1971 and ended in 1978, after the delivery of 398 locomotives. This year also marked the completion of the dieselization program at DB.

The transmissions had to keep pace with engine developments. After the 1 800 HP rated L 218 rs, Voith produced the reinforced L 821 rs for an input power of 2 000 HP, both with turbo couplings in the upper speed range. In 1966 work began on the new large L 820 rs two-converter transmission, designed for an input power of 2 350 HP. It acquired a welded-steel housing for weight reasons.

In order to achieve good part-load efficiencies in the upper speed range, converter II was only utilized up to half its normal speed range. This resulted in the tractive effort characteristics of a $1^1/_2$-converter transmission.

Large Diesel-Hydraulic Locomotives for China

At the end of the sixties, German locomotive exports, too, experienced a revival. While the operation of the Krauss-Maffei locomotives in the USA was nearing its end, the People's Republic of China expressed an interest in high-performance diesel-hydraulic locomotives. This time the partner was Henschel Werke in Kassel, which, under its chief design engineer Siegfried Kademann, had attracted much attention with its modern six-axle diesel-hydraulic loco-motives with outputs of 4 000 HP for DB and the Soviet Union. As part of a pre-liminary contract, Henschel delivered four, two-engine DHG 4000 CC locomotives with L 830 rU + KB 510/1 transmissions in 1967. Designated the Class NY 5, they were destined for applications on heavy passenger and goods trains.

In 1970, a follow-up order for 30 even more powerful NY 6/7 locomotives was placed. Kassel produced the most powerful diesel-hydraulic locomotives until then ever built. Sub-suppliers for engines and transmissions contributed with experiences gained with Deutsche Bundesbahn.

The new NY 7 had two power drive systems, each rated at 2 500 HP, driving one of the two three-axle bogies via a single-stage transfer gearbox. With 113 km/h its maximum speed was equal to that of the USA locomotives built by Krauss-Maffei, while its weight of 138 tonnes was significantly lower due to the Chinese not permitting wheel loads exceeding 23 tonnes.

By the end of 1972, all of the 30 locomotives had been delivered and were immediately put into scheduled service by Chinese State Railways. Applying German standards, the Beijing repair works with its 2 500 employees carried out

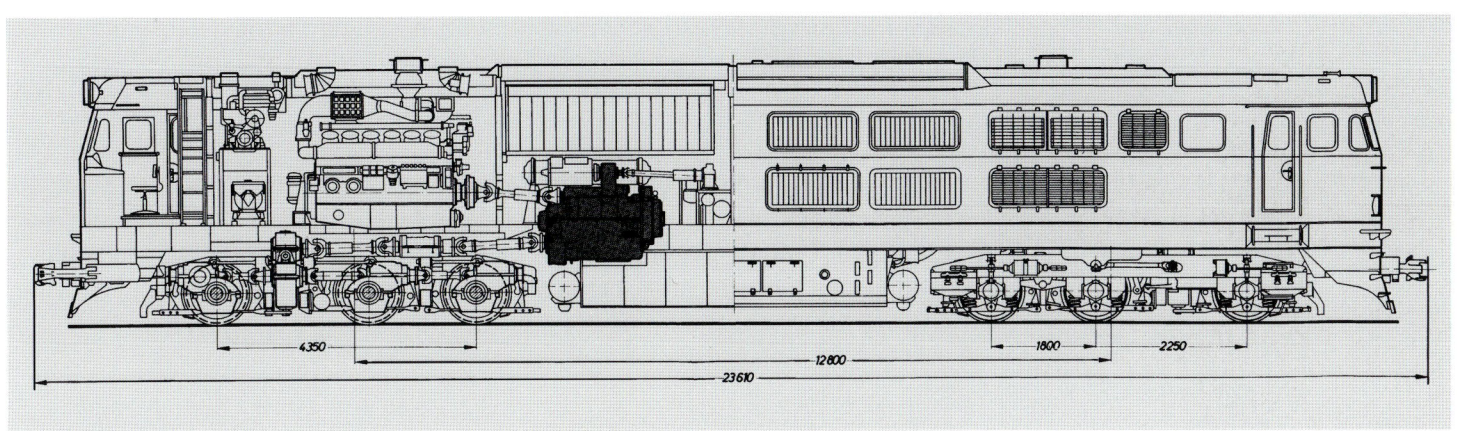

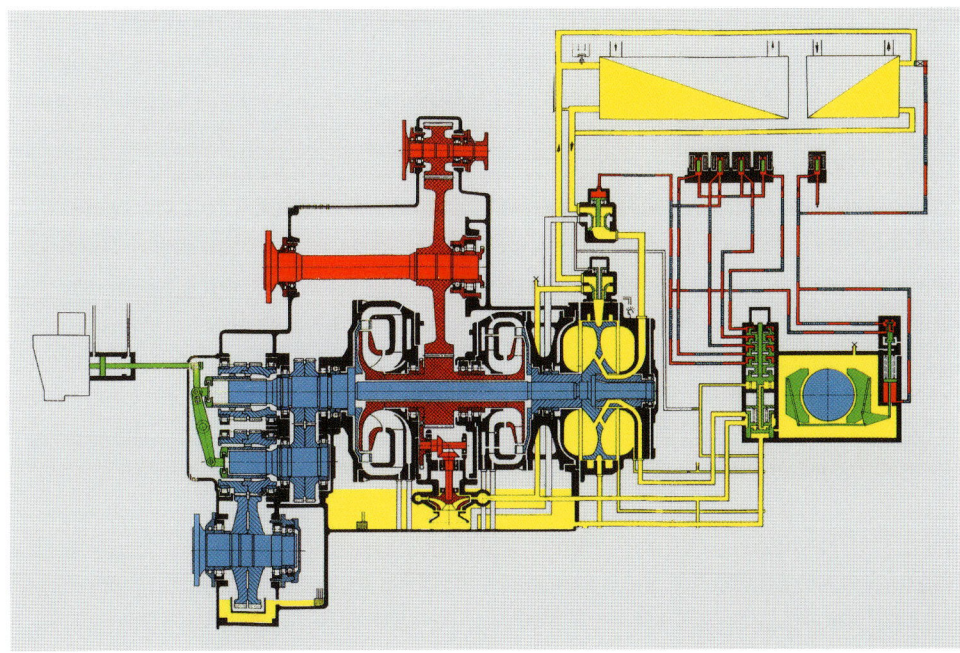

5 000 HP Class NY 7 diesel locomotive built by Henschel, Kassel, for the People's Republic of China.

Voith L 820 rU + KB 510 turbo transmission in braking mode with the traction circuits drained. The two transmission oil heat exchangers are separately installed in the locomotive.

maintenance and major overhauls meticulously. During the first ten years, the mileage of the locomotives amounted to an average 200 000 km per year; they were primarily used in combination with heavy long-distance passenger trains.

In 2004, six of the 34 locomotives were still in operating condition. They had completed an average of 3 million km. 27 locomotives were scrapped.

Despite the excellent performance of the 34 large diesel-hydraulic locomotives, there were no follow-up orders, as they were too expensive compared with the American diesel-electric locomotives. Consequently, for a number of manufacturers in the People's Republic in China, the technology of these American locomotives formed the basis for the series production of large CoCo diesel locomotives. However, with 3 600 HP, their outputs were significantly lower.

L 411 r Special Transmissions for SNCF Gas Turbine Trains

In 1965 the utilization of light-aviation gas turbines from helicopters marked the beginning of a new development in the design of high-speed railcars. SNCF, the French State Railways, and the manufacturer of the gas turbines, Turbomeca, were the pioneers of these new rail vehicles that allowed high traction power with low wheelset loads.

Compared to a diesel engine, a gas turbine requires less installation space and is also lighter. However during part-load operation and idling, it has a lower efficiency, as a result of which fuel consumption rises. At 4 000 to 6 000 hours, major overhaul intervals were far shorter than those of diesel engines, which are not less than 10 000 to 15 000 hours.

The advantages of higher power density were thus offset by higher costs. The new gas turbine trains were nevertheless a great success for SNCF – at least before the main routes had been electrified. The gas turbine trains could not keep up with the electric trains that were put in service in the mid-eighties on the newly built and electrified mainlines.

After some failed attempts with mechanical transmissions, SNCF selected Voith hydrodynamic power transmission. Voith consequently developed a

The first Voith L 411 rU turbo transmission for gas turbine drives, installed into the four-car ETG gas turbine train of SNCF with a maximum speed of 178 km/h.

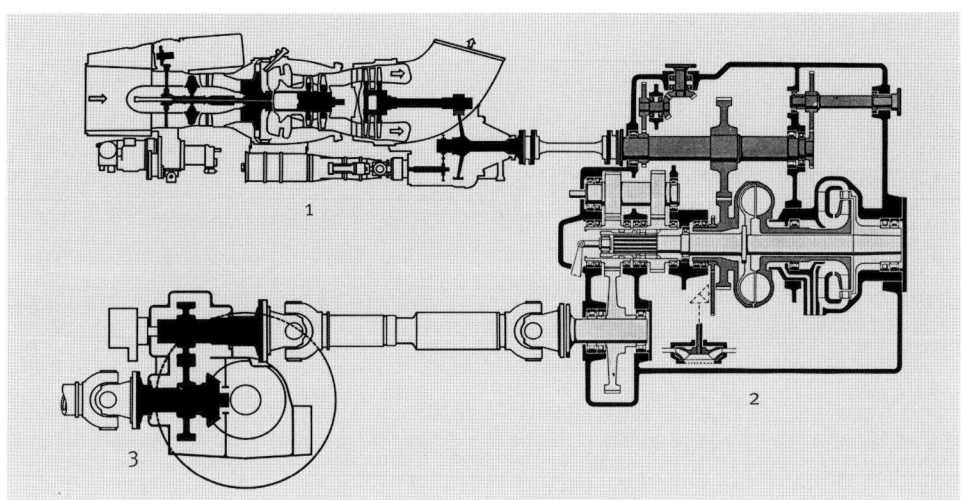

L 411 rU transmission with converter-coupling configuration and a maximum input power of 1 500 HP that was specifically adapted to the requirements of SNCF and the characteristics of the two-shaft Turbomeca gas turbine. The first ten transmissions were delivered at the end of 1969 and installed in ten four-car ETG (Elements à Turbine à Gaz) trains. They entered service in spring 1970 on the Paris – Cherbourg route.

The new transmissions were an instant success. The smooth running of the high-speed gas turbines, free from the torsional vibrations that are unavoidable with diesel engines, had a highly positive effect on the gears and the roller bearings of the transmission. The behavior normally encountered when shifting from one range to the other was adapted by the introduction of the double-filling principle to the hydraulic circuits, thus avoiding the situation in which the gas turbine may be unloaded.

Due to the high popularity of the new trains among travelers, SNCF had the RTG (Rames à Turbine à Gaz) developed in 1970. It was a five-car train with two gas turbines for traction, each rated at 850 kW, and two ancillary gas turbines with 290 kW each for the electrically driven auxiliary machines and air conditioning systems. The former seating capacity of 188 of the ETG was increased to 280 for the RTG.

For the RTG, the L 411 rU were fitted with a new KB 275 hydrodynamic brake. For cost reasons, the brake had a single flow design. The control and shutter principle had been adopted from the large KB 510. The cooling of the transmission oil occurred in the same way as in the ETG via a Voith oil to air cooler installed in the side of the vehicle with the fan driven by the bevel gear PTO of the transmission via a variable-speed coupling. An additional 230-liter oil tank was installed for buffering short-term braking power up to a maximum of 1 400 HP at high speeds.

The first gas turbine train, RTG 01, was tested in November 1972. With two gas turbines, each with an output of 1 000 kW and special final drives, it reached a speed of 260 km/h and thus proved its excellence for non-electrified high-speed traffic.

The procurement of gas turbine trains ended in 1976, by which time SNCF owned 14 ETG and 39 RTG trains. Then the electrification of the French mainlines and newly built routes began. The opportunities for gas turbine trains shrunk and they were forced onto branch lines. Owing to the lower speeds on these routes, the gas turbines ran mostly at part-load, while the transmissions operated primarily in the converter range. In both cases this led to increased fuel consumption and did not permit extensions to the gas turbine overhaul intervals. In the nineties, they were gradually phased out of service and replaced by modern diesel railcars.

Turbo Reversing Transmissions

The first Voith turbo reversing transmissions were developed for industrial applications. The RS 24 for the drive of a fish net winch on a trawler was delivered in 1950. The transmission had a torque converter with a slide valve for each direction of the winch's rotation. As a result, output torques could be steplessly controlled at constant input speeds.

The R 24 K was a variant of this first turbo reversing transmission, with two scoop tube-controlled turbo couplings driving windlasses in coal mines via electric squirrel cage motors. 19 such transmissions were delivered.

In the middle of the sixties, the subject of the "Turbo reversing transmissions" became topical again. Industrial operators of shunting locomotives, especially in steel mills where they required frequent changes of direction, complained about the poor availability of their mechanical reversing jackshaft gearboxes. Voith was urged to redress the problem.

Derived from a series unit, Voith built a test transmission at low expense in 1964. The L 2r 3 transmission was installed into a 240 HP Henschel locomotive – Voith works locomotive no. 4 in Heidenheim – and proved to be an instant suc-

RS 24 turbo reversing transmission for a winch drive with 180 HP diesel engine. June 1950

cess. Features such as the smooth and wear-free change from forward to reverse, reduced standstill periods during reversing maneuvers and the use of the contra-rotating converter as a dynamic brake; i.e the converter for the opposite direction of travel, were demonstrated to many visitors from home and abroad.

In 1969, the L 2r3 U transmission was, for the first time, also delivered to state railways. Swiss Federal Railways purchased 90 transmissions for their 380 HP Tm IV shunting tractors. In 1973, the transmission was reinforced and installed into 500 HP shunting locomotives destined for export (Norway, Kenya, Ivory Coast).

The development of a larger turbo reversing transmission started in 1967. While industrial companies observed the characteristics of the L 2r 3 with interest and were quite impressed, they asked for a turbo reversing transmission suitable for three-axle 500/700 HP locomotives with cardan shaft drives and maximum speeds of 40 km/h.

The basis of the new transmission was the two-converter L 420 r transmission that had proven itself in the new industrial locomotives equipped with three wheel sets driven via cardan shafts. Voith adopted the rotors with the starting and cruising converters and arranged both rotors side by side. An input shaft drove the latter via a trio of gears that formed the step-up arrangement.

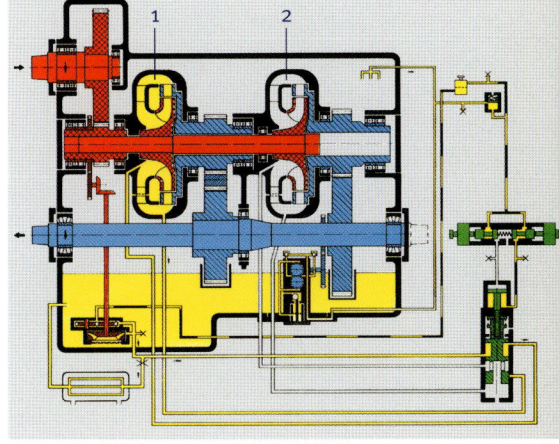

L 2r3 U turbo reversing transmission with reversing converter 1 and forward converter 2. April 1966

L Locomotive transmission
2 Number of circuits
r Hydrodynamically reversible
3 Code for circuit size
U Design with output below input

The DH 240 Henschel locomotive was the works locomotive no. 4 at Voith in Heidenheim; from July 1965 it was fitted with a turbo reversing transmission.

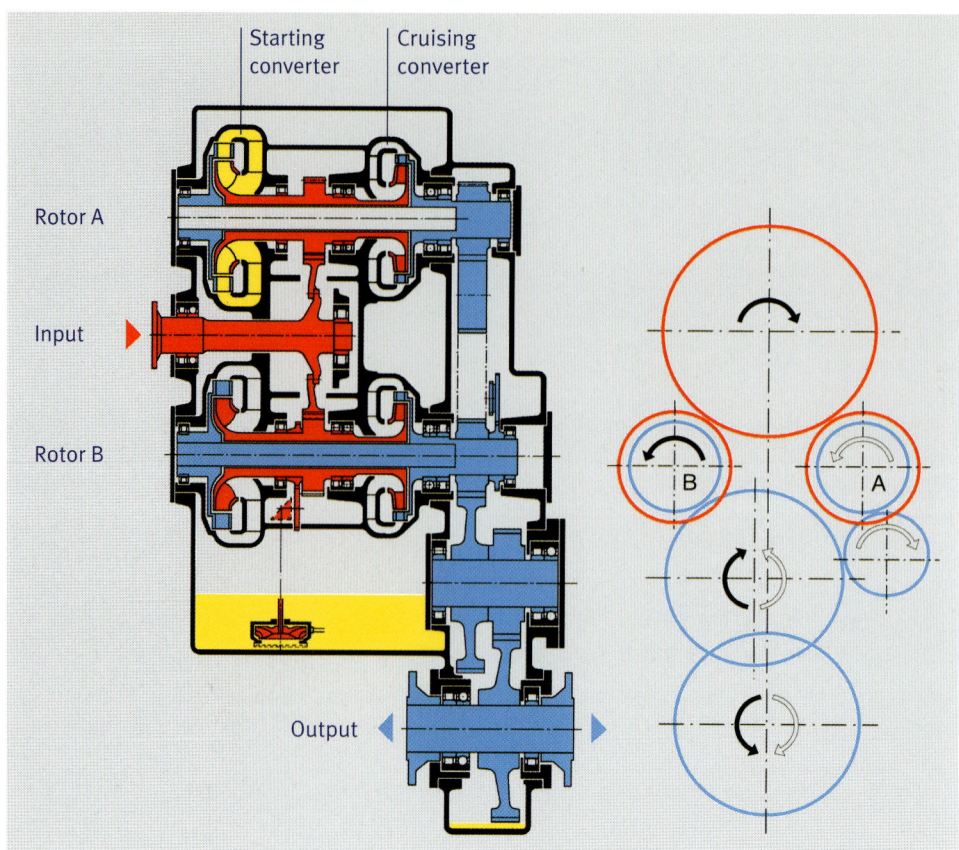

The L 4r4 for Australia was the first transmission with integrated braking effort limitation.
May 1970

Right: L 4r4 V2 turbo reversing transmission and gear schematic.

Tractive effort diagram of the L 4r4 for traction (2) and braking (1) at different engine speeds.

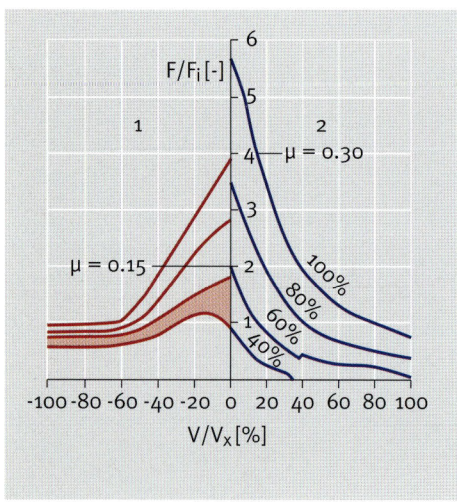

The first L 4r4 transmission was delivered in 1968 to Gmeinder, Mosbach, for installation into a three-axle industrial locomotive. Other transmissions of the first production series went to Henschel, Kassel. Over the years, the latter delivered a total of 70 three-axle locomotives to industrial companies at home and abroad. These were equipped with L 4r4 turbo reversing transmissions.

The first major order for the new turbo reversing transmission came from Australia at the end of 1969. A regulating valve supplemented the transmission control system, in order to prevent the contra-rotating converter over-braking the locomotive wheels and causing sliding during hydrodynamic braking. The valve achieved this effect by limiting the engine speed automatically to 60% of its rated value. The loco builder Walkers, based in Queensland had received an order from the New South Wales Government Railways (NSWGR) in Sydney. The order specified 20 bogie-locomotives with an output of 710 HP. These locomotives proved so popular that another 30 were ordered in late 1971.

NSWGR's 50, Class 73, diesel-hydraulic locomotives constituted the first locomotive fleet bought by a state railway company that featured turbo reversing transmissions. These operated successfully and reliably for some 25 years. In the middle of the nineties, the demand for shunting locomotives drastically declined. Walkers converted some of the locomotives for the sugar plantations in North Queensland, some of them were shipped to Vietnam. The rest was scrapped.

In order to round off the delivery program of turbo reversing transmissions, Voith also developed a larger Size 5 transmission. Relevant work began in 1969, approximately two years after the L 4r4, so that the operating experiences gained with the latter transmission could be exploited in the new L 5r4. A new starting

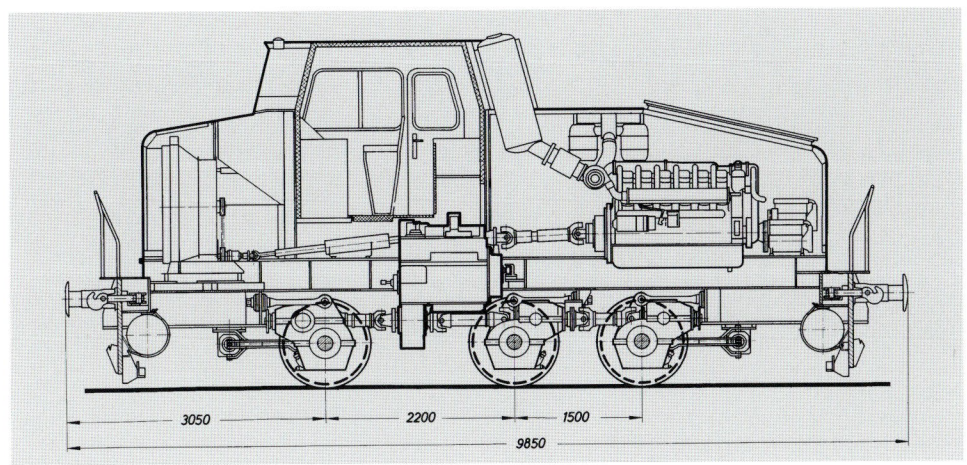

DHG 700 C industrial locomotive built by Henschel, Kassel, with L 4r4 V2 turbo reversing transmission in its original design from 1968.

and cruising converter combination was used. Another step forward was the installation of a rotary slide valve to reduce the flow of air, present when the two starting converters were emptied of oil. This prevented undesirable overheating of the converter housings during longer periods of operation in the high-speed range.

The L 5r4 U2 turbo reversing transmissions were installed into MaK industrial locomotives. For the first time in 1972, Krauss-Maffei launched a new bogie locomotive with the L 5r4 U2, which included an electronic control system. This system led to the development of the "Krauss-Maffei-Direct" electronic control, monitoring the adhesion between wheel and rail and adjusting the amount of power or braking effort delivered at the wheels thus preventing spinning and sliding during traction and braking.

Two M 1200 BB locomotives built by Krauss-Maffei in double-heading a heavy goods train near Hechingen, Baden-Württemberg in 1978.

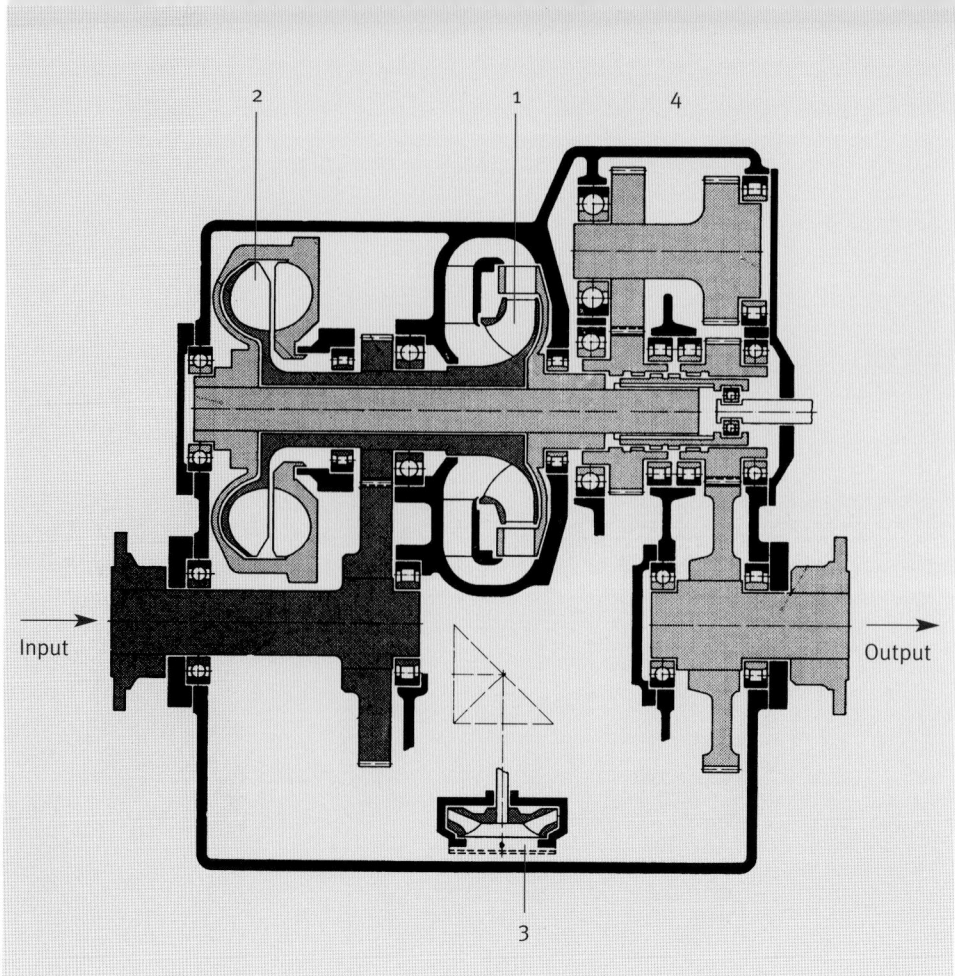

The first T 211 r.
December 1971
Mass: 700 kg
Oil filling: 65 kg

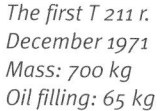

1 *Converter*
2 *Coupling*
3 *Filler pump*
4 *Reversing section*

T 211 r – the first Railcar Transmission of the New Generation

The T 211 r was developed in 1969 as an alternative to the DIWA 501/506 U+S transmissions for light underfloor drive systems rated up to 200 HP. It was intended to be suitable for engines between 200 and 300 HP. It was to be capable of flange fitting to the engine in a manner similar to the DIWA transmission and, most of all, it was expected to prove as reliable as the previous railcar transmissions. Voith decided in favor of a two-speed transmission design with a converter and a coupling, as well as an integrated mechanical reversing section and an optional hydrodynamic brake that could be fitted on the front face of the transmission, where it would be driven by the secondary shaft. For the first time a vertical joint face connected the hydraulic and the mechanical drive sections.

After a number of small orders, the first large order arrived in 1973 from Macosa, in Spain. ZTP Skopje in Macedonia required a three-car diesel train with a maximum speed of 120 km/h. It was to be suitable for service on mountainous routes with upward gradients of up to 25 ‰. MAN in Nuremberg designed the cars and their drive systems in co-operation with Macosa, Barcelona.

The three-car train had a power car at each end and a trailer car in the middle. Each of the underfloor powerpacks drove the two wheel sets of a bogie, in order to obviate wheel-spin when climbing gradients, especially during unfavorable weather conditions. For the first time, the engine was flange-fitted to the transmission by means of a bell housing and the engine output connected to the transmission input via a torsionally flexible Vulcan coupling. The Vulcan coupling was later replaced by a Voith spring coupling. The engine-transmission unit was installed in the car body via a three-point system using rubber mounts.

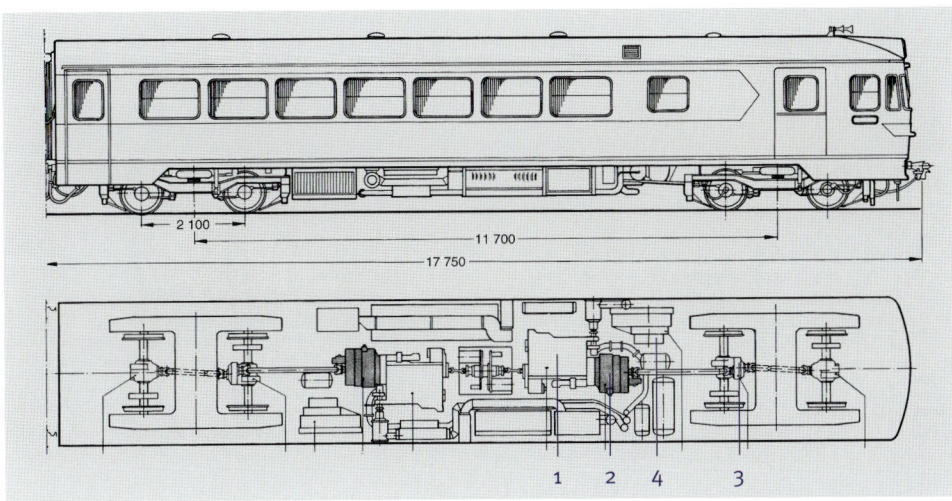

The Macosa DMU power car
for Yugoslavia with 2 underfloor power units, each rated
at 280 HP.

1 Engine
   (Büssing-MAN D 3256 BTXU,
   280 HP at 2 100 min⁻¹)
2 Transmission
   (Voith T 211 r, 250 HP)
3 Final drive
   (Voith V 13/15 + E 13/15)
4 Cooler Group

The three-car DMU built in 1975.

Macosa delivered another 10 three-car sets to Skopje and 10 to Croatia. During the eighties, Djuro Djakovic in Yugoslavia built 15 diesel multiple units under license. As a result, 180 underfloor Büssing/Voith powerpacks, were in operation in Yugoslavia.

T 320 R – the Second Railcar Transmission of the New Generation

In 1960, a new T 420 r two-converter transmission had been developed for underfloor installation. It was destined for the VT 23/24 railcars of Deutsche Bundesbahn and outperformed the competing EMG transmission. With its maximum input of 650 HP and its legendary reliability it marked the breakthrough for diesel railcars with underfloor drivelines and engine outputs up to 700 HP.

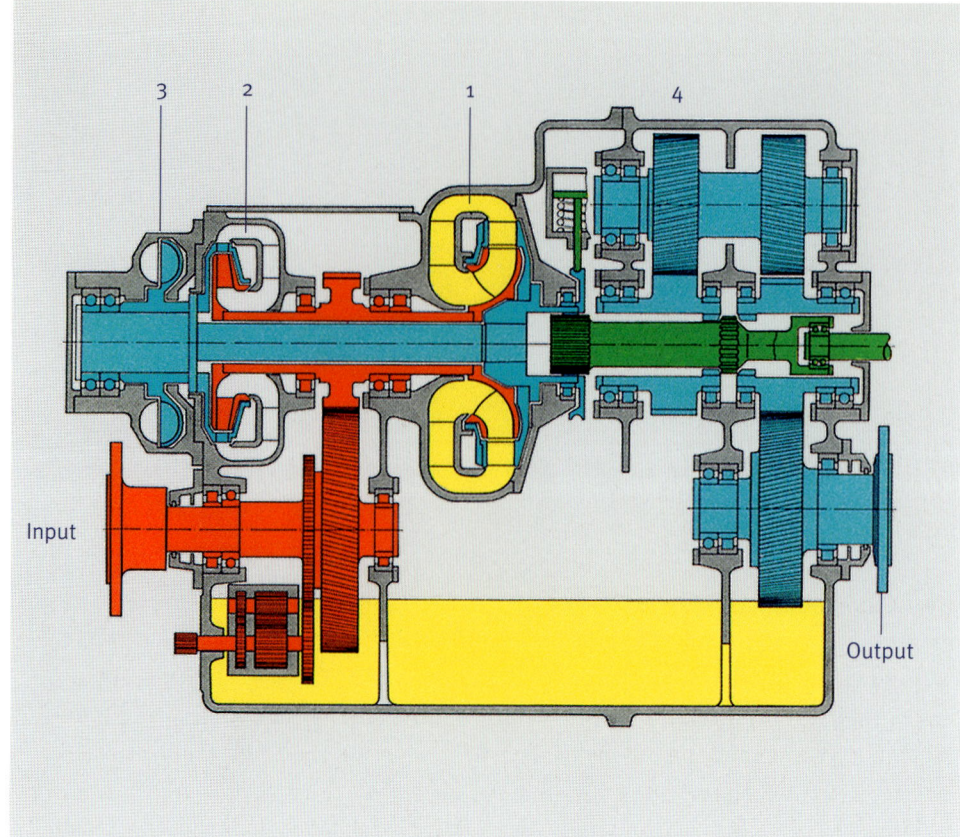

Input

Output

T 320 r two-converter transmission with KB 260 hydrodynamic brake and added hydrostatic pump (a) for driving the cooling fan.

1 Starting converter
2 Cruising converter
3 Hydrodynamic brake
4 Reversing section

VT 628.0 two-car DMU for DB with two 280 HP underfloor power units.

In the early seventies, the time had finally come to replace the older railcar transmissions with more modern designs. This materialized in the form of two new developments: the T 211 r for the lower power range up to 300 HP and the T 320 r for the medium range up to 400 HP.

The reason for this development was a need to replace the Deutsche Bundesbahn VT 98 railbuses with a large number of VT 627/628 light railcars. These first began to appear in 1971. The T 420 r would have been too heavy and too expensive for the new vehicles. The Central Office of Deutsche Bundesbahn (BZA) in Munich therefore gave Voith an order to develop a light underfloor transmission whose basic design was to be very similar to the T 211 r built two years earlier, but was to have a new transmission suspension with two flanged cross members.

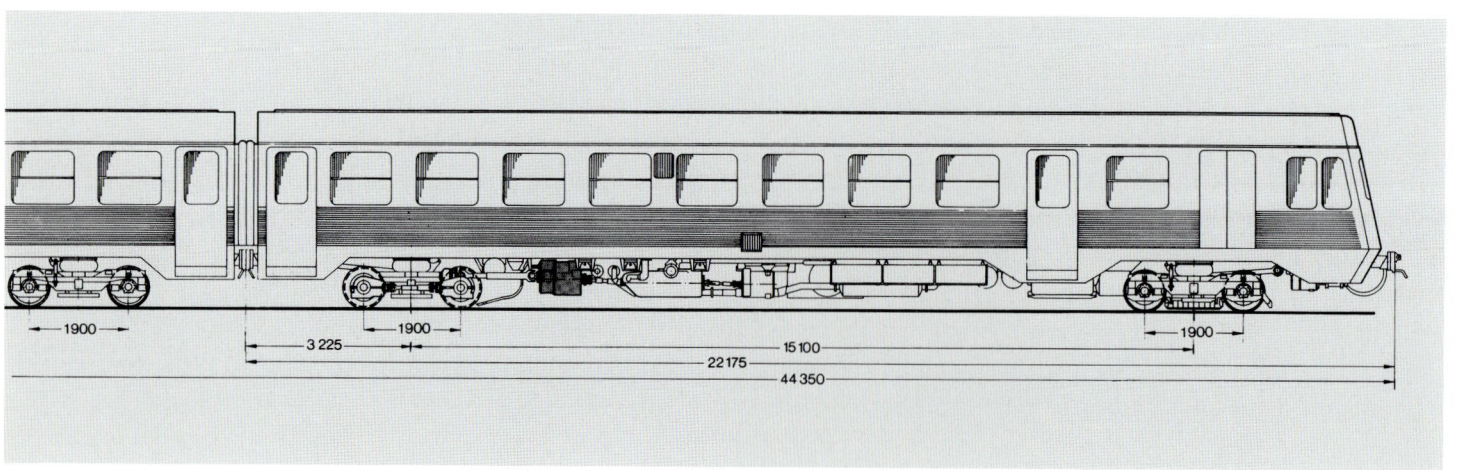

The first 32 transmissions were installed into 12 two-car VT 628.0 and 8 single car Class 627 DMU's in 1974. Despite being successful in service, series-deliveries of two-car VT 628.2 DMUs did not start until 1987, due to a shift in priorities within the procurement program of DB.

DIWAbus Transmissions for Overseas Light Railcars

In the early sixties, a version of the DIWAbus 200 S experienced a break-through in rail applications. As it did not require a reversing final drive, it was particularly suited for installation into light diesel railcars. The version in question was type U+S whose mechanical section included two planetary gear sets for both the forward and the reverse directions. It became a standard transmission for light railcars and was produced until the early nineties, although in road vehicles the D 851 had succeeded the original DIWAbus transmissions as early as 1974. A total of 1 070 type U+S DIWAbus transmissions were delivered for rail applications.

Danish Private Railways

In the course of modernizing their extensive rail network, a number of private Danish rail operators had Waggonfabrik Uerdingen develop a light, three-car DMU, with a maximum speed of 100 km/h. These DMU's were fitted with two light power units suspended underfloor via two cross members on metal-elastic elements.

The DIWAbus transmissions were hung off the engine and had an electro-pneumatically actuated reversing section.

501-380/U+S DIWAbus transmission with solenoid valves and pressure sensors for reverse actuation.

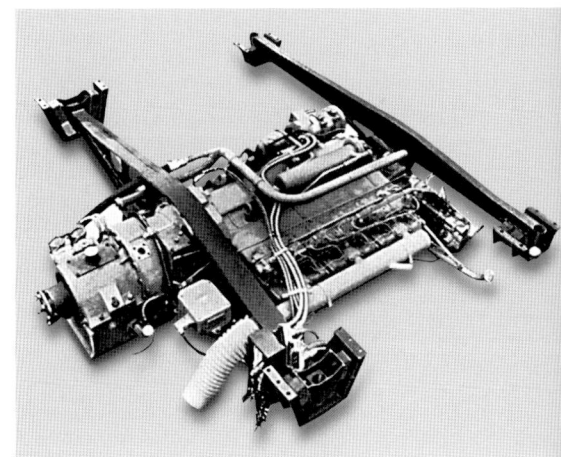

Uerdinger light DMU for Danish private railways and its 185 HP power unit and suspension.

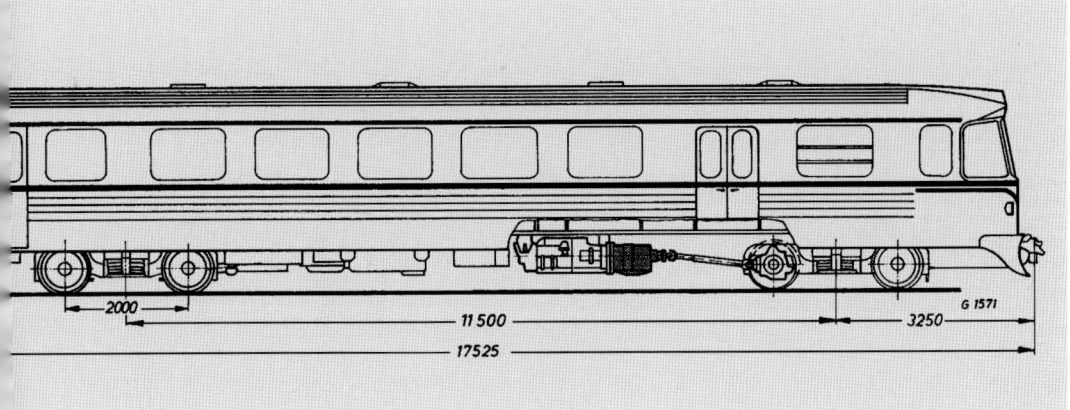

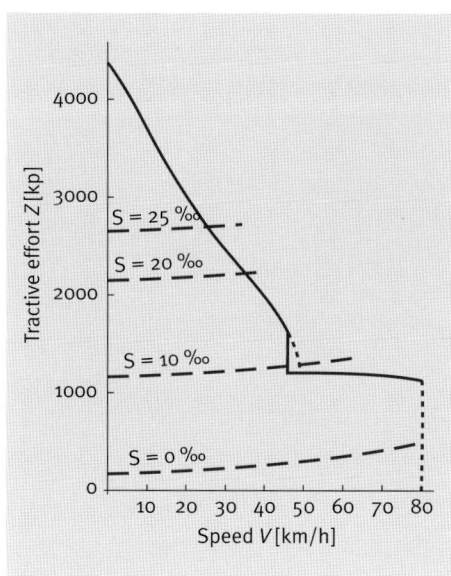

Tractive effort diagram of the 420 HP three-car Feve DMU.

With their smooth-running characteristics and high levels of comfort, the trains exceeded the expectations of railway operators and passengers alike. In the following years, another 65 trains were purchased, mostly in two-car configuration. From 1973 they were fitted with 506/U+S DIWAbus transmissions in which the reversing was accomplished by the adoption of multi-plate clutches rather than the earlier brake bands. With Büssing engines no longer being available, Daimler-Benz OM 407h engines were fitted instead.

Depending on the railway operator, annual mileages differed greatly and could range from 80 000 to 200 000 km. By mid 1988, the leading DMU had achieved three million kilometers.

MAN Railcars for Spanish Meter Gauge Railways

In 1966, MAN received a joint order from five Spanish railway operators for 20 railcars to be built in Nuremberg, as well as "knock-down" kits for another 43 railcars to be assembled in Spain under MAN license. For this major order, Voith delivered seventy, 501 – 380/U+S DIWAbus transmissions worth half a million deutschmarks. Due to the high engine output of 210 HP, a 436 TD turbo coupling was fitted in between the engine and the transmission to protect the planetary input gearset from torsional engine vibrations.

The tractive effort diagram of a 64-ton three-car DMU shows that the speed maintained on the steepest ascents never exceeded 25 km/h. Within the speed range up to 80 km/h, there was only one fully automatic shift at 45 km/h. With an input power of 193 HP, the transmissions operated at their permissible limit. Particularly on mountainous routes, the trains had to run continuously at full load for long periods of time. Although the trains were geared to 80 km/h and

Single-engine Meter gauge diesel railcar by MAN with two driven axles. Output 210 HP.

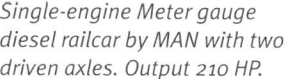

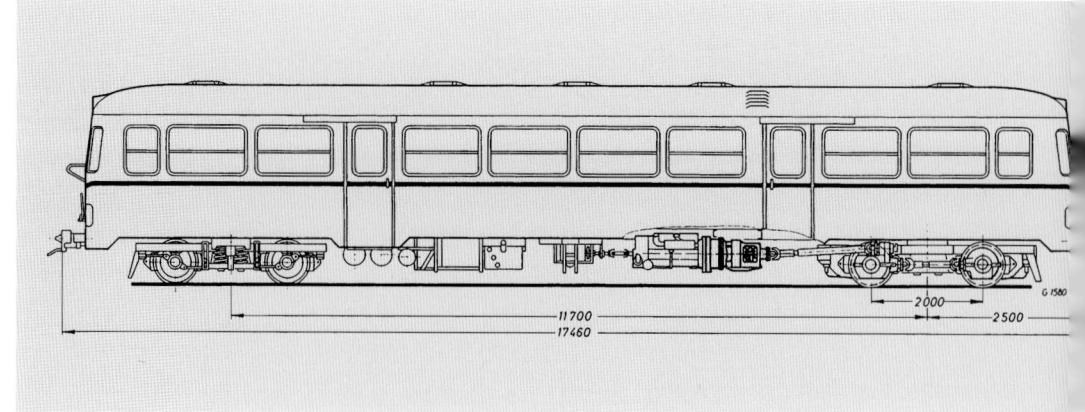

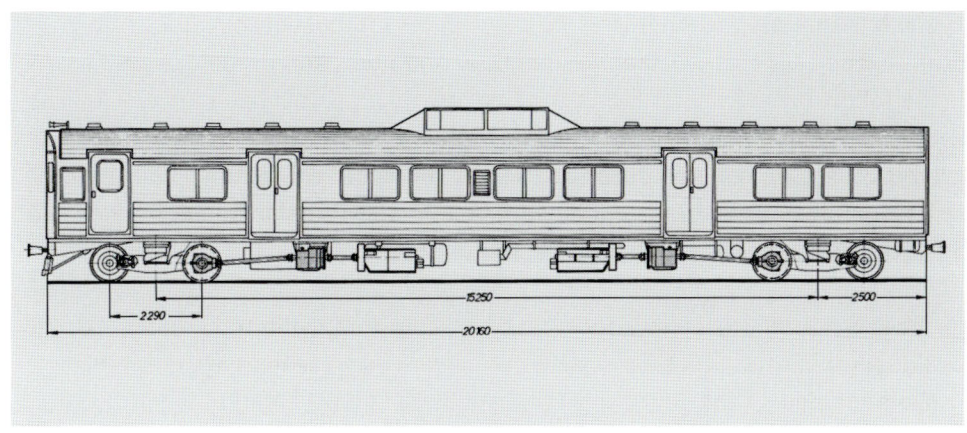

Powercar from the three-car
ADK multiple units of WAGR.
Output 2 x 185 HP.
Empty weight: 38 t.

had sufficient power for applications over flat terrain, they had insufficient power for mountainous routes where upward grades could reach 25 ‰. For the engine and transmission, this resulted in a very tough load duty-cycle that could not be compared with that of the city bus.

Diesel Railcars for Western Australia

In 1968, Western Australian Government Railways (WAGR) purchased five three-car DMUs from the ADK series for unelectrified suburban routes outside Perth.

The trains with a 1 067 mm track gauge were supplied by the Australian company Comeng located in Granville, New South Wales. As a result of the efforts of Voith's Australian subsidiary, instead of the usual, Wilson type, SCG transmissions they were fitted with 501 – 380/U+S DIWAbus transmissions with single-stage Voith E13/15 final drives.

Three-car ADK DMU for suburban routes outside Perth.
The center car does not have
power equipment.

Transmissions and engines were separately suspended underfloor and connected by a cardan shaft. In view of the frequent starts typical of suburban commuter traffic, a maximum speed of 50 mile/h, or 81 km/h, was sufficient. The combination of the Cummins engine and DIWA transmission was such a success, that only one year later the SCG transmissions fitted to the nine older ADG DMU's and two ADH sets were replaced by 44 DIWA transmissions.

The ADK railcars remained in service until 1992. Owing to the electrification of the suburban routes around Perth in the early nineties, they were eventually made redundant. During 24 years of operation they had achieved 2 million kilometers per train. In early 1993 they were sold to New Zealand, where, by mid 2003, they had, on average, covered another 440 000 km.

New Beginnings –
The End of German Dieselization
1978 – 1986

General Situation

For two decades, the transmission plant had been working to capacity producing large turbo transmissions and final drives for diesel locomotives. Railcar transmissions only took up a 20% share of the manufacture, while transmissions for Deutsche Bundesbahn dominated with approximately 50% of total sales; consequently machine tools in the workshop were set up for the production of large transmissions. This led to the manufacture of smaller hydrodynamic units beings moved to new plants; turbo couplings in 1957 and DIWA transmissions in 1963.

In the middle of the seventies, the German rail vehicle market began to alter. DB's demand for diesel locomotives was accounted for, and the industrial locomotive sector had also reached saturation point. Additionally, diesel hydraulics faced a new rival in the shape of electronics in power transmission, i. e. new three-phase current technology. It offered advantages in terms of traction and was regarded as a solution for the future by many buyers of industrial locomotives. Hence the situation for locomotive transmissions in the German and export markets became increasingly difficult, despite the decline in orders being less pronounced for railcar transmissions. This resulted in a change of production priorities in the workshop: railcar transmissions prevailed and large locomotive transmissions became the minority. Converting the production lines to build smaller turbo transmissions was difficult and could not take place overnight. Only after the introduction of modern machining centers in the eighties was it possible to produce smaller transmission housings economically.

The technological development of turbo transmissions continued. New and more affordable transmissions were added, and existing units were uprated. In co-operation with Krauss-Maffei, Munich, the traction behaviour of diesel-hydraulic locomotives was brought to the equivalent level of three-phase technology. This period formed the foundations for an upswing in diesel-hydraulic transmissions in the nineties.

L 3r4 U2 for Three-Axle Industrial Locomotives

In June 1977, while Rolf Keller was still in charge, work began on a new 500 kW*) power class turbo transmission for industrial locomotives. The new L 3r4 U2 was intended to replace the previous L 4r4 U2/V2, offering significantly lower production costs yet without compromising the reliability or the service life.

*) From this chapter, all outputs will be stated in kW rather than HP and all tractive efforts in kN rather than tonnes. The gradual transition to ISO units began in 1975; internally, the original units remained in use for quite some time.

The attempt to achieve particularly high starting tractive effort using two-stage starting converters failed due to their lower efficiency and unsuitable braking characteristics. For this reason, Voith returned to the single-stage standard starting converter. Like the L 5r4 U2 it was fitted with rotary slide valves using a simplified design.

The L 3r4 became the standard transmission for three-axle diesel-hydraulic industrial locomotives and hence competed directly with the three-phase diesel-electric locomotives entering the market at the same time.

In combination with high-speed diesel engines, the new transmission allowed an extremely compact drive unit and excellent accessibility. Over the course of extended operation, the trouble-free service and the low maintenance costs demonstrated that diesel-hydraulics combined with electronic control systems were still more economical for industrial locomotives.

The Reinforced L 4r4z U2

After the introduction of the L 3r4 U2, the L 4r4 was reinforced in 1975, to accommodate an input power of 660 kW. As a result of this upgrade, the ratio of cost to power compared to three-phase technology was improved further. The L 4r4z U2 was a new design employing rotary slide valves in the starting converters (see page 98). This led to an increase in the starting tractive effort from 5.6 to 6.9, as the cruising converters could now be utilized up to higher turbine speeds. The rate of heat dissipation from the transmission was improved by increasing the converter oil flow while at the same time changing from an open to a closed oil circuit.

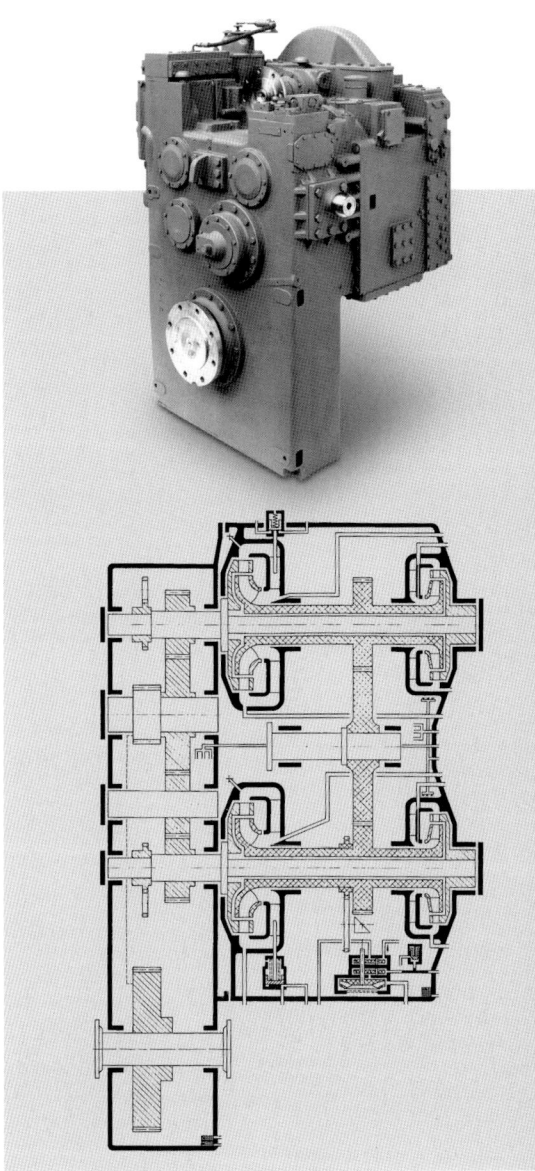

L 3r4 U2 turbo reversing transmission in the design adopted after 1982.

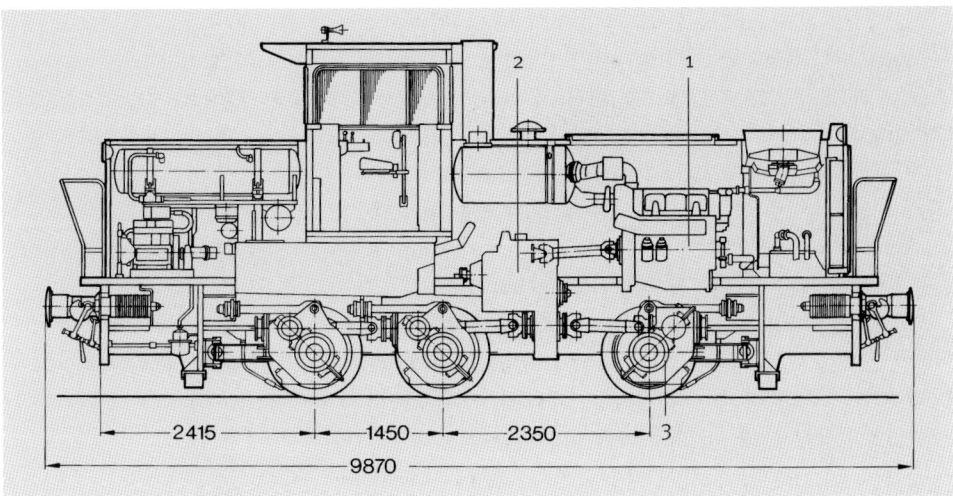

G 763 MaK industrial locomotive built in 1982.
1 MTU 6 V 396 TC13 engine
2 L 3r4 U2 turbo reversing transmission
3 MaK final drive

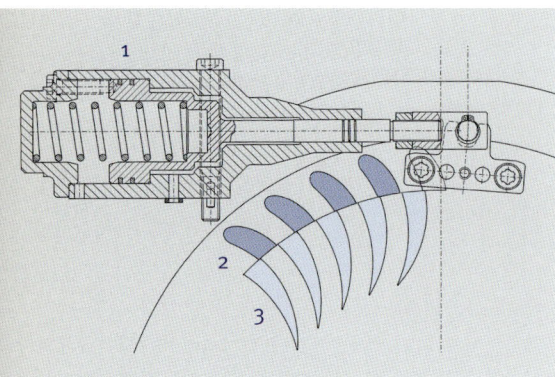

The L 4r4z U2 rotary slide valve actuation
1 Adjusting cylinder
2 Stationary outer section
3 Rotary inner section

L 4r2 U2 – the Export Transmission for India

In addition to the L 4r4z, another turbo reversing transmission was designed in 1978. Primarily intended for applications in non-European countries, it had just one converter for each direction of travel; its function, design and manufacture were simple.

In the seventies and eighties, Indian State Railways built diesel locomotives in their CLW works in Chittaranjan, where they replaced the steam locomotives previously used on its narrow gauge, 610 and 762 mm networks. This took place in technical co-operation with MaK, Kiel. Initially, the locomotives were fitted with Suri transmissions that did not prove successful. From 1982, the narrow gauge locomotives were fitted with Voith L 4r2 transmissions produced under license at Kirloskar Pneumatic in Pune. Between 1982 and 1991, CLW built a total of 50, ZDM 4 A locomotives for use on 762 mm tracks, and 45, NDM 5 locomotives for the 610 mm gauge, all fitted with L 4r2 U2 transmissions.

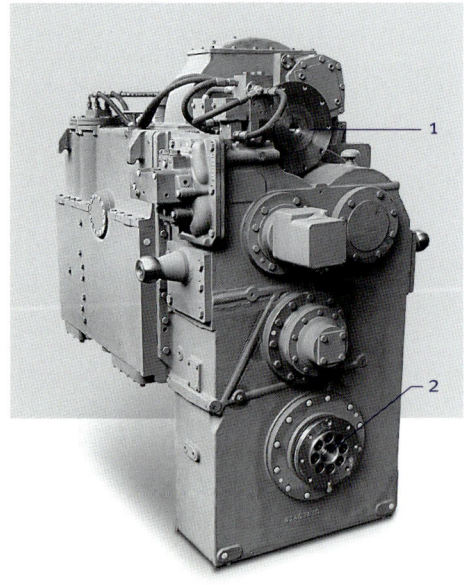

L 4r2 U2 turbo reversing
transmission from 1978
1 Input
2 Output

Indian State Railways' ZDM 4 A
locomotive used on 762 mm gauge.
1   MaK 6M282AK engine
2   L 4r2 U2 transmission
3/4 Driving axles
5   Pony axle

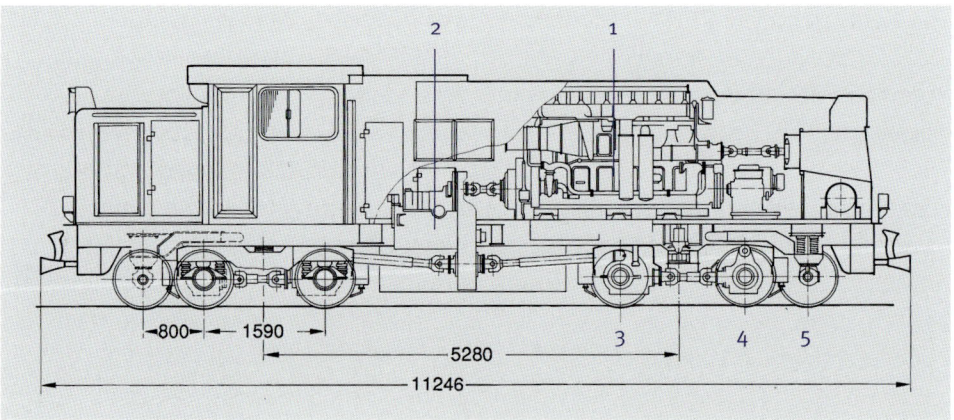

The L 4r2 U transmission variant developed for the Indian State Railways' WDS 4 D broad gauge shunting locomotives was built under license by Kirloskar Pneumatic. The production of hydraulic locomotives for the State Railways ended in April 1993. Afterwards, SAN, Bangalore only built diesel-hydraulic locomotives in limited numbers, for the private sector.

The Reinforced L 520 rzU2 Transmission

In 1968, Voith launched the compact L 520 rU2 two-converter transmission, which replaced the older three-converter transmissions. Over 300 of these transmissions were installed in export locomotives built by Henschel, Kassel, and supplied to Indonesia and East Africa.

The Indian WDS 4 D standard
shunting locomotive for
1 676 mm gauge was fitted with
the Voith L 4r2 U transmission.
m = 60 tonnes
V_x = 25/60 km/h

In 1981, the transmission input power was uprated from 1 100 kW to 1 400 kW and a value analysis program undertaken to reduce production costs.

With improved hydrodynamic brakes, the reinforced transmissions were installed in eight, two-engine high-speed locomotives built by Krauss-Maffei for Spain's TALGO trains. The basic design of these locomotives derived from DB's V 200, however the transmission oil sump had to be adapted to the pivotless bogies, and the output shaft of the transmission had to be lowered (Variant "spez").

The locomotives demonstrated both their riding capability during high-speed operation and their high availability. On average, trouble-free distances of 740 000 km were achieved during the first five years of operation.

Krauss-Maffei's M 4000 BB
built for the Spanish broad
gauge TALGO trains.
Locomotive mass 82 t
Maximum speed 180 km/h

L 520 rzU2 + KB 380/11 designed
for an input power of 1 400 kW.
Transmission mass 3 300 kg
Oil filling 215 kg

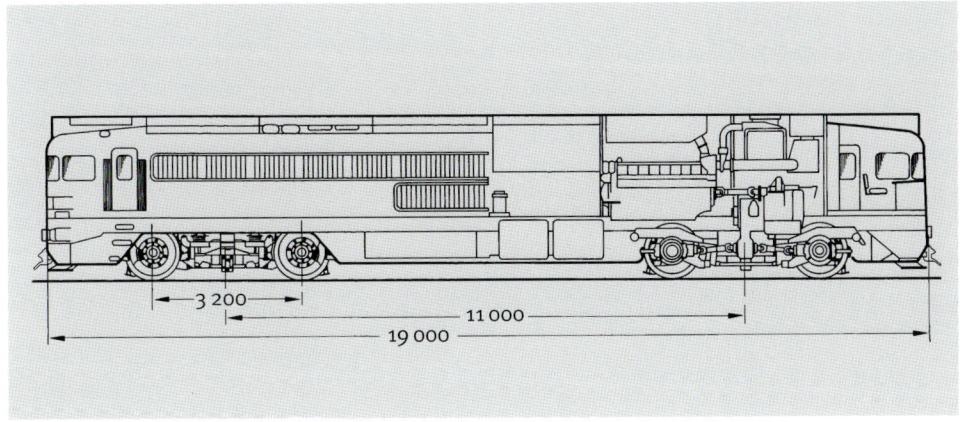

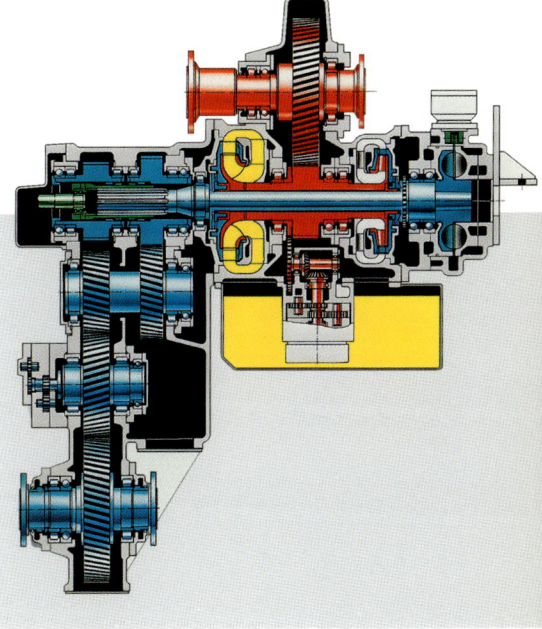

T 320 r Two-Converter Transmission for Denmark

Towards the end of the seventies, the T 320 r developed for DB, also gained a foothold in the export market. In order to maintain pace with the development of commercial vehicle engines that were also used for underfloor installations, the T 320 r was upgraded in several stages.

During this time and in subsequent years, Danish State Railways (DSB) placed a number of orders with Voith. The new two-car MR trains built by Waggonfabrik Uerdingen had two power units each rated at 288 kW, which were introduced on low-density services that operated at hourly intervals. As a result of their success, manifested by a significant increase in demand, DSB had ordered a total of seventy-two MR trains by 1983.

The trains were built under license by Scandia in Randers. The service performance of the Voith transmissions in the MR trains was extremely good. By the beginning of the nineties they had reached distances of 2 to 3 million km without major overhauls.

The International Breakthrough of the T 211 r

The period until 1986 was primarily characterized by the success of the smallest turbo transmission – the T 211 r. Like the non-federal railways of southern Germany, overseas state railways had to provide modern light rail cars for branch lines at a time when passenger volumes had fallen. Despite this higher comfort levels were expected. Over time and after several design improvements the T 211 r became the most successful railcar transmission and the mainstay for the transmission plant.

MR Class, two-car DMU for
Danish State Railways.
Built in 1978
m = 2 x 40 t
V$_x$ = 120 km/h

Among the numerous orders received during this time, two contracts from Spain and the United Kingdom are referred to as typical examples.

Three-Car Diesel Multiple Units for RENFE

After years of project work, Spanish State Railways placed a major order for 134 three-car diesel multiple units with two Spanish consortia:

– 70 trains Class 592 (diesel-hydraulic) from Macosa, MAN, Ateinsa
– 62 trains Class 593 (diesel-mechanical) from CAF, Babcock & Wilcox, Fiat.

All trains had four engines rated at 4 x 169 kW, a non-powered center car, air conditioning and 228 seats. This large fleet provided the first opportunity to compare two transmission systems: the modern, hydrodynamic two-speed transmission with converter and coupling and the older concept of a mechanical, semi-automatic five-speed gearbox with a turbo coupling and reversing final drives. An assessment of the two systems shows that the hydrodynamic two-speed transmission was able to span the tractive effort offered by the five speeds of the mechanical gearbox.

The order for 280 transmissions had a total value of DM 16 million and was the largest export contract for Voith transmissions since the war.

Power car of the three-car Class 592 DMU of RENFE.
1 Engine (MAN-Büssing, 169 kW)
2 Air conditioning system
3 Transmission (Voith T 211r, 149 kW)
4 Underfloor cooling system (Voith)
5 Final drive (Voith E 15/19)

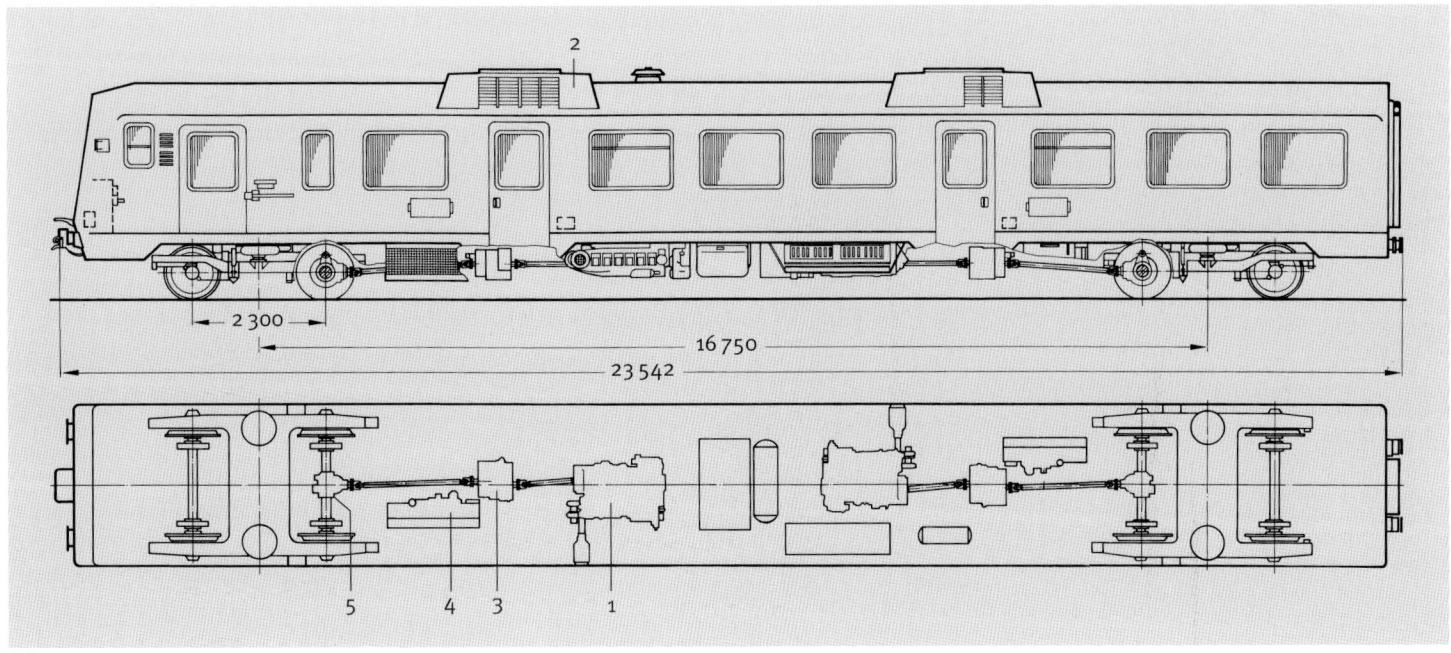

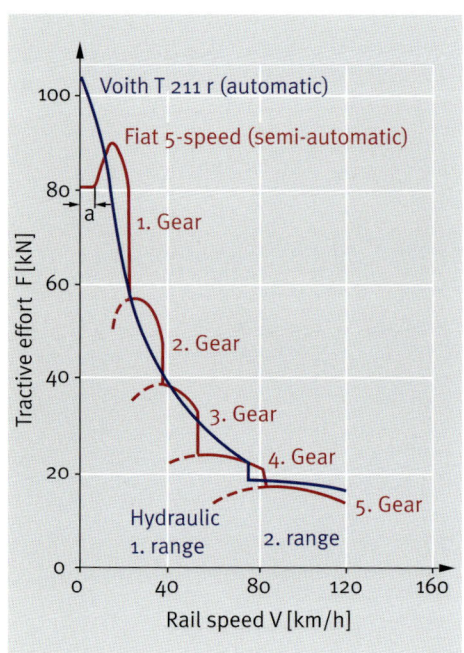

 (chart labels)
Voith T 211 r (automatic)
Fiat 5-speed (semi-automatic)
1. Gear
2. Gear
3. Gear
4. Gear
5. Gear
Hydraulic
1. range
2. range
Tractive effort F [kN]
Rail speed V [km/h]

Comparison of tractive effort of the RENFE trains with different power transmission systems.

The 70, S 592 diesel multiple units were delivered by Macosa between 1980 and 1982. In the middle of 1987, the front runner had achieved 550 000 km; the average annual distance was 138 000 km in commuter traffic and 180 000 km for trains on long-distance routes.

From 1993, all 280 T 211 r transmission underwent a major overhaul after distances of 1.2 to 1.5 million km. These were initially carried out by Voith, Madrid, and later in RENFE's own workshops. In order to ensure availability of the railcar fleet during overhaul, Voith supplied 32 spare transmissions.

Return to Diesel-Hydraulics in the United Kingdom

In the United Kingdom, a traditional stronghold of diesel-electric mainline locomotives Voith and Maybach transmissions provided a diesel-hydraulic interlude in the sixties for the Great Western Region.

Starting in the fifties and continuing into the sixties, some 2 000 diesel railcars and 1 000 trailer cars were purchased for BRB's branch lines services. The power cars were fitted with SCG, Wilson type transmissions and reversing final drives. By the beginning of the eighties, this fleet was obsolete, caused high maintenance cost and was increasingly unpopular with the traveling public.

The British Rail modernization program began with two developments: a two-axle Class 140 railbus by Leyland and a diesel-electric three/four-car Class 210 multiple unit. These vehicle types represented the lower and upper limit of the profile for regional traffic. The optimum solution was somewhere in the middle, and eventually aligned itself to the then new two-car DH2 diesel railcars of Netherlands Railways, supplied by Duewag, Uerdingen. These railcars had two underfloor power units with 212 kW Cummins engines and T 211 r turbo transmissions.

With different engine/transmission combinations, the British Railway Board ordered four, three-car prototype DMU's from two suppliers in March 1983. The Class 150 was delivered by BREL in Derby and the Class 151 from Metro-Cammell in Birmingham.

The BREL DMU, 150 001, was equipped with the same drive components as the NS trains that had already proven themselves during two trouble-free years of service.

Locomotive Transmissions

Further Developments of Turbo Reversing Transmissions

These developments were characterized by the transition from electro-pneumatic to electronic control systems with hydraulic cylinders. This meant that no air was required to actuate the transmission. The two speed, range-change gearbox that is shifted during standstill is actuated electro-hydraulically and automatically moves into neutral when the engine is switched off.

The range of turbo transmissions in sizes 2 to 5 remained unchanged, while their power capacity was significantly increased by reinforcement, so encompassing the entire range of industrial locomotives from 300 to 1500 kW. Intensive efforts were needed, in order to develop the electronic control to be as reliable as the electro-pneumatic system, with its mechanical governor.

Turbo Reversing Transmission Size 2

The smallest turbo reversing transmission was developed as a modular concept for installation into a wide variety of small locomotives, in the early seventies. By now, only the most successful type, the L 2r4zseU2 with range-change gearing and electronic control is being produced, with over 540 such units having been delivered to SNCF. In 2004, further deliveries went to Geismar, France, for seventeen, 348 kW-locomotives of Algerian State Railways.

Turbo Reversing Transmission Size 3

In 1978, Voith launched the compact L 3r4 U2 turbo reversing transmission as an answer to the three-axle AC-AC locomotives offered by a number of German manufacturers. Combined with the electronic locomotive control system developed by Krauss-Maffei it set new standards for the traction behavior of diesel-hydraulic locomotives. The main reasons behind this were the improved anti-slip and slide device with rapid transmission disengagement and an automatic limitation of tractive effort at low speeds. Additional features were the constant-speed control, as well as enhanced braking effort characteristics delivered by the counter-rotating converter.

In 1998/99, a reinforced version with electronic control was developed for the multi-purpose Rh 2070 locomotive of Austrian State Railways. The range-change gearing was derived from the size 4 transmission.

VTL 500 shunting locomotive for Algeria with turbo reversing transmission.
Locomotive mass 34 tonnes
Maximum speed 32/62 km/h

L 3r4 zseU2 with range-change gearing suited for an input power of 660 kW. Built in 2000

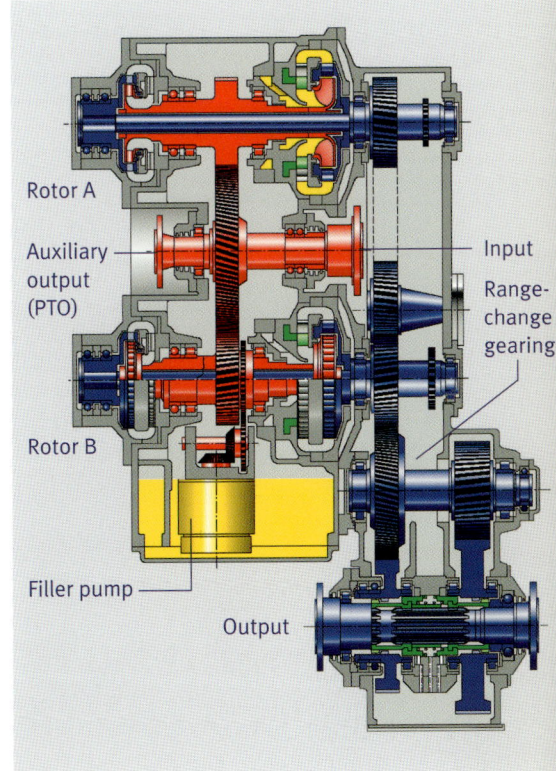

Rotor A

Auxiliary output (PTO)

Rotor B

Filler pump

Input

Range-change gearing

Output

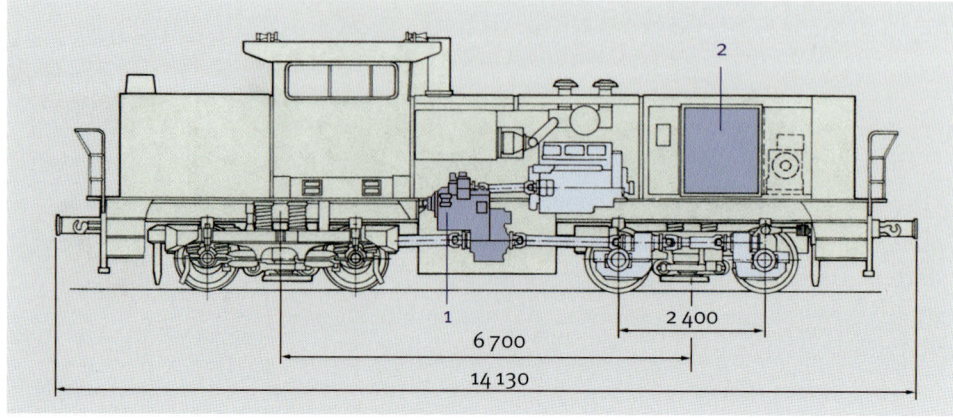

The order for 60 locomotives went to Siemens Schienenfahrzeugtechnik in Kiel and was increased by 30 locomotives in early 2003. The 90 new locomotives replaced a series of older classes owned by ÖBB. In the middle of the eighties, Voith in St. Pölten had already delivered a special version L 3r4 U2 spez whose distance between input and output had been increased. This special transmission was required for the rebuild of various locomotive fitted with L 26 and NG 600 transmissions and built by Voith St. Pölten. At the Hungarian State Railways central Szombathely workshop, 24 four-axle, class M 47 locomotives were modified and fitted with turbo reversing transmissions and 600 kW-MTU engines. By the end of 2004, a total of 500 size 3- turbo reversing transmissions had been delivered.

Turbo Reversing Transmissions Size 4

During the eighties, the permitted input power of the reinforced and improved L 4r4 zU2 transmission was gradually increased from 575 kW to 1 040 kW. At the same time, mechanical section with its range-change gearing was enhanced and changed to electro-hydraulic actuation.

In 1997, Vossloh placed a major order with Voith for 90 special transmissions for use with medium-speed diesel engines rated at 1 000 min^{-1}. Regarding the engine, the operator Belgian State Railways decided in favor of the robust 6-cylinder engine manufactured by ABC of Gent. This special transmission was designated L 4r4 zseU2a.

Similar to the German V 90 locomotives, they were destined for heavy shunting duties in hump yards and mainline service, hauling light goods trains. Auxiliary machines such as cooling fans, compressor and air conditioning system in the driver's cabin were driven by an AC alternator. The first locomotive was

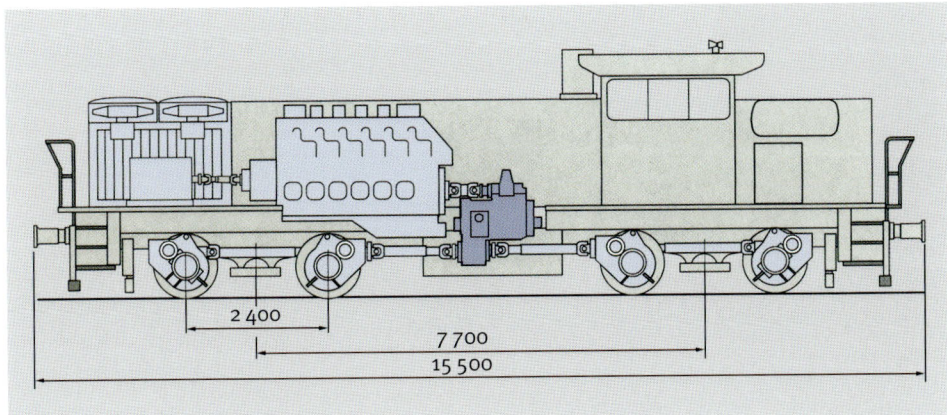

*Belgian State Railways
1 150 kW-multi-purpose locomotive, class HDL77.
Locomotive mass 90 tonnes
Maximum speed 60/100 km/h*

handed over to SNCB in October 1999 and immediately entered scheduled
service. In 2002, another 80 locomotives were ordered, and by mid-2005 all of
them had been delivered.

Turbo Reversing Transmission Size 5

For services similar to those of the SNCB locomotives, Vossloh in Kiel also
modified a modern multi-purpose locomotive from its standard program for
Swiss Federal Railways. As this locomotive had to haul goods trains on long
mountainous routes, an engine output of 1500 kW was required. The low-speed
range of the L 5r4 zseU2 turbo reversing transmission with range-change gearing
had a particularly high speed reduction. This made it suitable for service on
construction sites where vehicle speeds of 2 to 5 km/h are frequently required.
An important feature for SBB was the possibility of braking the train dynamically
with the counter-rotating converter on longer downward gradients. In this way,
the locomotive can continuously hold trailing loads of 600 t at 50 km/h on 10‰

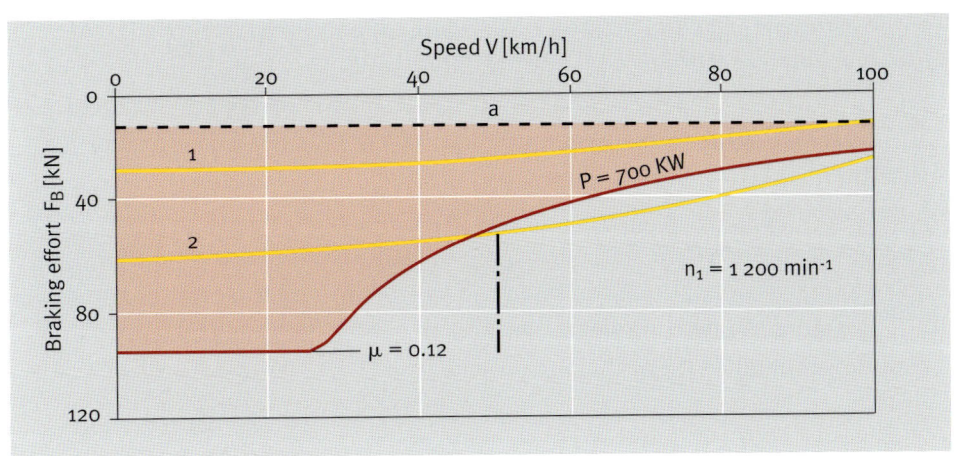

*Dynamic braking of the Am 843
locomotive fitted with a size 5
turbo reversing transmission in
the high-speed range.*

*a Lower control limit
1 250 t trailing load
2 600 t trailing load on
   10 ‰ downward gradient*

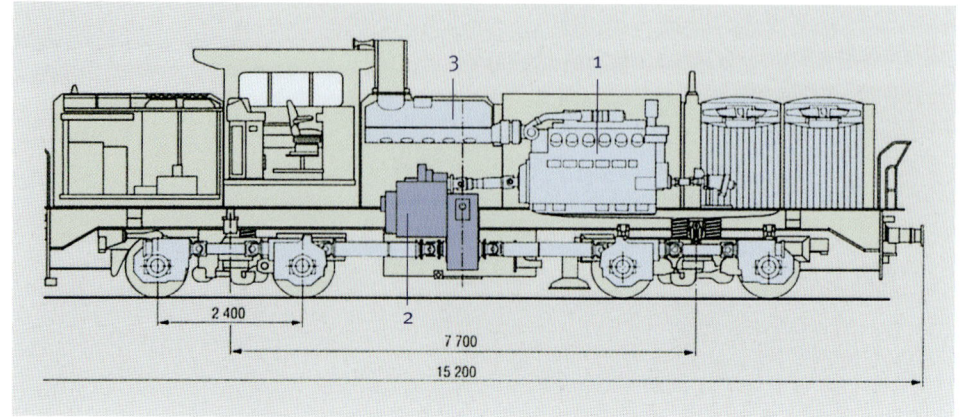

Am 843 locomotive of Swiss
Railways (SBB).
Locomotive mass 80 tonnes
Maximum speed 40/100 km/h
1 Engine (Caterpillar, 1 500 kW)
2 Transmission (Voith, 1 400 kW)
3 Particulate trap for exhaust

L 620 reU2 + KB 385 turbo transmission for inputs up to 2 700 kW.
1 Input
2 Step-up gear
3 Step-up pinion on primary shaft
4 Starting converter
5 Cruising converter
6 Combined filler, control and
 scavenger pump
7 KB 385 brake
8 Output

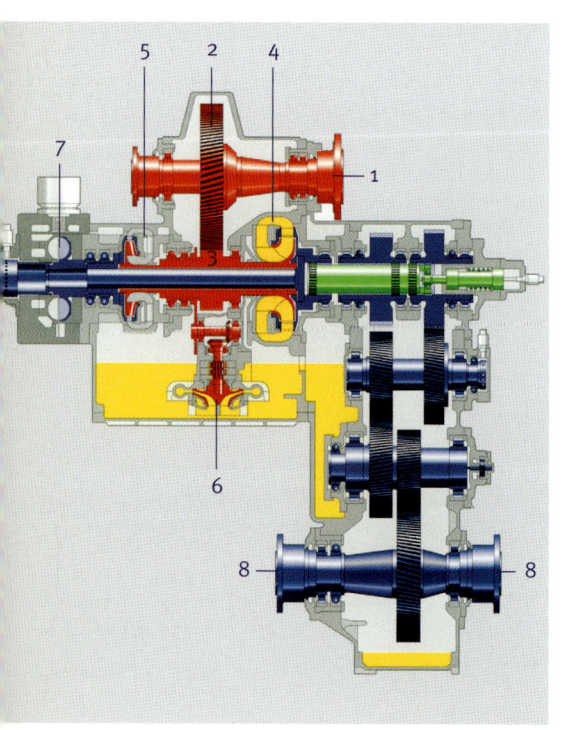

downward gradients, without wear and without the risk of the transmission oil or
the engine cooling water overheating.

Turbo Transmissions for Maximum Power

In 1999/2000 Voith developed a completely new two-converter turbo transmission with electronic control. The L 620 reU2 transmission was intended
for the MaK 2000 mainline locomotive of Vossloh in Kiel whose rated power of
2 200 kW represented a further extension of the power range of four-axle diesel-
hydraulic locomotives and allowed heavy freight trains to operate at 120 km/h.

The design of the transmission corresponds to that of the L 520 rzU2.
To enable a light and compact unit, the improved 52 k starting converter with a
profile diameter of 525 mm was selected, while the profile diameter of the
cruising converter measured 434 mm.

In 2003, the permissible input power was raised from 2 200 kW to 2 700 kW
and a hydrodynamic KB 385 brake was added. Compared with the L 520 rzU2,
the new turbo transmission excels in two outstanding aspects:

– Low weight to power ratio: without brake 1.76 kg/kW, with brake 1.98 kg/kW.
– Higher efficiency due to higher power density.

As the diagram for the MaK 2000-4 with its 2 700 kW-MTU engine shows,
trailing loads of up to 1 000 t can be towed at 110 km/h on level routes. On downward gradients of 10‰, the same train can be kept constantly at the relevant
permissible speed.

In order to dissipate the required 2 000 kW continuous braking power of the new KB 385 brake, without exceeding the maximum oil temperature of 130 °C, the flow cross-sections of the pipes and valves had to be adapted to flow rates of 24 to 33 l/s. This resulted in a brake weighing 600 kg.

By the end of 2005, 69 of the new MaK mainline, Caterpillar engine locomotives rated at 2 250 kW had been delivered. Three locomotives rated at 2 700 kW and fitted with MTU engines have hydrodynamic brakes fitted. Another 21, G 1700 locomotives with central driver's cabins and an engine output of 1 700 kW were fitted with L 620 reU2 transmissions, too, so that the total number of L 620 reU2 ordered or delivered units amounts to 102 at the of end 2005.

Railcar Transmissions

Apart from the L 620 r, further development of locomotive transmissions focused on turbo reversing transmissions. However the changing market of the railcar sector required a number of completely new designs.

Two-speed transmissions were no longer sufficient for the ever-increasing vehicle speeds. The new three-speed transmissions returned to the converter-coupling-coupling design that Voith had adopted in the thirties. The new engines with their wide speed range and high torque multiplication thus allowed a come-back for the high efficiency turbo coupling. Combined with an integrated brake

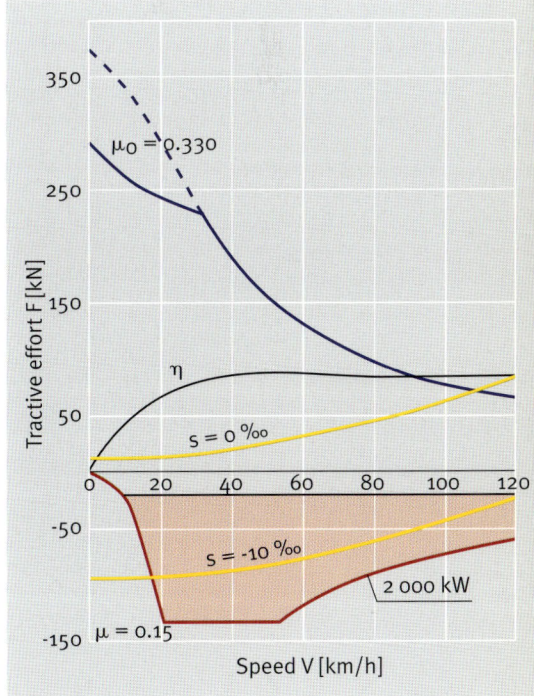

*Tractive/braking effort diagram
of MaK 2000-4 rated at
2 700 kW and 2 000 kW braking
power.
Locomotive mass 90 tonnes
Maximum speed 120 km/h
Resistance to motion for 1 000 t
trailing load*

and an electronic transmission control, diesel-hydraulic transmission systems became even more attractive for modern railcars.

The reinforced version of the T 211 r and the T 311 r were supplemented by a three-speed version. The rapid developments of the last 15 years eventually leading to complete drive systems in the shape of Voith powerpacks are demonstrated by the examples of Germany and the United Kingdom, complemented by the new generation of DIWA transmissions for light railcars and rail buses.

The Success of the Regio Shuttle in Germany

At the beginning of the nineties, the new DIWA transmissions again became attractive for rail applications. Existing two-axle rail buses in the new Federal States of the reunified Germany were modernized using the DIWA transmission since they already had reversing final drives. The same applied to a larger fleet of rail buses owned by Hungarian State Railways. These vehicles, capable of 100 km/h maximum speed and rated at 200 kW were equipped with the new D 863.2 three-speed transmissions.

For the new RS1 Regio Shuttle built by Waggonunion/Adtranz/Stadler, Berlin, the four-speed version D 864.2 was chosen. The RS1 has a large low-floor area between the two bogies and is significantly more comfortable, efficient and faster compared with the old rail buses.

Voith D 864.2 DIWA transmission, rail version.

1 Input with torsionally flexible coupling
2 Input planetary gear
3 Torque converter
4 Output planetary gear
5 Reinforced output bearings
6 Secondary lube oil pump
7 Heat exchanger

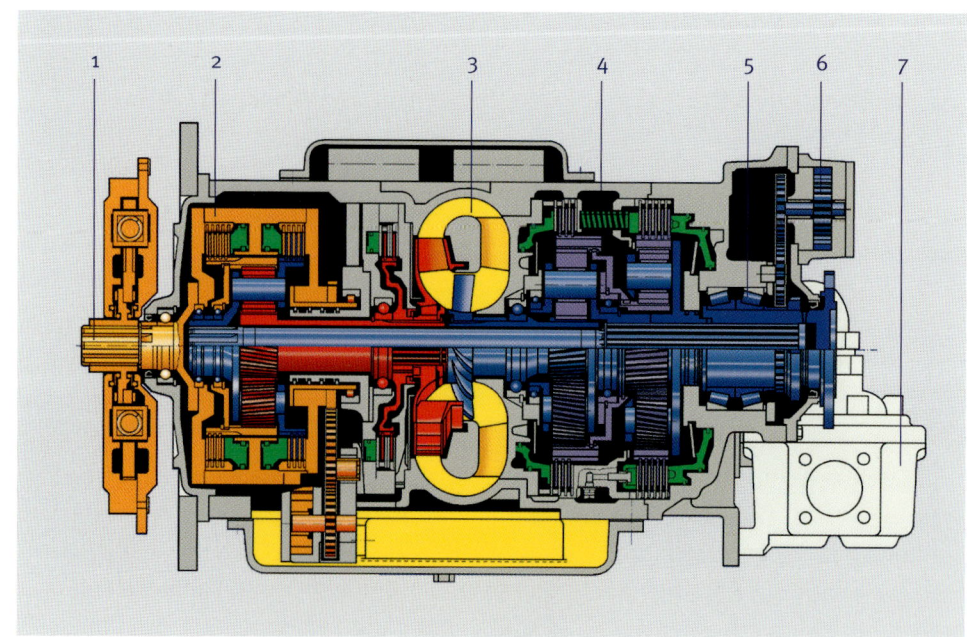

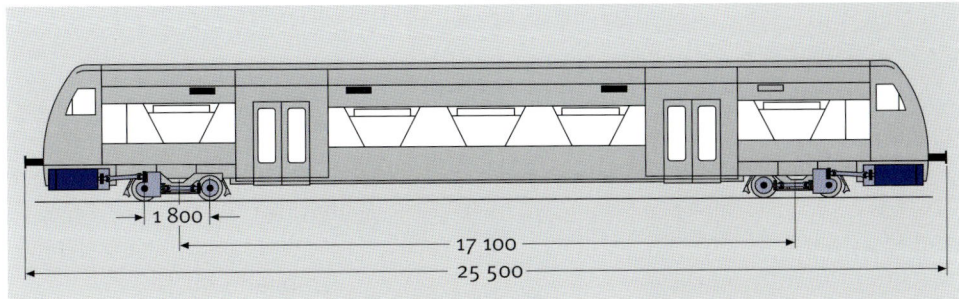

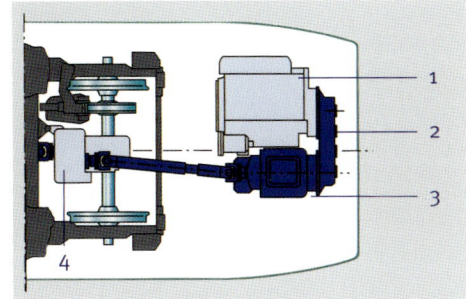

During the nineties, the Regio Shuttle successfully represented the new generation of light railcars procured in the course of a modernization campaign by regional railways in Germany.

Due to the low-floor section, the two 257 kW drive systems had to be accommodated in the front and rear area of the car. Engine and transmission were arranged next to each other and connected via a mechanical transfer gearbox. The underfloor cooling unit was integrated into the engine-transmission assembly, which was mounted in the car body by flexible rubber elements.

By 2005, 325 Regio Shuttles had been delivered and are now in service all over Germany. Major applications can be found with Deutsche Bahn as VT 650 (73 units) and with South-West German Regional Railways with its mountainous routes (100 units). The average mileage of the railcars amounts to 110 000 km per year with top runners up to 250 000 km.

The T 211 r becomes the Standard Transmission of the New British Diesel Railcars

The small T 211 r railcar transmission developed in 1969/70 is an example for the power reserves of a hydrodynamic transmission. Whilst retaining its main dimensions, the transmission was continuously adapted to the required higher engine outputs. Hand in hand with the increase in the engine output from 213 kW to 315 kW, the peripheral speed in the circuits rose by 17%. The oil press fits and bearings were strengthened.

The first 571 British railcars had a maximum speed of 120 km/h and were built as two-car sets. The BREL Class 150 was built with 20 m car bodies, while the Metro-Cammel Class 156 and Leyland Class 155 were built with 23 m car bodies. All units were delivered between 1985 and 1988 and had the same driveline: 213 kW NT855R5 Cummins engines and Voith T 211 r turbo transmissions.

The two-engine RS1 Regio-Shuttle built by Adtranz/Stadler, Berlin with four driven wheelsets and 2 x 220 kW for traction. Vehicle mass 42 t (empty) Vehicle mass 48 t (with passengers) Maximum speed 120 km/h

1 Engine (MAN)
2 Mechanical transfer gearbox (Voith)
3 DIWA transmission (Voith)
4 Reversing final drive

Tractive and braking effort diagram for the Regio-Shuttle with two engines, each rated at 257 kW and 220 kW transmission input power

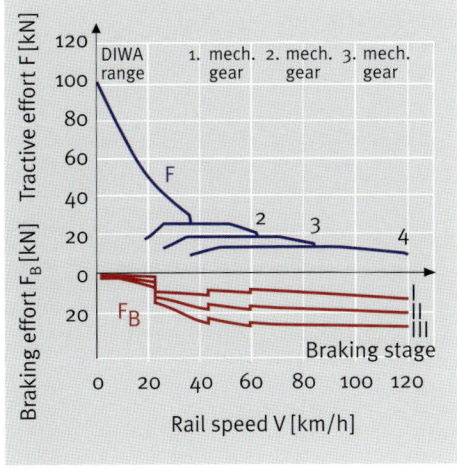

Three-car DMU Class 158 in Edinburgh. October 1990

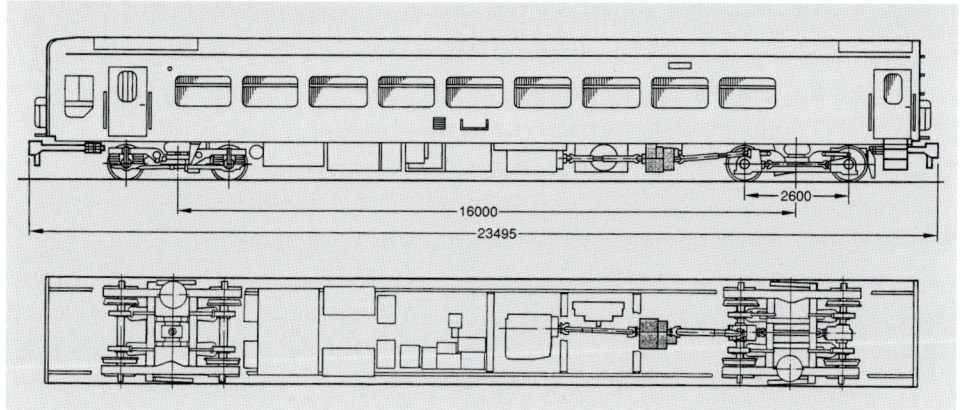

All of the Class 158 cars were fitted with 261 kW underfloor power units.

In 1989, the second generation of the new British railcars arrived in the form of Class 158. The 400 railcars were delivered by BREL, Derby, as three-car units in 1989/1990. At 145 km/h, they were faster than the first generation, had 23 m aluminum car bodies and a loading gauge that allowed them to operate on all routes in the United Kingdom.

The second series was concluded in 1991/92 by 22, three-car Class 159 sets, for Network South East (NSE) – the "South Western Turbo". Similar to the Class 158, they had higher engine outputs.

At the same time, BREL in York worked on a new family of railcars, the Class 165/166 "Networkers". These trains had wider bodies and had no through access, when run in multiple formations. The driveline was adopted from Class 158, but Perkins engines with an identical power output replaced the Cummins engines.

Class 165/1 DMU driving
towards London Paddington.
July 1992

T 211 rzze, the first transmission
of the T 211 r family equipped with
an electronic-hydraulic control.

1 Control unit with solenoid
   valves

Including spare units, Voith delivered 778 T 211 rz transmissions for the second generation. In the middle of 1994, the manufacturer – BREL had meanwhile been taken over by ABB Transportation Ltd. – stated that the operating cost of the new vehicles were comparable to those of the electrical railcars run by Network South East; fuel consumption and maintenance cost had declined drastically.

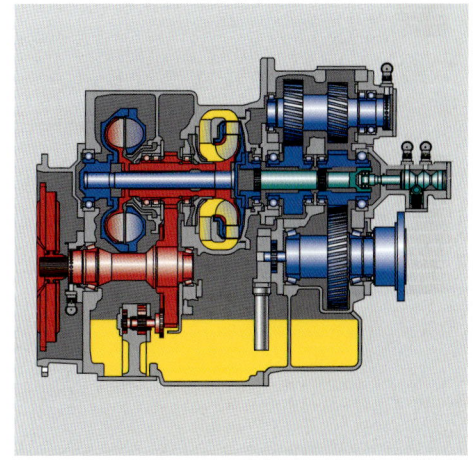

By 1997, British Rail had been privatized and divided into 25 "TOC's" (Train Operating Companies).

For the British railway industry, privatization was a very difficult process, and there were no new orders between 1993 and 1997. During this time, Adtranz as the successor of ABB, developed a new railcar concept, the Turbostar. On the outside, Class 168 for Chiltern Railways was still a Networker-like DMU, but the technology was new. For service on Intercity routes, the maximum speed had to be increased to 160 km/h. With its 315 kW MTU underfloor engine, a three-car train had an output of 945 kW. Another new feature was that engine and transmission were flanged together into a powerpack in order to provide space for the numerous auxiliary machines including air conditioning.

The turbo transmission was again redesigned, and the control system was converted to Voith Turbo Control. This control processes the driver's commands and the signals from the sensors fitted to the transmission and forward the output to the electro-hydraulic solenoid valves on the transmission. They open and close the hydraulic control lines leading to the reversing cylinder and the

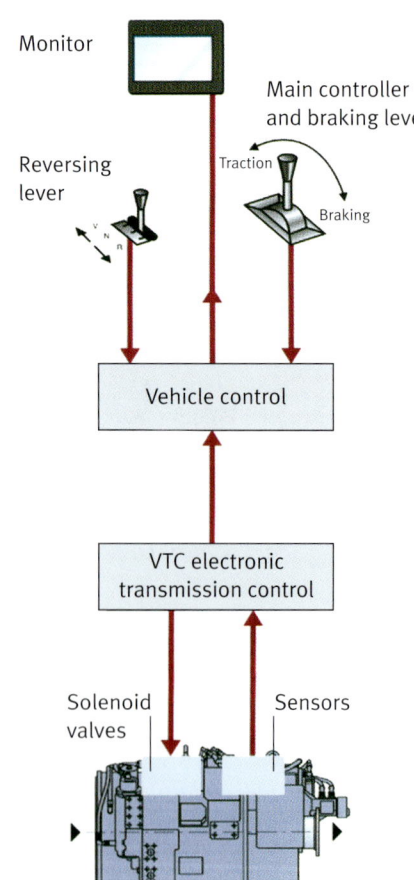

Monitor

Reversing lever

Main controller and braking lever

Traction

Braking

Vehicle control

VTC electronic transmission control

Solenoid valves

Sensors

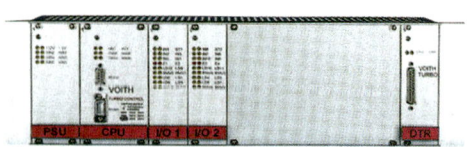

Linkage of transmission and vehicle control. With Voith micro-processor rack system in the control cabinet.

main control cylinder. As a result, the transmission can be actuated without the need for control air.

Class 168 was not suitable for universal service on all routes of the privatized British railway companies, as its car body was too wide. For this reason, Adtranz in Derby developed the Turbostar family mentioned earlier, which was fitted with the drive components from the Class 168, but had a far more attractive and aerodynamic appearance. Compared to the equally freely usable Class 158/159 it also had no end-gangways.

The maximum speed of 160 km/h of the Turbostar was very high for a two-speed transmission. However the engine output of the three- and four-car trains ensured good acceleration so that the vehicles could primarily run in the coupling range between 100 km/h and 160 km/h. This was assisted by the favorable torque characteristics of the MTU engine, resulting in a tractive effort characteristic that was similar to that of a two-converter transmission.

From 1999/2000, numerous private rail operators purchased the Turbostar as the third generation diesel-hydraulic multiple unit. After overcoming initial difficulties, by the beginning of 2003 it had developed into a highly successful universal vehicle, with a casualty rating of 23 000 "miles per casualty". By then, 359 Turbostar vehicles had been ordered from Adtranz and its successor Bombardier, Derby; all of them were fitted with Voith turbo transmissions and MTU engines.

As key supplier of diesel-hydraulic railcars apart from Adtranz/Bombardier in Derby, GEC-Alstom the successor of Metro-Cammell in Birmingham, also offered a new range of high-speed railcars that complied with the more sophisticated demands made by Intercity traffic.

In early 1999, First North Western ordered from Alstom 11 two-car and 16 three-car Class 175 diesel multiple units capable of 160 km/h. Unlike the Turbostar, a 335 kW Cummins engine was flanged to the T 211 re.3 transmission, which was, for the first time in the UK, fitted with a hydrodynamic brake. As a result, 50% of all wheelsets of the multiple units could be dynamically braked. The continuous braking power per drive unit was 220 kW; while 390 kW was permissible over short periods, depending on the oil temperature.

Between January 1999 and March 2000, 70 turbo transmissions were delivered to Alstom; the commissioning of the trains was completed by mid 2001.

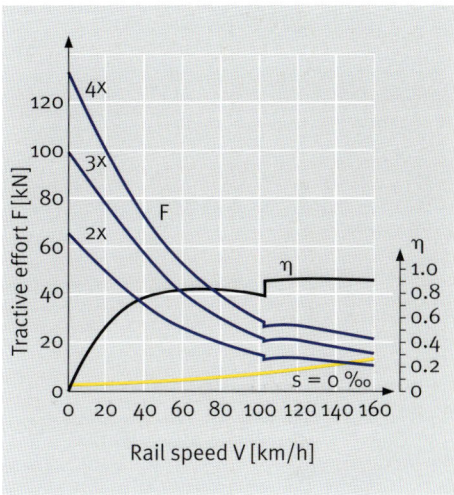

By the middle of 2004, i.e. three years later, the entire fleet was covering an average of 200 000 km per year. From the middle of 2004, many of the trains were used on the Transpennine route and, from 2006, replaced by the new Class 185.

The New Diesel-Hydraulic Class 185 Multiple Units for the Transpennine Express

From 1991, intercity traffic had been served by Class 158 diesel multiple units running from Liverpool via Manchester to Leeds and Newcastle across the Pennine hills in the north of England. From 2001, Class 175 took on more of the task. In 2006, the higher-powered Class 185 high-speed trains replaced them. In August 2003, First Group and Keolis had placed a major order worth £ 250 million with Siemens for 56 three-car diesel multiple units. The vehicles were developed and built by Siemens/Duewag in Uerdingen.

The new trains are the diesel-hydraulic version of the electric Desiro trains built by Siemens. The cars measuring 23 m in length are fitted with new bogies from Siemens, Graz, with Voith final drives and a 559 kW underfloor Voith Turbopack drive system consisting of three modules:

– Cooler Group with three hydrostatically driven fans and cooling elements for the charge air and the cooling water of the engine. The 70 kVA on-board alternator is driven at constant-speed by an hydrostatic drive system.
– Cummins QSK19-R750 diesel engine
– Voith T 312 bre turbo transmission.

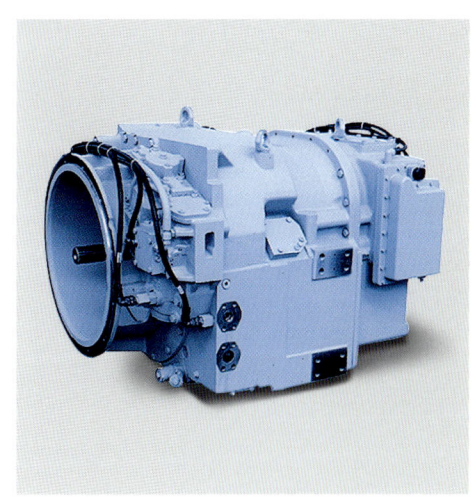

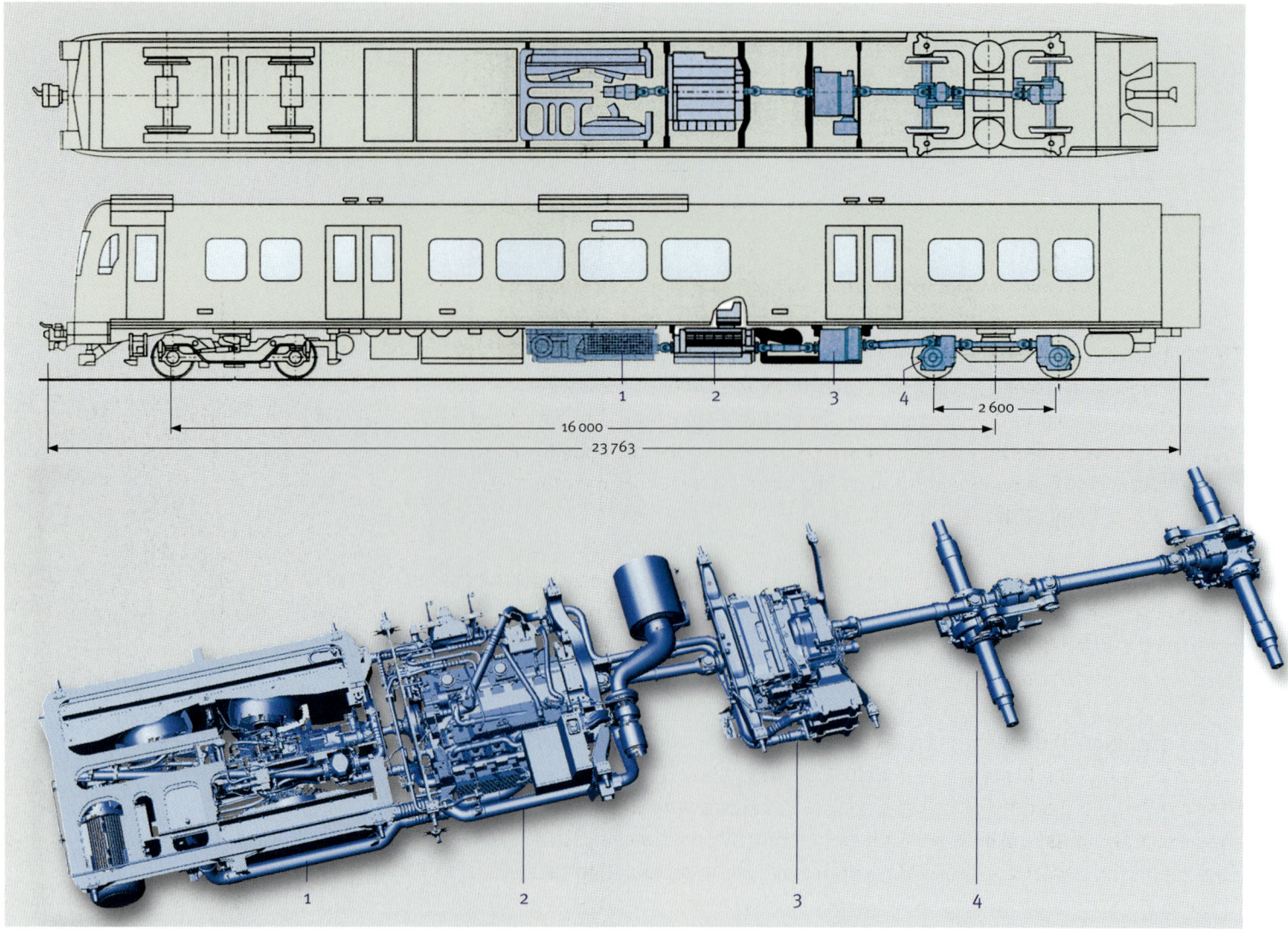

Class 185 diesel multiple unit with 559 kW underfloor drive unit.

1 Cooling unit with hydrostatic triple pump driven by the engine via a cardan shaft
2 Cummins diesel engine with air filter
3 Voith T 312 bre turbo transmission
4 Voith SK 485 final drive

The Voith scope of supply also includes the two double reduction final drives, the cardan shafts and the driveline control. The vehicle manufacturer supplies the integrated Cummins diesel engine.

This order is not only the largest in the history of the Voith's Rail Division, but also is an example of the changed procurement strategies of rail vehicle manufacturers. Until then, vehicle manufacturers purchased individual components separately from sub-suppliers and were hence responsible for their ultimate interaction. Against the background of ever more complex vehicles, shorter development times and the need for standardization, the late nineties were characterized by an increasing demand for deliveries of complete packages. As a systems supplier, Voith was now in charge of the entire driveline including the integration of the diesel engine. This development strategy allowed intensive trial runs of the complete drive system on the test bed before delivery.

The T 312 bre three-speed transmission with the converter-coupling-coupling design was developed in 1994/95 for transmission inputs of up to 650 kW for the Class 611/612 tilting trains of Deutsche Bahn. In 2000 Alstom, Birmingham also employed it for the first time in a British application, the five-car Class 180 diesel multiple units, built for First Great Western. The four hydrodynamic circuits are driven by the diesel engine via a trio of gears divided into two rotors.

The hydrodynamic brake is integrated, as its application in high-speed railcars offers economical advantages. The reversing gear is actuated via two hydraulic "on/off" cylinders and does not use air. The transmission and the engine are mounted via two cross members that are connected to the car body by means of rubber elements. The cooling module is mounted on a steel frame mounted below the car body at the four points.

Compared to the Class 158 and 175 high-speed trains previously used on the Transpennine route, the new trains are heavier yet significantly more powerful. Their acceleration is better and they are capable of maintaining their maximum speed of 160 km/h on the 5‰ uphill gradients.

Construction of the vehicles at Siemens/Duewag in Krefeld started in the middle of 2004. The first three-car train was completed in May 2005 and thoroughly tested on the Wildenrath test circuit. Further trial runs followed in the Eifel mountains in Western Germany, until finally the train was shipped to England at the end of 2005. Regular service at TPE is scheduled to take place in 2006.

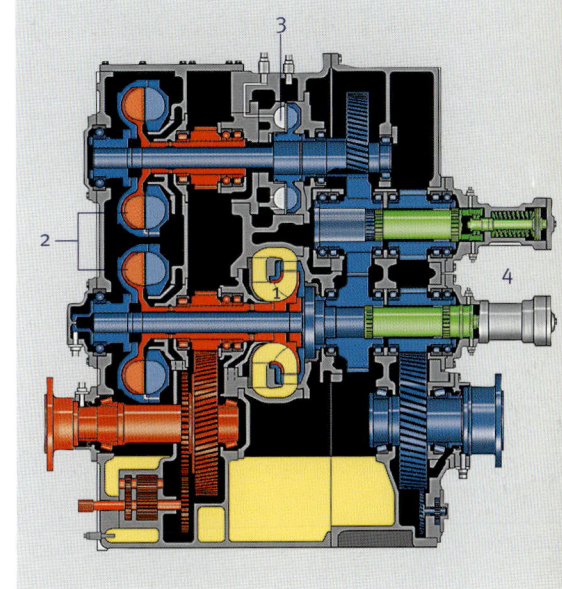

T 312 bre turbo transmission
1 Converter
2 Couplings
3 Dynamic brake
4 Reversing section with
  operating cylinders

Three-car Class 185 DMU
during test runs in Wildenrath.
August 2005

The Voith T 212 bre Railpack, four-station, continuous assembly line with air cushion transport system, destined for the Minuetto diesel multiple units purchased by Italian State Railways.
In the background a transmission and engine module, part of the Voith Turbopack, for the Class 185 DMUs of First Group/United Kingdom.
November 2004

Outlook

New Strategies in a Changed Environment

Voith hydrodynamic power transmission for rail vehicles has a long tradition, spanning almost 75 years. During this period, a large number of turbo transmissions have been developed for the most diverse applications and installation conditions, many of them still proving themselves in daily operation. They always represented the latest state of technology. Yet this remarkable technical development is not completed but continues and reacts to the changing conditions of the market.

In Germany and Europe, many of the traditional vehicle builders no longer exist and have made way for large groups such as Siemens, Bombardier and Alstom. Outside Europe, the market for diesel locomotives is still dominated by General Motors and General Electric.

Equally far-reaching changes have been experienced by state railways. In Germany and the United Kingdom, they underwent partial or complete privatization in order to be able to react more flexibly to the redefined structures of the rail traffic.

For diesel locomotives, Voith continues to offer modern turbo reversing transmissions for the lower and medium power classes. This is complemented by new developments for medium and high power locomotives dedicated to mainline operation. Voith is currently in the process of developing an attractive alternative to the single-engine, six-axle diesel-electric locomotives that dominate the world especially in the higher power range.

As far as diesel railcars with underfloor drive systems are concerned, diesel hydraulics still offer numerous advantages compared to diesel electrics. By evolving into a systems supplier and owing to large overseas orders, Voith has maintained its leading position. The new British Class 185 high-speed DMU built by Siemens for First Group is an example of this. Three new developments intended to secure the future transmission sales of the Voith Rail Division are introduced in the following text.

L 311 reV2 for Special Vehicles

Voith has been delivering turbo transmissions for track construction/maintenance vehicles for many years. As early as 1970, the company delivered a series-manufactured two-converter transmission for the drive of a track-bed-cleaning machine built by Plasser & Theurer, Linz, with an engine output of 735 HP/540 kW.

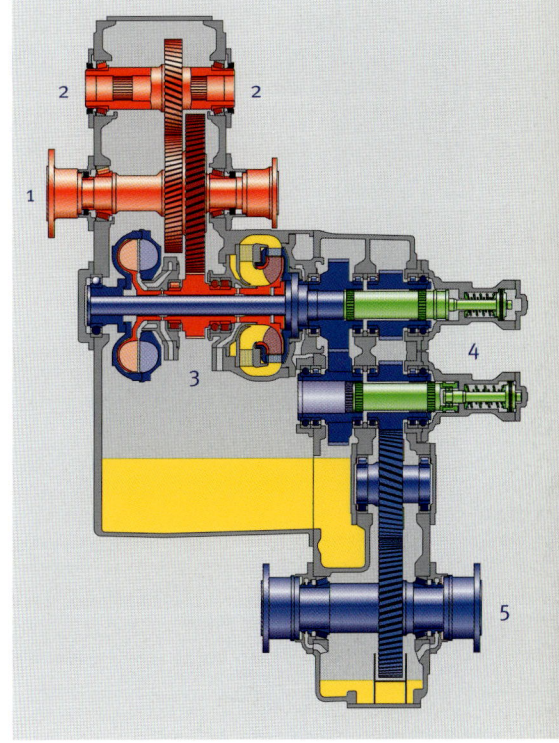

L 311 reV2 special vehicle transmission for an input power of 600 kW.

1 Input
2 Primary auxiliary drive (PTO)
3 Rotor
4 Hydraulic cylinder of reversing section
5 Output

Plasser & Theurer 09-4X Tamping machine, Linz, June 2005.
Total mass 126 t
Engine output 2 x 440 kW
Maximum vehicle speed 115 km/h

Modern track construction machines such as tamping machines with track stabilization require increasingly higher outputs. They have to be able to drive quickly from their depots to the construction site and back, which can only be done with high traction power. On site, the entire output is transmitted hydrostatically to drive the tamping machinery.

The division between hydrodynamic and hydrostatic power transmission is a feature of these special rail vehicles, one in which the high-performance hydrodynamic traction drive is augmented by hydrostatic drives for lower power at lower speeds.

New Transmission Concept for High Power

With the high-performance, 2 700 kW input rated, L 620 reU2 unit, the development of large turbo transmissions for diesel locomotives has reached its current zenith. The transmission allows the construction of four-axle, single-engine 3 000 kW diesel locomotives. For higher outputs, six-axle diesel locomotives are mainly used.

With a maximum power rating of 3 600 kW they would have the potential to move high-speed cargo and passenger trains on electrified lines right across Europe. Irrespective of the system supply, they would be able to offer the same journey times as multi-system electric locomotives without the need for locomotive changes at borders. Voith has developed a transmission concept that is based on the principle of the double turbo transmissions of the thirties. At the time, Voith hydrodynamically decoupled the two wheel sets of a power bogie

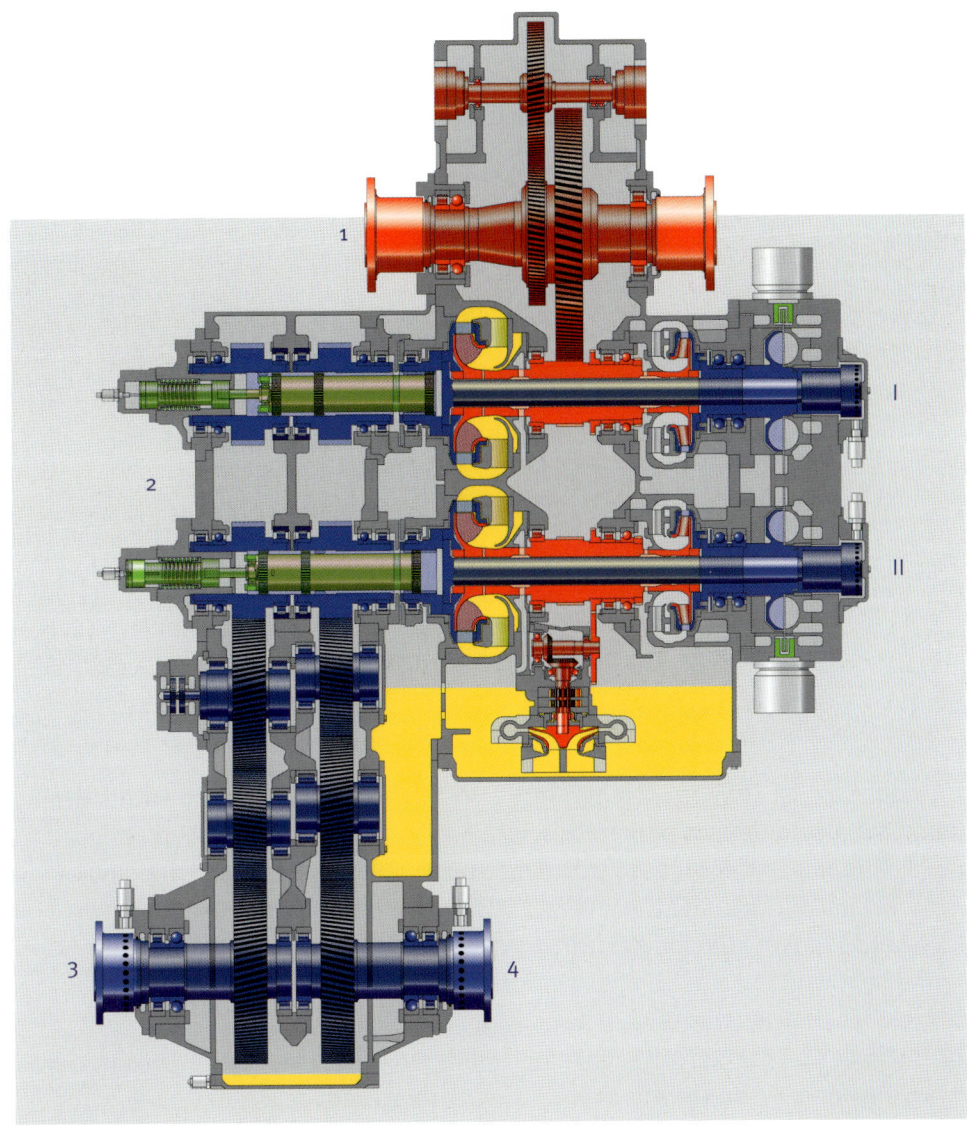

Voith LS 640 reU2 + KBD 385
turbo split transmission.
Mass with double brake 9 200 kg
Oil filling 400 l

1 *Input*
   I Rotor for bogie 1
   II Rotor for bogie 2 (both with
   hydrodynamic brake optional)
2 *Reversing gearbox, separate*
   for both rotors
3 *Output to bogie 2*
4 *Output to bogie 1*

using two separate torque converters. In the new turbo split transmission, two separate rotors hydrodynamically decouple the wheelsets of the two three-axle bogies. This results in a four-circuit transmission with two converters for each of the two bogies.

The turbo split transmission is designed for medium-speed or high-speed engines with speed ratings of 1 000 min^{-1} and 1 800 min^{-1} and with a maximum input power of 4 200 kW. Each of the two rotors can be equipped with a hydrodynamic brake.

During mainline service the two torque converters for the particular direction of travel are filled, and during shunting operation turbo reversing at half power is possible. In this case, the two rotors of the reversing gearbox are engaged for different direction of travel, and only one of the two starting converters is filled with oil.

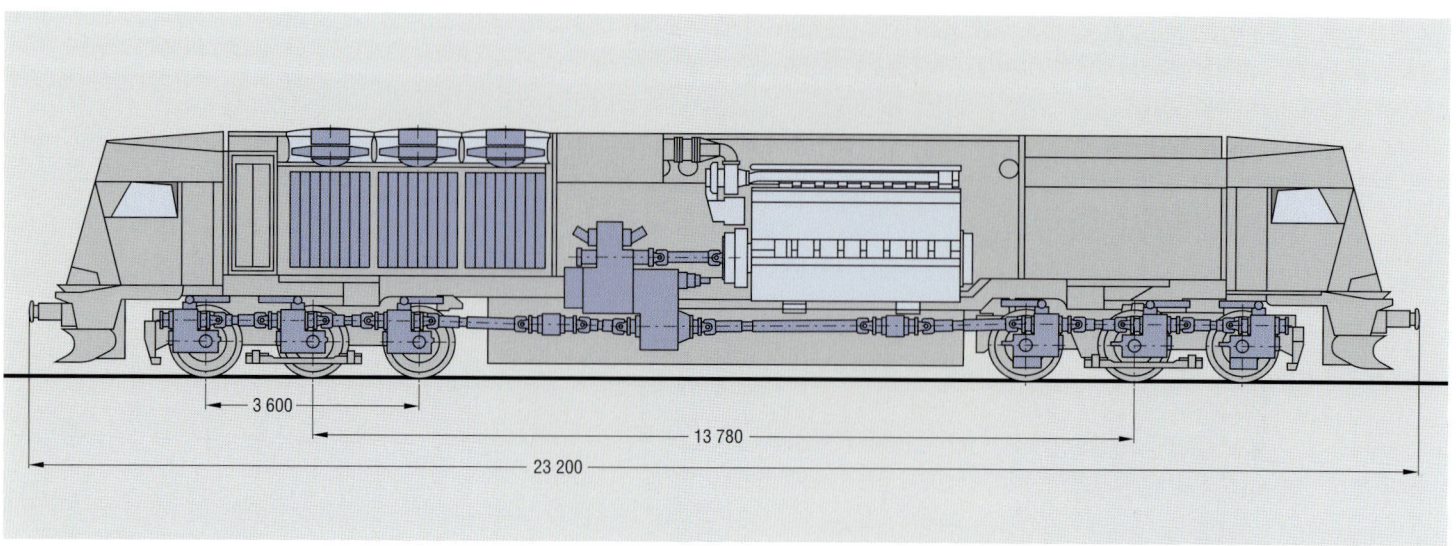

3 600

13 780

23 200

Diesel-hydraulic 3 600 kW locomotive for high-speed goods and passenger traffic with decoupled bogie drive.

In the case of Central European and Non-European railways Voith at its own risk is developing the complete locomotive concept up to the state of series-build. This will enable Voith to prove the suitability of the drive concept as an alternative to electric and diesel-electric high-performance locomotives. The locomotive is available for construction under license by any interested party around the world.

TurboFlexx – a Modular Transmission Family for Diesel-Hydraulic Locomotives

In the 800 kW to 1 400 kW power class, turbo-reversing transmission with and without range-change gearboxes have been dominant. In some cases, locomotives exclusively used for mainline duties were fitted with the L 520 rzU2 two-converter transmission.

In German goods traffic, railway operators primarily purchased new electric locomotives for electrified mainlines. Refurbishment and upgrades, carried out in railway workshops have extended the service life of diesel locomotives on branch lines and in shunting duty. As a result of the growing emphasis on main-line operation compared to shunting, private railways are procuring new loco-motives: this naturally has an effect on the transmission type required. Turbo transmissions with mechanical reversing and hydrodynamic brakes are regaining popularity. Unlike comparable turbo reversing transmissions, their design is fairly simple and they offer better efficiencies, especially at high rail speeds.

To cover the range of modern diesel locomotives it is therefore planned to develop a new transmission family with input powers up to 1 700 kW. The basic transmission has two converters with two outputs situated below the main body of the transmission. The optional hydrodynamic brake is integrated and, as with the T 312 br railcar transmission, arranged on a second rotor. With the three-converter type, this rotor also contains an additional cruising converter III which is mechanically staggered in respect to the converter II.

The four possible variants
– Two-converter transmission without and with brake
– Three-converter transmission without and with brake
have the same transmission housing. Each converter has its own main control valve with optimized channels incorporated in the compact housing.

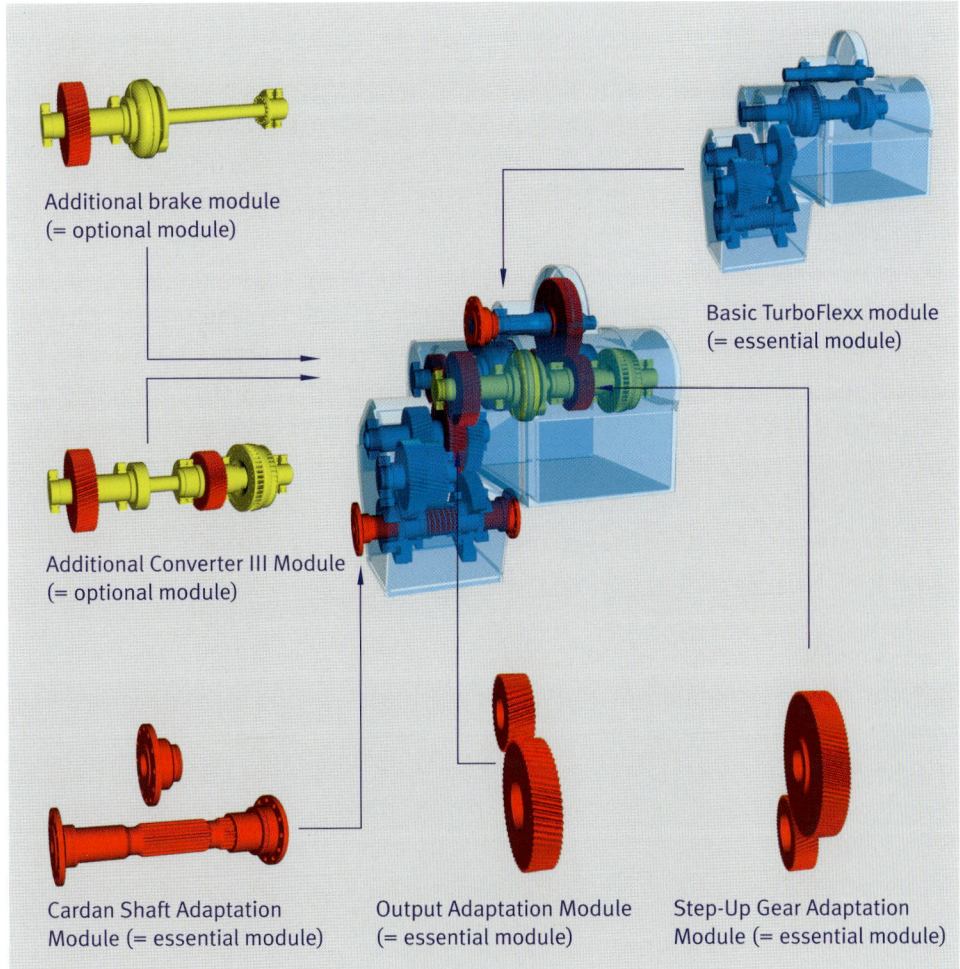

Additional brake module
(= optional module)

Basic TurboFlexx module
(= essential module)

Additional Converter III Module
(= optional module)

Cardan Shaft Adaptation
Module (= essential module)

Output Adaptation Module
(= essential module)

Step-Up Gear Adaptation
Module (= essential module)

Overview of the new modules, with the option of two or three converters and hydrodynamic brake. For an input power of 1 700 kW.

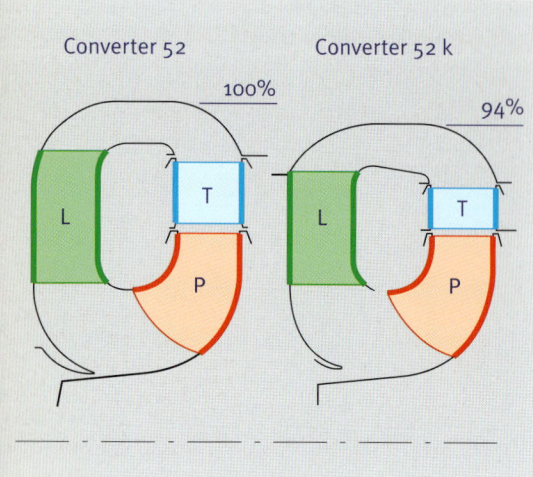

Converter 52   Converter 52 k
            100%
                              94%

Blade arrangements of starting
converter type 52 and improved
version 52 k.

P Pump
T Turbine
L Guide wheel

Further Development of Hydrodynamic Torque Converters

The torque converter is the heart of any turbo transmission. Over the last few decades, it has been continuously adapted to the traction requirements of diesel railcars. The development programs were aimed not only at maximum efficiency, but also at high stall torque multiplication combined with the broad speed range for the starting converters and, as far as possible, constant power absorption of the cruising converters. Among the numerous converter types, the single-stage model with centrifugal turbine flow proved to be the most suitable. It had a simple design and was ideal for high speeds owing to the high ring strength of its radial turbine. It became the standard converter for Voith turbo transmissions.

In the early seventies, two-converter transmissions gradually replaced the three-converter design for diesel locomotives. At the same time, turbo reversing transmissions became increasingly important for industrial locomotives. Both applications required hydraulically staggered starting/cruising converter combinations. Relevant development activities were based on the converter types of the fifties and continued over many decades. In countless field tests, new variations with a wide variety of stall torque multiplication, power absorption, a broad operating speed range, and different speed ratios were examined.

An example is the development of the type 52 starting converter. After numerous measurements of different versions, the Voith design department decided in 1967 that test number 199 provided the best results for the new L 520 rU2 turbo transmission. A series of further tests finally resulted in type 52 k with a six percent-reduced outer profile diameter. This gave the same power absorption and demonstrated a broader efficiency curve. These advantages were gained at the expense of lower stall torque multiplication. By improving the guide wheel blades, the latter was eventually raised to the higher value of the type 52 and, for the first time, realized for L 620 reU2 and T 212 bre transmissions.

Structured grid of a turbine
wheel with centrifugal flow as
part of the overall model of a
torque converter.

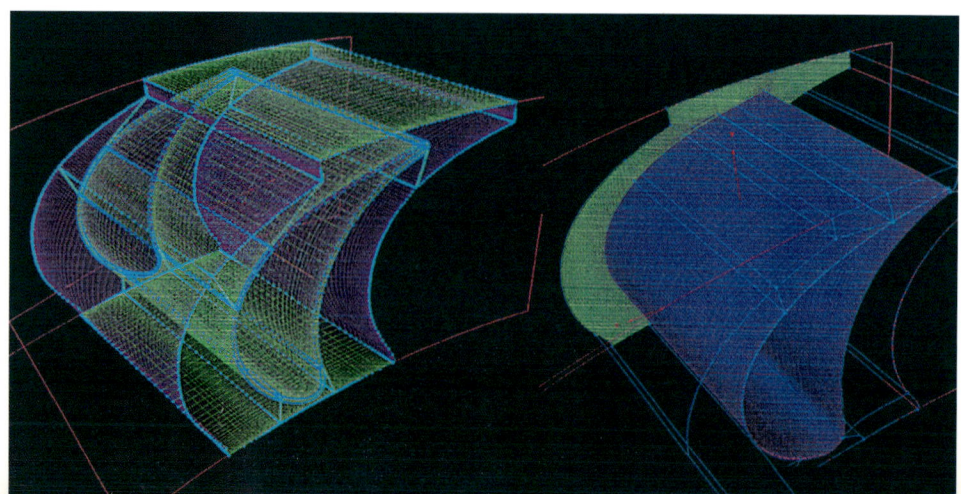

124

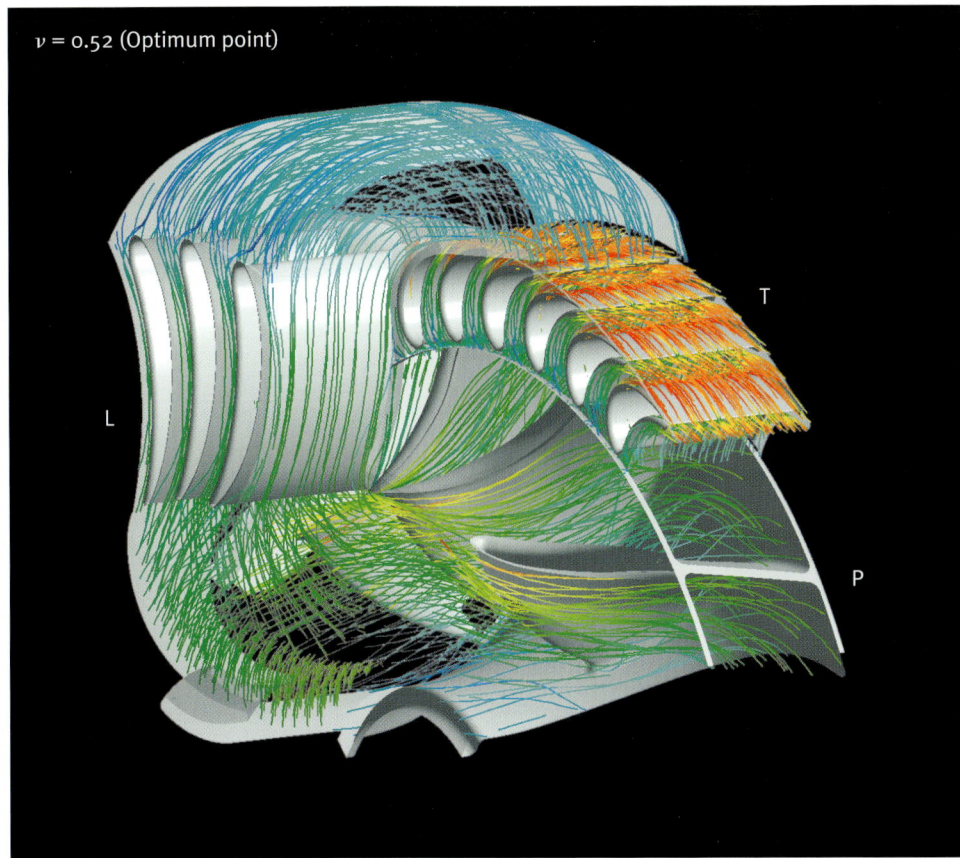

ν = 0.52 (Optimum point)

L

T

P

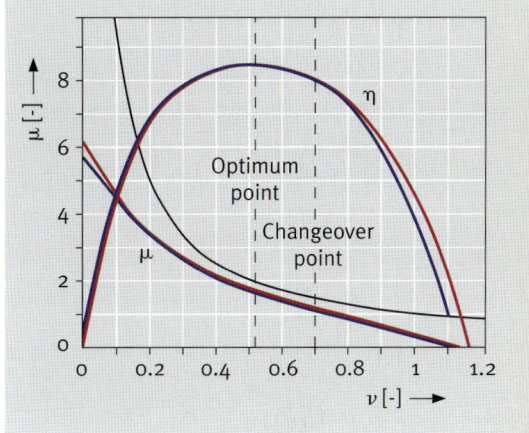

The successes in developing modern hydrodynamic torque converters for vehicle drives and industrial applications are closely associated with the names of the heads of the hydraulic R & D department, Rolf Keller, Norbert Pfisterer and Elmar Rohne.

Although it has reached a high level, the development of converters still continues. Conventional research methods have meanwhile been enhanced by computational fluid dynamics (CFD). Testing an individual converter at the end of a development chain is simply carried out as a means of checking results. In the first instance it is necessary to establish basic patterns of the flow through the blade-grid of existing converter types, targeted improvements would otherwise be impossible. Generating the grid of the converter turbine is a building block of the overall model "Torque Converter". The complete model of the blade grids for pumps, turbine and guide wheels allows the flow processes to be simulated inside the converter during different operating conditions. The optimum point at a speed ratio of *ν* = 0.52 is characterized by a balanced flow field with minimal secondary flows.

The comparison of CFD simulation with measured test bed values displays good conformity. On this basis, computer simulations can be carried out in order to identify the effects of modified grid geometries on the converter characteristics thus removing the need for time and cost consuming test bed measurements.

Computer-generated flow field of the 52 k starting converter at optimum point.
Red: high relative speed at the turbine outlet (approx. 25 m/s)
Green: medium absolute speed between guide wheel and pump (approx. 14 m/s)
Blue: low absolute speed between turbine and guide wheel (approx. 7.5 m/s)

Comparison of efficiency η and torque multiplication μ based on the example of starting converter 52 k.
Red: numeric simulation (CFD)
Blue: measured values

Heinz Höller, Wolfgang von Berg, Helmut Fleuchaus

Starting and Controlling

Voith Turbo Couplings and Torque Converters in Industrial Plants

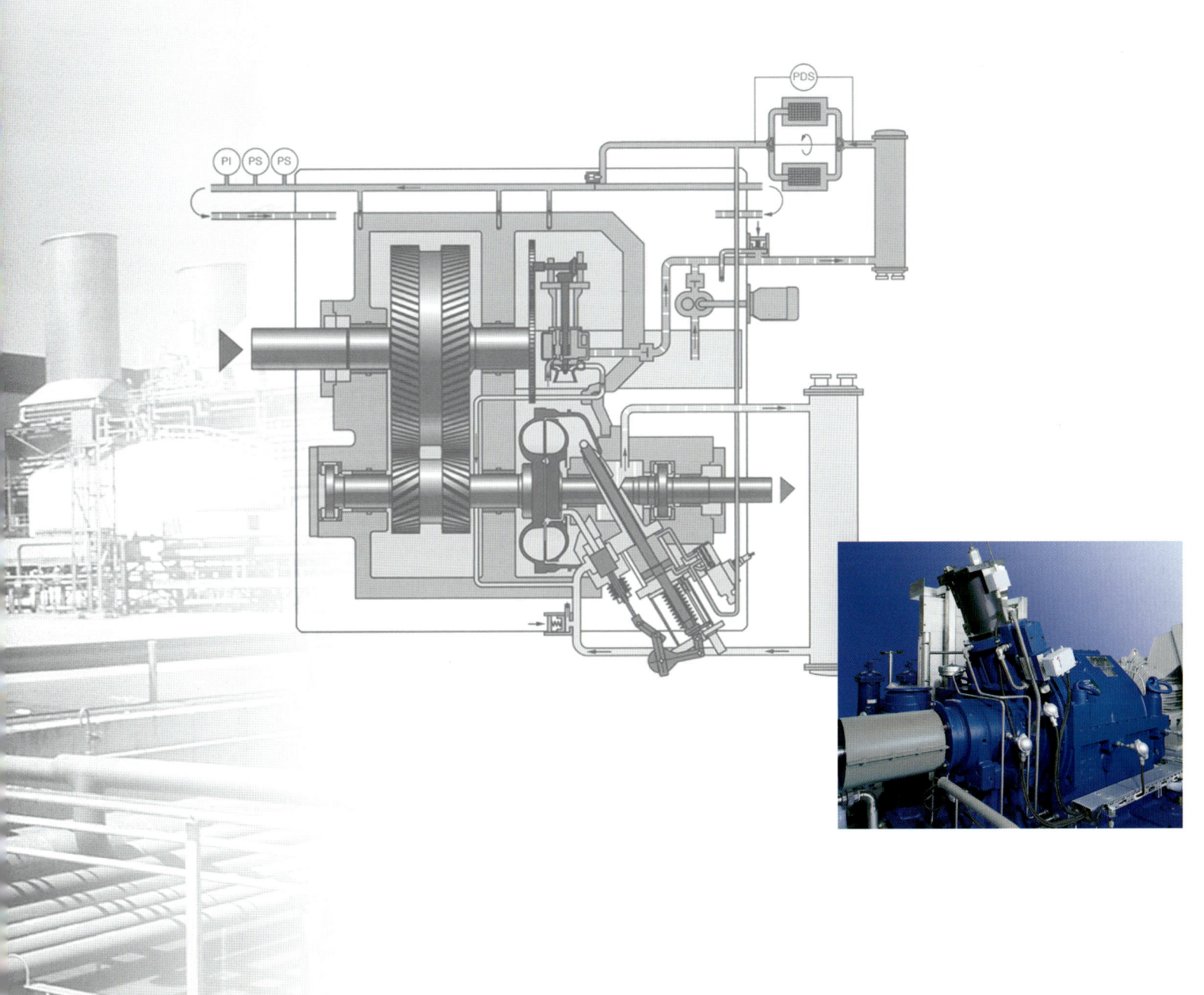

Development of Turbo Couplings and Converters

Industrial applications of hydrodynamic couplings and converters differ from those in rail and road vehicles in many respects. They are characterized by a much greater product variety and, as far as certain types are concerned, also by high unit sales. The following chapters show the wide range of applications in which these components are used. Their product development, initiated by a concrete requirement, can therefore be beneficial to an installation in a totally different field, either directly or after small adaptations. Additionally, close cooperation with end users soon revealed that it was commercially advisable to divide the market into a section for variable-speed couplings and one for constant-fill couplings.

Given these two conditions, the wide variety of similar applications and the existing market division, it appears practical to describe the past and future technical development of the products concerned in a summarized, more theoretical chapter. In contrast, the two chapters about applications are based on actual installations in the market and refer directly on individual products. Shared scientific and constructive details will be pointed out accordingly.

The further development of Föttinger units such as torque converters, couplings and brakes in the Market Division "Industry" from 1934 until today can be roughly divided into three periods with different focal points:

The years until 1949 were largely spent on establishing the **basics**. The findings from independent coupling developments for turbo transmissions and feedback from the actual application of Voith Sinclair couplings revealed weaknesses in the progression and the stability of characteristic curves. This resulted in the development of new design features, control methods and -devices.

From 1950 to 1969, numerous **market-driven functions** were added. Hydrodynamic drive components appeared in the market in new applications. An increased performance characteristics, the integration of gear drive and other new functions had a decisive impact on the path of developments.

After 1970, the range of tasks began to expand and is still expanding. The **process integration** of converters, couplings and brakes has significantly increased. Some main functions require **numeric models** to enable process studies. New applications and the **optimization** of efficiencies have come into focus. There is an increasing demand for environmentally friendly operating media.

Development Period Until 1949

Start-Up Couplings and Couplings with Switch-On Function

The development of this coupling type is strongly influenced by the research carried out in context with the "Herdecke" start-up and synchronization coupling, as well as turbo transmissions for rail vehicles.

In fill-controlled start-up couplings, the operating medium flows in at the hub area, while draining takes place via nozzle boreholes at the outer diameter. This results in a relatively slow draining process, which is disadvantageous especially during speed shifts in turbo transmissions.

For the new series of T-couplings in turbo transmissions, Voith therefore developed a membrane quick-draining valve that works automatically, depending on centrifugal forces and operating fluid pressure. Harold Sinclair later acquired the license for this development and used it in his fluid couplings. In a lecture at the "Institution of Mechanical Engineers" in April 1938, he made special reference to this Voith invention.

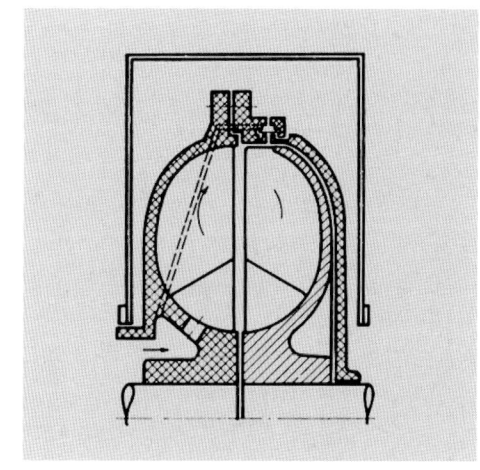

Föttinger coupling further developed by Voith with quick draining valve.

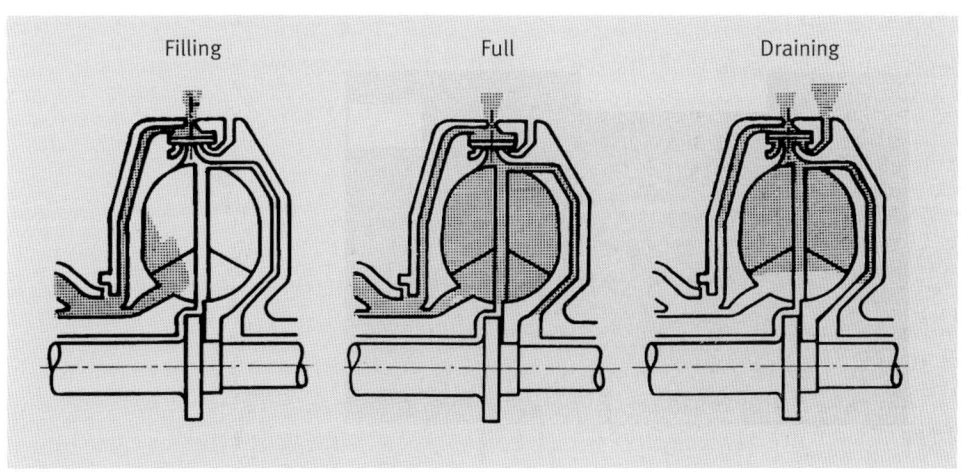

Filling	Full	Draining

Membrane quick draining valve. The pressure chamber of the membrane valve is connected with the filling side of the coupling via an outer pipe ("Hydraulic Piping"). The valve is activated by the volumetric flow.

Start-Up and Variable-Speed Couplings

The type S was the basic design of four types of start-up couplings from the licensing agreement with Harold Sinclair. The coupling profile with its closed core ring "f "complied with the design introduced by Vulcan in Hamburg ("Vulcan Profile"). A stationary scoop tube "d" acting as a static head pump collects the operating fluid flowing off from the inner to an outer shell via nozzles. The static

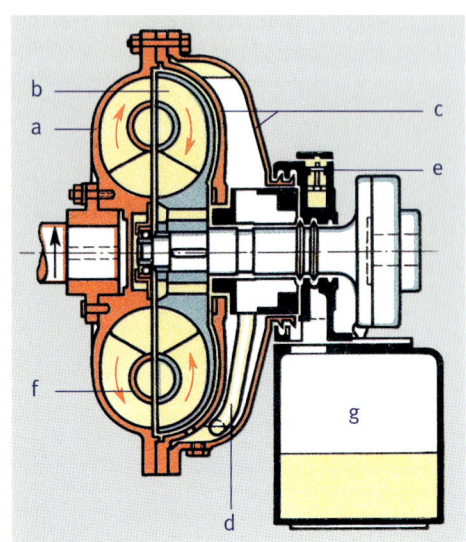

Voith Sinclair coupling VSK,
Type S...

a *Primary wheel (pump)*
b *Secondary wheel (turbine)*
c *Coupling shells*
d *Scoop tube*
e *Distribution housing*
f *Core ring*
g *Oil tank*

pressure pumps the fluid via the distribution housing and a heat exchanger into the hub area and through boreholes back into the operating chamber "a – b". A double-acting control pump is integrated in this closed cooling and control circuit, allowing exchange of volume between operating chamber and oil tank "g".

In the operating manual of 1934 the term "Variable-speed turbo coupling" was used for the first time. In 1938, the core ring in the bladed chamber was omitted as one of the measures undertaken to improve the characteristic curve.

As a result of the operating principle of the Sinclair series, the intended wide control range reaching as low as 20% of the engine speed, led, however, to thermal problems, especially with higher machine power. Additionally, the dependence of the external cooling flow on the filling level proved to be rather unfavourable for accurate speed control.

During talks with Voith about a hydrodynamic start-up and damping coupling for superchargers, the engineer Helmut Müller of Junkers in Dessau suggested a new way of controlling filling levels.

Due to the war, the patent was registered by Voith not before 1953, stating 23 July 1942 as its publication date. In this document, Müller who had meanwhile moved from Junkers to Voith is named as co-inventor.

Still working for Junkers at the time, Müller described the basic idea for this patent in 1944 as follows:

"The height of the fluid level in the working chamber and thus the setting of the slip is achieved in such a way that, following the principle of communicating tubes, a ring chamber is connected to the operating chamber, in which the height of the oil level can be changed by a movable scoop tube that is situated excentrically to the rotation axis of the coupling. The key advantage is that virtually any amount of oil can be sent through the coupling at any state of slip, i. e. any slip that might occur can be thermally controlled. A particular feature is that the connection between coupling and ring chamber is no longer designed as a nozzle, as a result of which the scoop tube aperture that is led to the outer diameter of the ring chamber allows a complete drainage or the working chamber. This means that nearly 100% slip can be achieved. In this case the control core in the working chamber must be omitted."

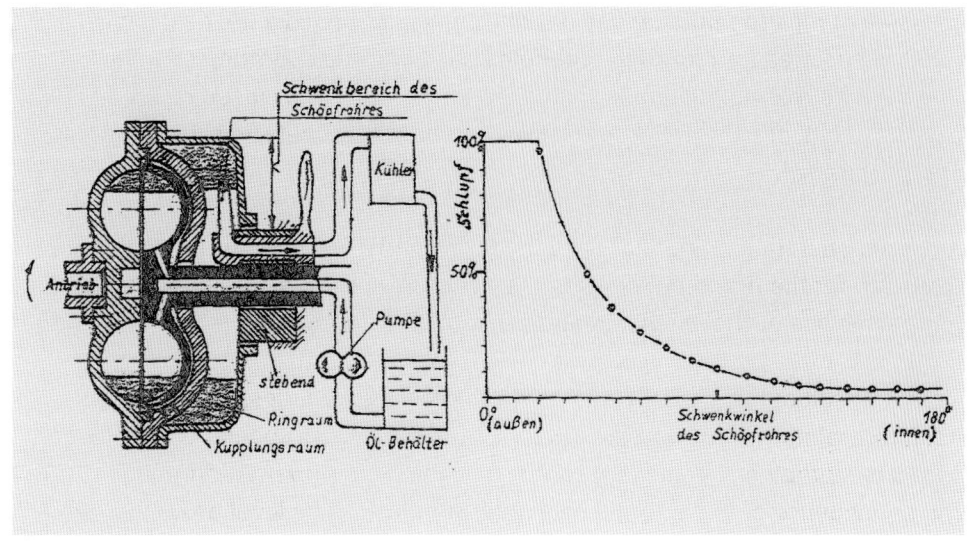

Filling control according to Müller.

Start-Up and Safety Couplings

During the years until 1949, the development of start-up and safety couplings took a rather complex course. Apart from a worldwide flow of information controlled by the licensor, experiences from turbo transmission design and feedback from practical applications had to be assimilated.

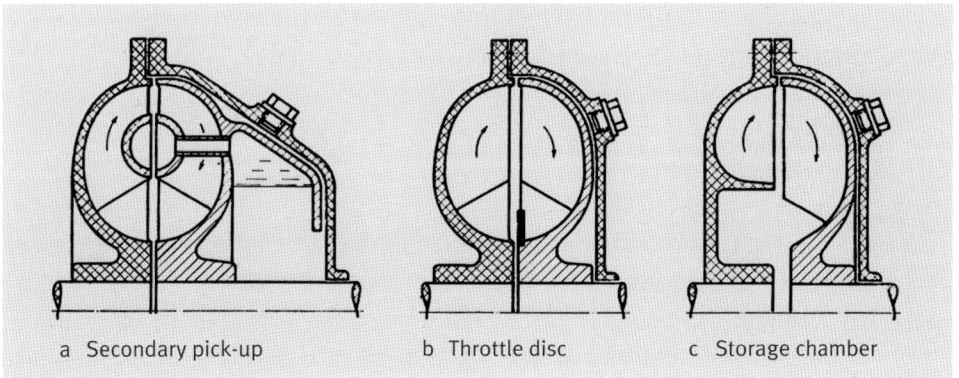

a Secondary pick-up   b Throttle disc   c Storage chamber

Development stages of constant-fill hydrodynamic coupling.

The cross section shows the design of the Voith Sinclair coupling "Type T" in its basic version from 1934, with a pick-up chamber on the secondary side. For many applications, the achievable torque reduction with a braked turbine wheel was, however, not sufficient. The coupling without a core ring and instead fitted with a throttle disc, as shown in cross section b, was again only partially successful. Cross section c shows a successful design with storage chamber developed by Voith in the forties, internally referred to as "stepped profile" or, in honor of its inventor Fritz Kugel "Kugel-profile".

Overload protection for the diesel engine of an icegoing vessel with double-engine single propeller drive.

Safety coupling and clutch with slide plate control, type TR.
1 Cooling oil nozzles
2 Slide plate
3 Standpipes

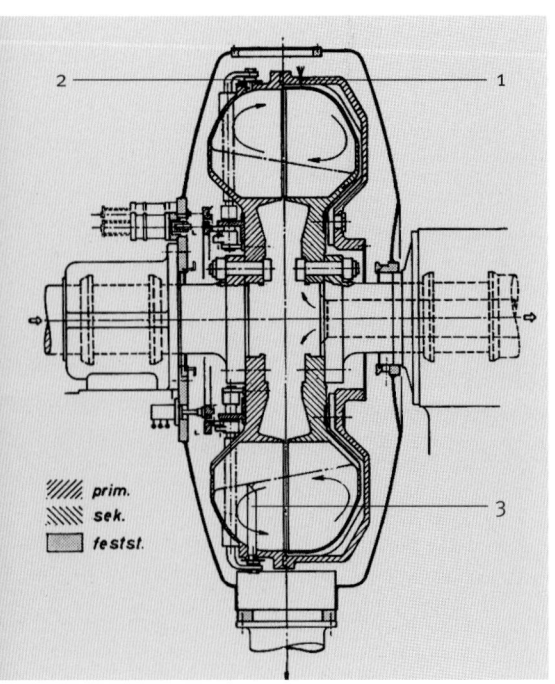

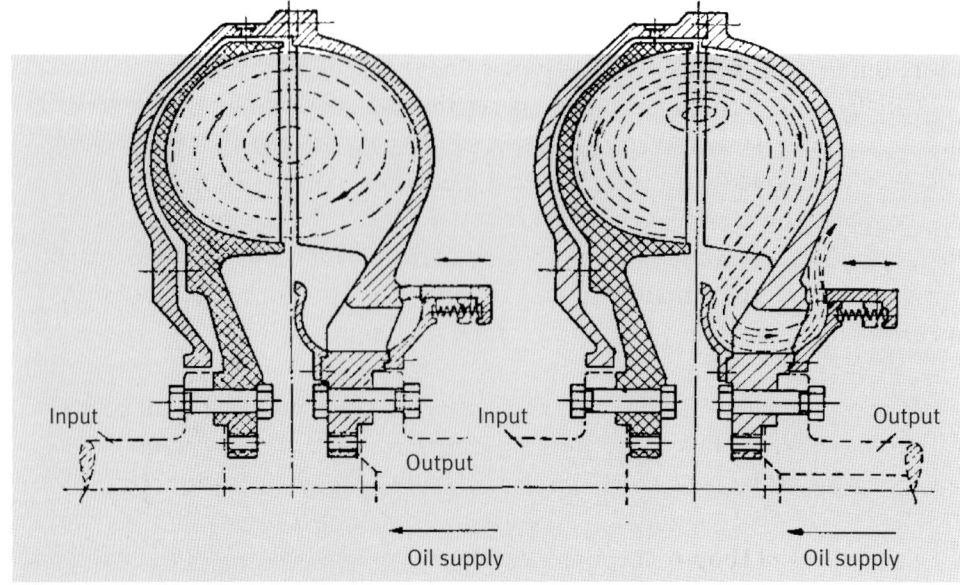

Start-Up Couplings and Couplings with Switch-On Function

During the first application of safety clutches in an icegoing vessel with diesel engine drive, the experiences from the storage chamber development for constant-fill couplings could be successfully realized for this coupling type, too.

When the output shaft is blocked, a diverting shell takes up the operating fluid that is dynamically guided towards the hub and directs it towards boreholes through which it can escape from the coupling chamber. This process provides quick and comprehensive engine relief.

A further development of this safety coupling and clutch is type TR with slide-plate-control. Apart from the cooling oil nozzles in the shell, the coupling also has a row of nozzles with different boring diameters in the primary wheel. This row of nozzles is covered by a steel band assuming the function of a slide plate. The slide plate can be put into three different positions via two adjustable control rings and control shafts:

Normal operation: all boreholes are tightly closed
Quick draining: large draining boreholes are uncovered
Partial filling: boreholes of the partial filling standpipe are open

With the partial filling, it was for the first time possible to prevent a predictable engine overload, for example during a reversing manoeuver or at a standing pull trial, by a targeted decrease of the transmission capacity of the coupling.

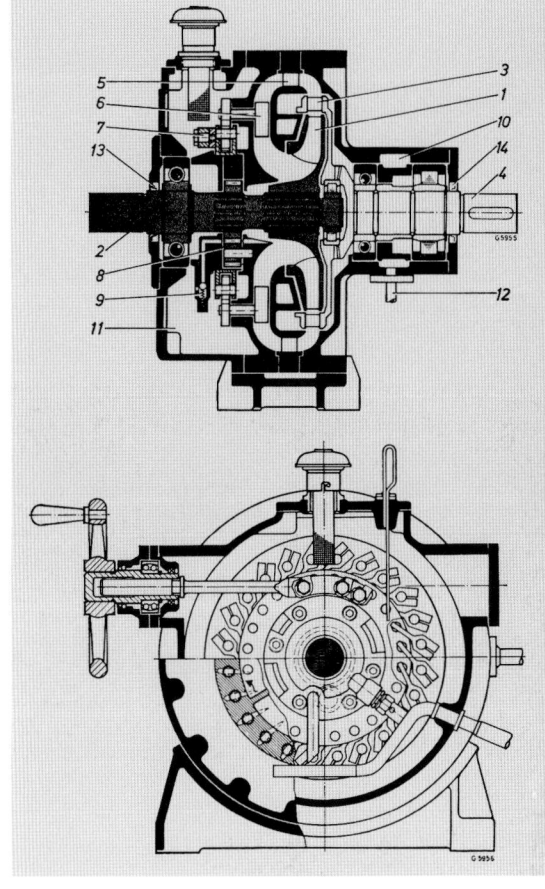

Start-Up, Open and Closed Loop Control Converters

Variable-speed couplings did not prove the ideal solution for the control of industrial processes with largely constant torques. In 1950, Karl Hanselmann therefore developed a controllable converter in Heidenheim with manually adjustable guide vanes. The typical characteristics of the converter type that became a "classic" were a centrifugal-flow turbine with short, onedimensionally curved "steam turbine blades" and centripetal-flow rotating guide blades with an adjusting device known from water turbine design.

At constant input speeds, this converter type has a nearly constant pump torque at a typically falling turbine torque vs. the speed ratio. By adjusting the guide blades, a highly diversified pump and turbine curve range is achieved.

Open and Closed Loop-Controlled Couplings

The new method of controlling the degree of filling by adjusting the fluid level in a communicating, rotating chamber was applied to all new designs in those years. The Voith Sinclair series was gradually replaced by further design developments.

At the time, the adjustable scoop tube of type 1 was used as a temporary solution in a few variable-speed couplings and later substituted by types 2 and 3. With current developments of environmentally compatible water couplings, the swivel scoop tube proved to be less sensitive to blockages caused by lime deposits and has recently been reactivated for those applications. Today, the diagonal scoop tube 3 is the preferred control element for all other applications.

Design of a guide vane controlled converter.

1   Pump wheel
2   Input shaft
3   Turbine wheel
4   Output shaft

5   Stationary guide vanes
6   Adjustable guide vanes
7   Control device

8   Gear pump
9   Relief valve
10   Oil channel
11   Oil tank
12   Connecting pipes
13   Seal
14   Seal

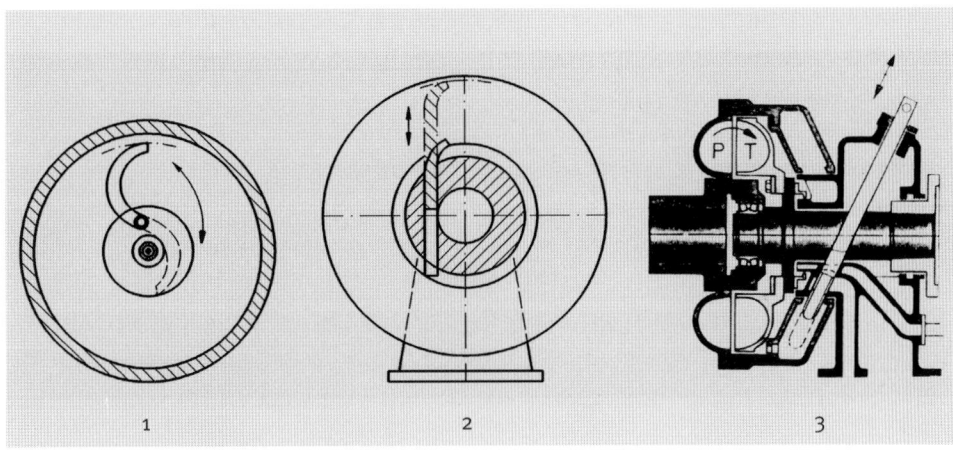

Voith-developed adjustment of scoop tubes.

1   Swivel scoop tube
    (adjustable lever)
2   Sliding scoop tube
    (gear segment)
3   Diagonal scoop tube
    (freely removable)

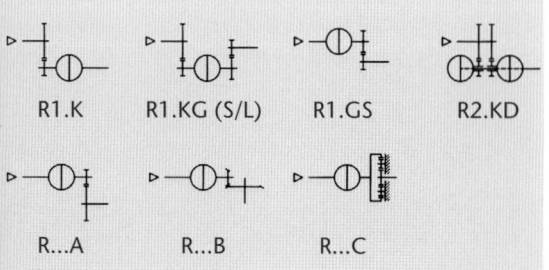

Geared variable-speed coupling – modular system.

During the period under review, there were other significant findings and developments: in the USA, the car builder Chrysler had successfully used couplings without core rings. In comprehensive test series, Voith recognized that this profile was also suitable for variable-speed couplings ("Chrysler Profile"). As a result, it became standard for variable-speed couplings.

Owing to a new fill-control method it became possible to adapt the outer cooling oil flow to the requirement determined by filling time and heat transport. A special filling and flow-control system was patented.

A suggestion from the market was, to adapt the operating curve of the specific speed levels of different applications by adding gear stages, as shown in the picture on the left. This approach resulted in a graded modular system for geared variable-speed couplings.

Start-Up and Safety Couplings

The period until 1969 is characterized by an expansion of the product spectrum sparked by new markets and applications, and the development of application-specific characteristic curves. Main applications are stationary machines driven by electric motors, as well as special machinery with combustion engines, for example tractors and construction machines. Especially for the lower power range of up to approximately 30 kW, a coupling range was developed that could be universally integrated into the driveline and was suitable for coaxial and axle-parallel installation.

Modular coupling series for applications with electric motors.

T   Stepped profile with storage chamber

TV With storage chamber on the primary side (delay chambers)

TVV With enlarged delay chamber

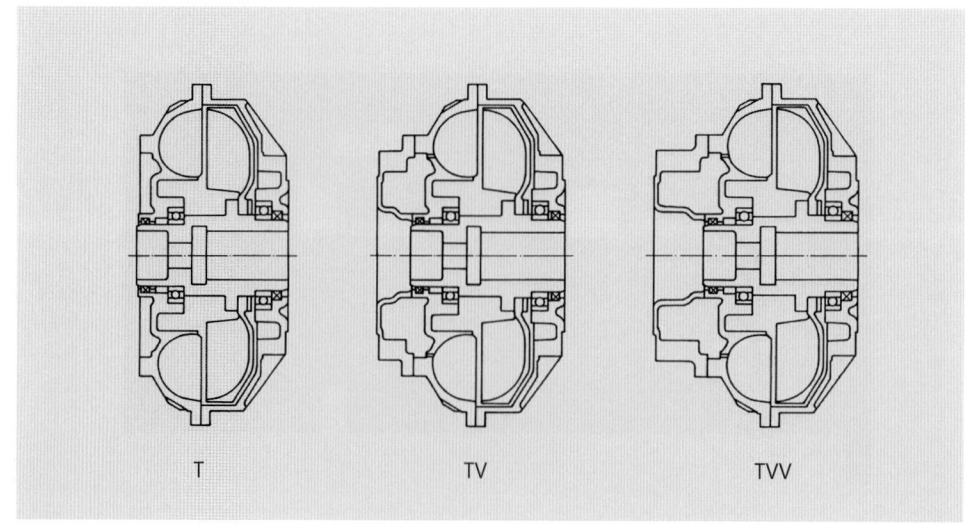

Interchangeable turbine wheels
Left: "Stepped" profile.
Right: Mixed profile.

For applications with **electric motors**, the possibilities of influencing the characteristic curve (storage room T and delay chamber TV) from 1947 were utilized in a modular coupling series. Couplings with enlarged delay chamber which further decreases the start-up torque transmitted by the coupling were designated TVV.

The solutions that had been realized so far were, however, still unable to fulfill particularly stringent customer demands for even further torque limitation. During nominal operation of a constant-fill coupling, the oil filling required for an acceptable nominal slip of S_N = 2 to 4% needs to circulate in a bladed operating chamber. With this filling, the coupling torque at stall condition is too high as described in the theoretical part above. A partial filling curve reduces this torque to the desired level, which is a horizontal curve slightly above the load (M_N). Further developments thus aimed at a better utilization of this partial filling effect.

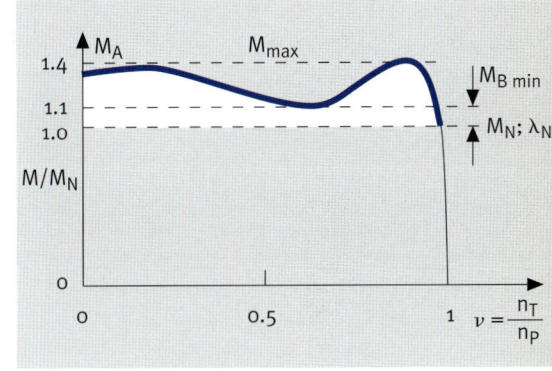

Start-up curve for mixed profile:
Acceleration torque at least
10% and torque imitation max.
40% above load torque.

The development result was that all torque levels of the coupling during the start-up phase are within the limits of 110% to 140% of the nominal torque. By changing the meridian contours in the turbine wheel ("Mixed profile"), an invention of the design engineer Siegfried Mlacker, a self-developing curve shape was achieved that fully met the demands of the belt conveyor manufacturers. A secondary goal of these development efforts was to be able to preserve the modular design of the series by exchanging just one component.

Below left:
Dynamic coupling curve for start-up protection of a supercharged diesel engine.
Below right:
Turbo coupling in a driveline with combustion engine.

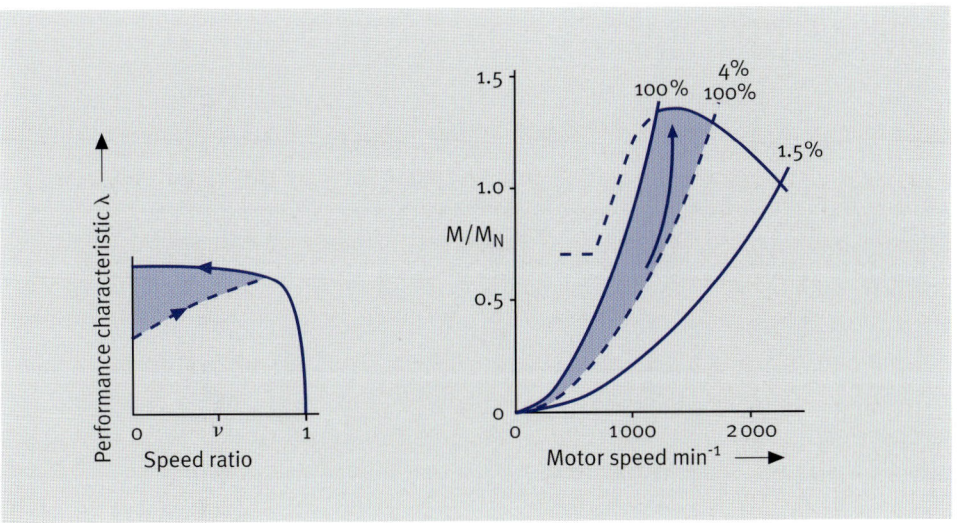

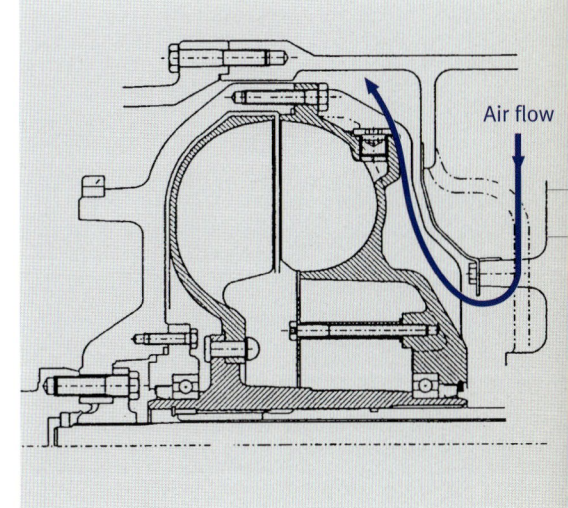

Largely inspired by demands for more effective run-up protection of charged diesel engines, the development efforts for couplings at **combustion engines** led to the same solutions. The goal was a reduction of the soot emission during starting. For the first time, the enlarged storage chamber is divided by a throttle disc into two differently functioning chambers. When the turbo-charged diesel engine is started, both chambers are active and the engine can run up at drastically reduced load. Above 1000 min⁻¹, the effect in the storage chamber at the back decreases, the engine reaches its maximum torque. If the engine is lugged down from nominal operation, only the storage chamber in the front is active; the characteristic curve of the coupling runs in its upper range and full engine torque can be utilized.

Typical torque sequence during start-up of a gas turbine (converter turbine torque) for three guide blade positions.
1 Break-away
2 Resistance without ignition
3 Ignition speed
4 Acceleration resistance
5 Self-sustaining speed

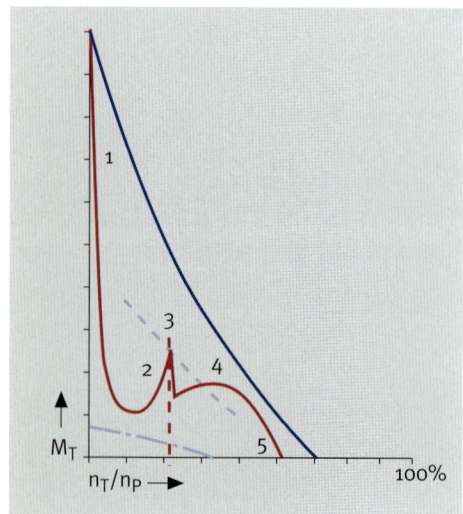

Development Period After 1970

This period is marked by new applications and the increasing integration of Föttinger units into the overall process. It proved to be imperative to create and utilize numeric models for simulation calculations.

Start-Up Fill-Controlled and Variable-Speed Converters
Gas turbines in peak load power plants often have to be started several times per day. In order to do this, the 20 to 270 MW single-shaft gas turbines used for this purpose require a starting drive. With guide blade control, a start-up drainable converter with high start-up torque ratio can carry out this start-up process in an ideal manner. For break-away (1) it provides a sufficiently high stalling torque and afterwards a large acceleration torque. After igniting (3), the run-up is supported up to self-sustaining speed (5). During nominal operation of the gas turbine, the starter motor is switched off and the converter is drained. It can then be decoupled or co-rotate on the turbine side with low losses. After shutting down the machine set, the turbine runner needs to be able to cool down at low speed. Ideally, this occurs through a turning device integrated into the converter housing, or via a free-running worm gear.

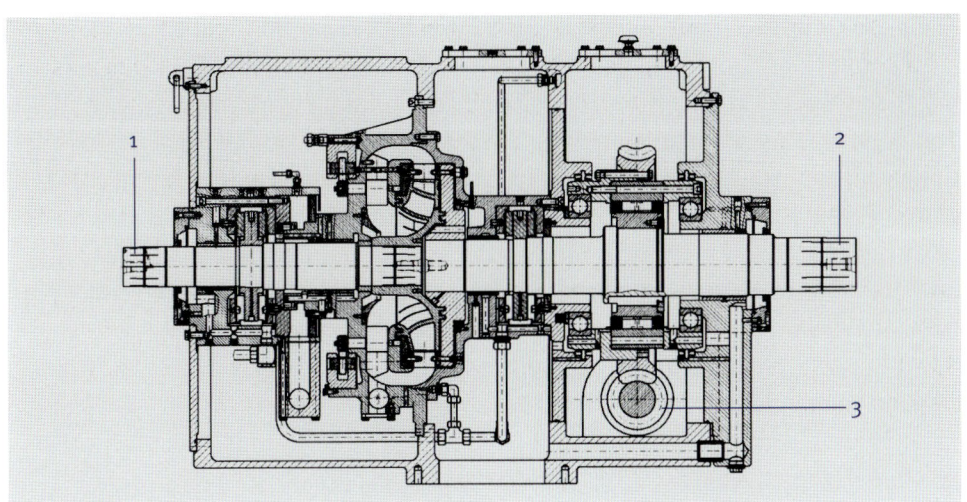

Guide vane controlled start-up converter for gas turbines with integrated turning device (worm gear).

1 Input
2 Output
3 Worm gear shaft

In many cases, large centrifugal compressors with power consumptions of several MW in LNG (Liquefied Natural Gas) plants cannot be started when directly coupled with the driving machine. This might be caused by the special operation of the gas turbine used in this process, or by the power output and the torque development of the electric motor.

Apart from starting assistance for the run-up of the driving machine (1) (single-shaft gas turbine or electric motor), a CSTC (compressor starting torque converter) can facilitate this starting process by decoupling the compressor (2). As soon as the driving machine has reached its nominal speed, the compressor, with its gas flow being throttled, is run up via guide blade control of the torque converter. As the compressor is operated at constant speed in nominal operation, it is recommended to eliminate the systems-inherent converter loss. Once it is locked up, the converter is therefore drained. As a result, the loss remains extremely low.

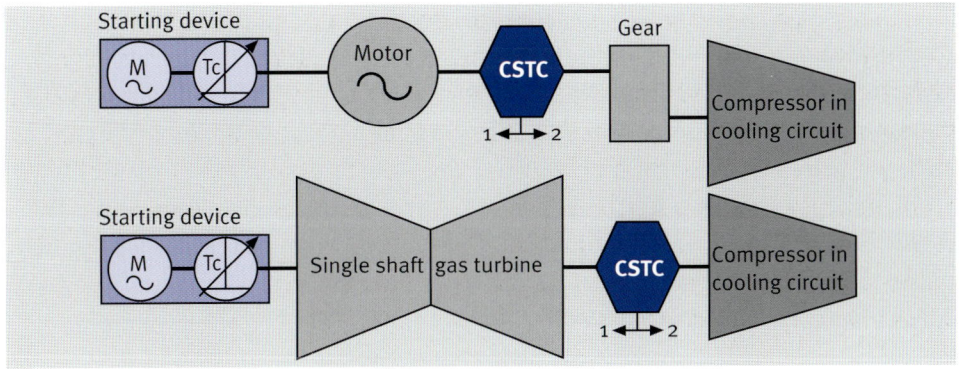

CSTC (VOSYCON – synchronizing converter) between main drive and driving machine (compressor).

Start-Up Couplings and Couplings with Switch-On Function

By now, the ongoing mechanization of coal mining requires input powers of more than 1000 kW for the head and tail drives of underground conveyors. This is added by aggravating conditions. While the conveyor cannot start up unless most of the load has been taken off the electric motor in order to comply with the frequently insufficient local electricity supply, the pull-out torque of the electric motor is expected to be available at short notice in the event of extraordinary loads. This scenario can occur, for example, when the conveyor has been buried under a collapsing coalface and has to be broken free.

With the coupling series DTPKW developed for such applications, the limiting curve of the coupling can be set to the respective pull-out torque of the motor. Its shape at 100% filling is described as "Longwall curve".

Systems characteristics of an armored face conveyor drive

– Largely load-free motor run-up (starting torque, left side of picture)

– After the run-up, the coupling is filled (secondary characteristic), right side of the picture. This allows short-time use of the pull-out torque of the motor.

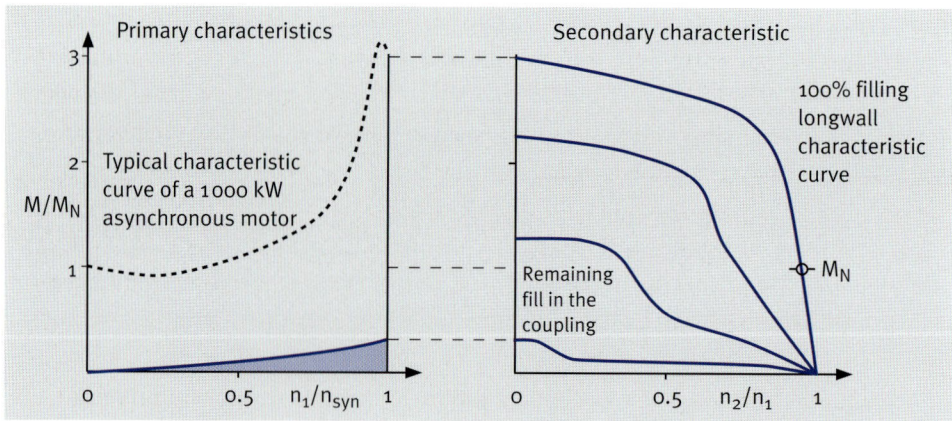

The height of the coupling must comply with the coalface and thus stays rather low. Complete drainage at standstill is therefore not possible. For this reason, the electric motor has to start with a low amount of fluid in the coupling – a situation that does not have a disruptive effect if the operating medium is oil. The demand for operation with water (pit water!) prompted the Voith engineers to design a solution that was largely influenced by specific applications. The double-flow, self-supporting coupling components supplied by Voith are connected with the motor and/or the conveyor.

If the conveyor is blocked, the high output provided by the motor very quickly leads to evaporation. In order to avoid cavitation damage, the runners must be made of propeller bronze. The outer (closed) cooling and control circuit is driven

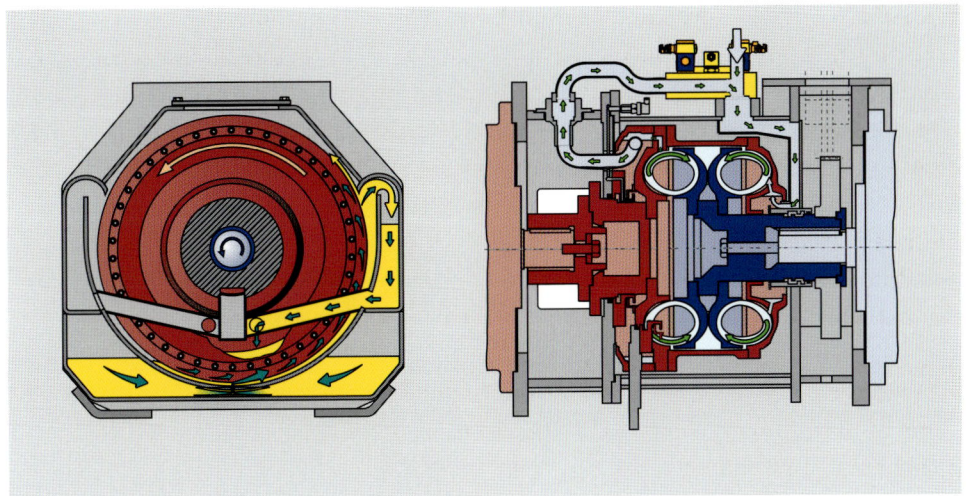

Self-supporting, double-flow
start-up and safety coupling
between e-motor (left) and
armoured chain conveyor (right)
– suitable for mining application due to low height
– suitable for pit water as
  operating medium.

by a static discharge pump that interacts with the external scoop device. An
upper valve plate controls the filling and the regular exchange of hot operating
water against cold pit water. If the motor shuts off very quickly, part of the water
remains in the housing and the runners are wading. At the restart, the water is
lifted by the outer contour of the impeller into a side chamber of the housing,
from where is flows to the filling side of the coupling near the hub. On site, the
double-flow self-supporting coupling with its stainless steel housing is integrated between motor and transmission of the driveline of the armoured chain conveyor. Sliding skids facilitate slow underground transport.

In order to start long belt conveyors smoothly and in a controlled manner, the
same operating principle (closed cooling and control circuit with static discharge
pump) was used for the self-supporting series TPKL with mineral oil as operating
fluid. As the pipes in the closed outer circuit stay filled, the coupling volume can
be changed indirectly via the control valves integrated in the feeding and return
line. Owing to the good dosing capability of the enclosed volume, the speed can
be adjusted more accurately compared to open designs that spray into the
housing. Yet this procedure does not reach the control accuracy of the SV series.

Increasing power requirements of single machines, for example mills or fans,
and the request of the operators for energy savings during continuous operation, made the developers return their minds to the construction principles of
the first Voith coupling when they designed the turbo coupling type 1210 TPL-SYN.

Fill-controlled start-up coupling
with closed cooling control
circuit.

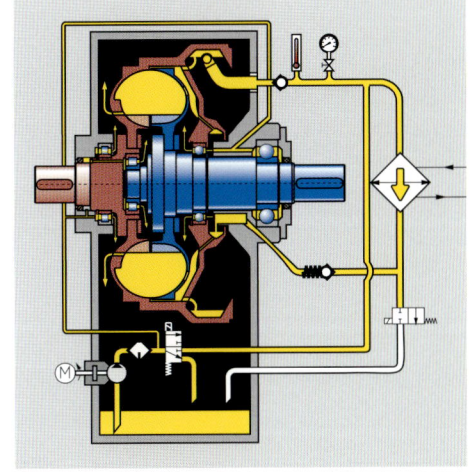

1210 TPL-Syn coupling
for 9 500 kW at 1 200 min⁻¹
– Start-up device for single
  machines with high outputs
– Slip-free nominal operation.

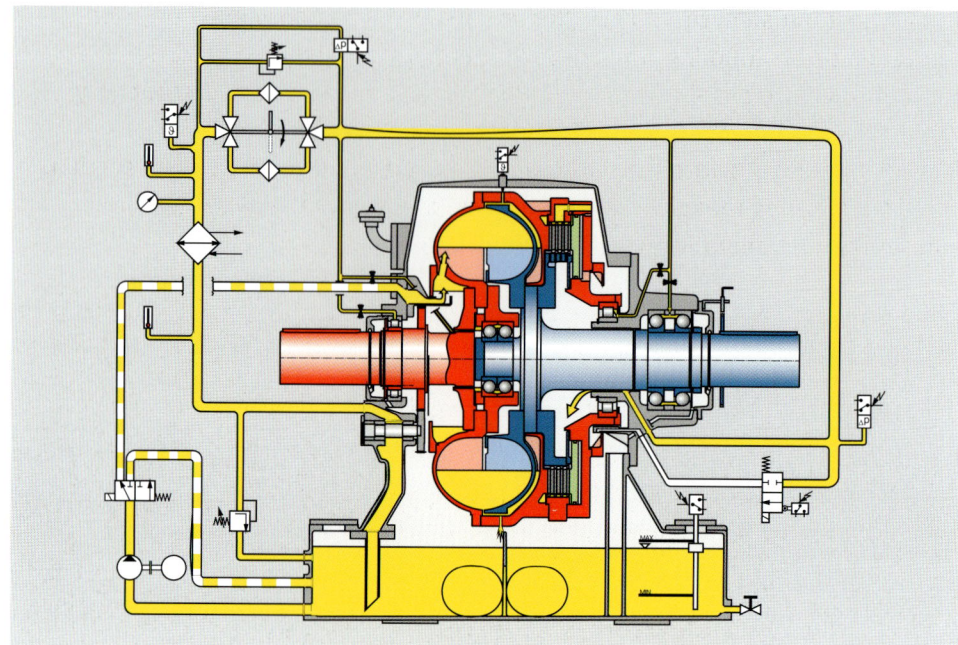

The operating fluid mineral oil flows in near the hub. Drainage into the housing via nozzles was retained, as there are no specific requirements to the accuracy of the speed development during the start-up process. Lock-up eventually occurs via wet-running clutch plates. For this purpose, oil is directed behind a ring piston that rotates with the shell. The ensuing rotation pressure causes a friction connection between the plates and thus shifts into synchronous operation.

Fill-Controlled and Variable-Speed Couplings

At the time, the expansion of the operating characteristics of couplings and converters by gears was well-known and had proven itself. In 1985, the first transmission series with hydrodynamic superimposing gear was developed for the Megawatt-class. The VORECON is used in plants, where pumps, beater mills or compressors have to supply large mass flows to a process, in accordance with the prevailing requirements and with stepless control. Depending on the process, the speed control range can be between 10 and 100%.

The basic type RW designated Long-VORECON is characterized by three Föttinger units coupled with mechanical gear elements that act in different operating phases and in line with their main operating range.

On the input side, the VORECON consists of a synchronous, scoop-tube-controlled turbo coupling. The converter that is arranged coaxially after the coupling on the main shaft 3 acts via the secondary shaft 4 and a stationary planetary gear E on the internal gear of the rotating planetary gear U (gear assembly). A hydrodynamic brake is integrated into this split-off section, practically turning the rotary gear into a stationary gear with a fixed ratio during coupling operation.

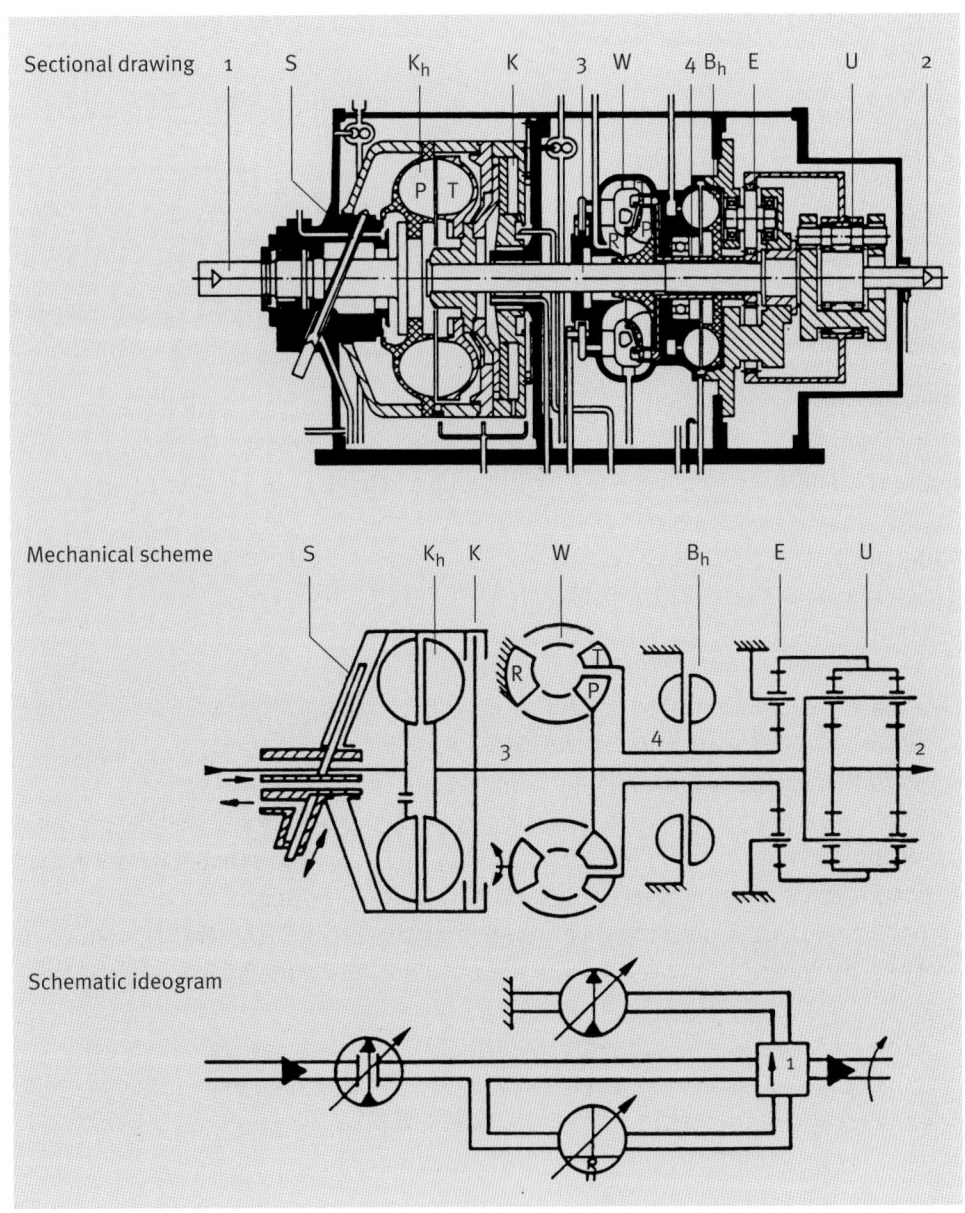

Sectional drawing

Mechanical scheme

Schematic ideogram

VORECON
Basic type RW with maximum
speed control capacity.

1   Input shaft
2   Output shaft
3   Main shaft
4   Superimposing shaft
W Converter
P   Pump
T   Turbine
R   Reaction device
B_h Hydrodynamic brake
K_h Hydrodynamic coupling
K   Clutch
E   Stationary gear
U   Rotating gear
S   Scoop tube

Type RWS (without coupling)
– Lower speed control range
– Reduced space requirements
– Brake with adjustable guide
  vanes assumes speed control
  in operation mode one

Type RWE (guide vane
controlled converter only)
– Lowest speed control range
– Lowest construction
  requirements
– Starting with converter in
  contra-rotating braking range
– Short VORECON for gas
  compressors and pumps

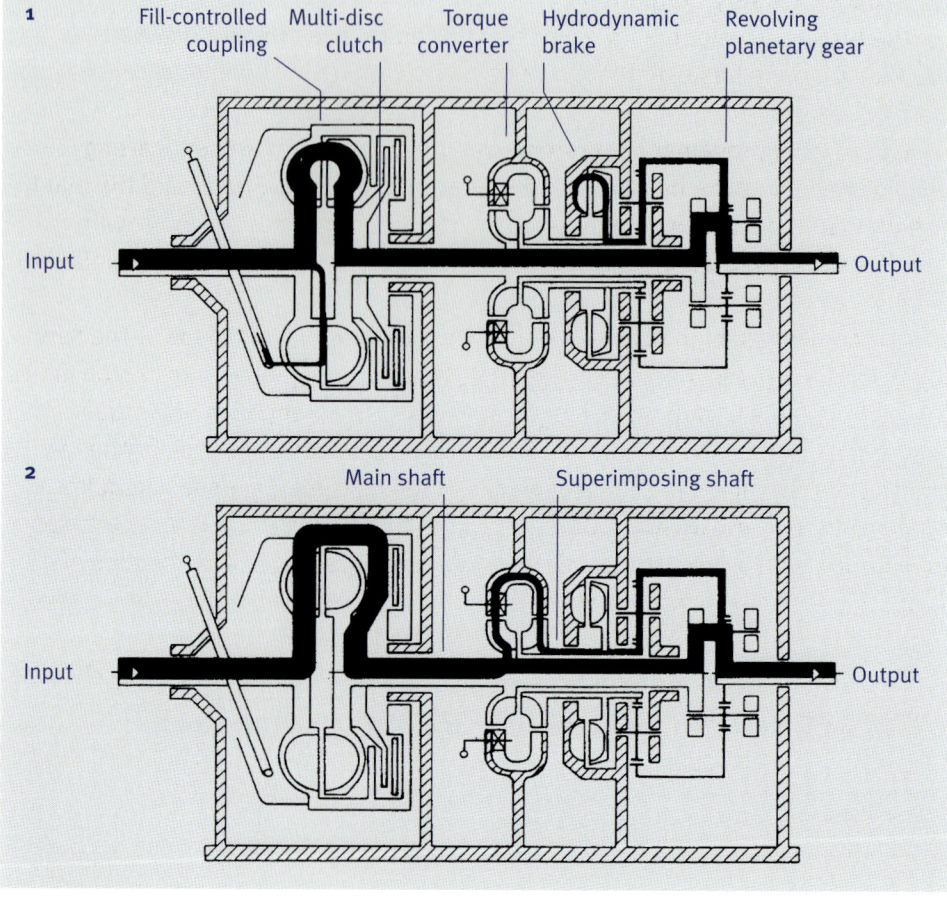

VORECON multi-circuit variable-speed drive, operating modes 1 and 2.

1   Fill-controlled   Multi-disc   Torque   Hydrodynamic   Revolving
   coupling   clutch   converter   brake   planetary gear

Input   Output

2   Main shaft   Superimposing shaft

Input   Output

VORECON prototype 1985 (internally referred to as "Blauer Klaus"). Built for test purposes and customer demonstrations on the test stand.

– In operating mode **1** (coupling operation, speed range 0 – 75% of the nominal value), speed control occurs via the scoop tube controlled coupling, the converter is drained. The hydrodynamic brake supports the annulus gear torque towards the housing.

– In operating mode **2** (converter superimposition, speed range 75 – 100%), the turbo coupling is synchronized via a multi-disc clutch. Speed control occurs via a guide vane adjustment of the converter. From main shaft 3, some of the output (maximum 30%) is split off, its power factors are converted and then redirected to the rotary planetary gear. Now, the converter efficiency is only linked with the split-off partial output, while the main output with its good efficiency is transmitted. With process-specific converters and planetary stages, efficiencies of up to 95% can be achieved by the converter range of the VORECON.

Start-Up and Safety Couplings

The further developments of constant-fill couplings after 1970 can be illustrated on the basis of three major categories:

1. Further improvement of systems behaviour, especially when starting belt conveyors, as the belt is often the most expensive component of the plant. The drive is expected to provide smooth introduction of tractive effort with simultaneous adaptation to the load condition, as well as close torque limitation.

In recent years, high standards have been achieved in this respect. The new type TVVS meets the market requirements to a large degree. Alongside the motor run-up, it creates a coupling torque of only 60 – 80% of the nominal torque. Compared to type TVV, the torque development time has been tripled. The low starting torque can be used to safely start an empty belt that usually requires only one third of the nominal torque. Carrying full load, the belt is smoothly set into motion with a time delay, the maximum torque remains below a value of 140% of the nominal torque.

This operating behaviour was achieved by an additional external ring chamber – the annular chamber – for which Heinz Höller was granted a patent.

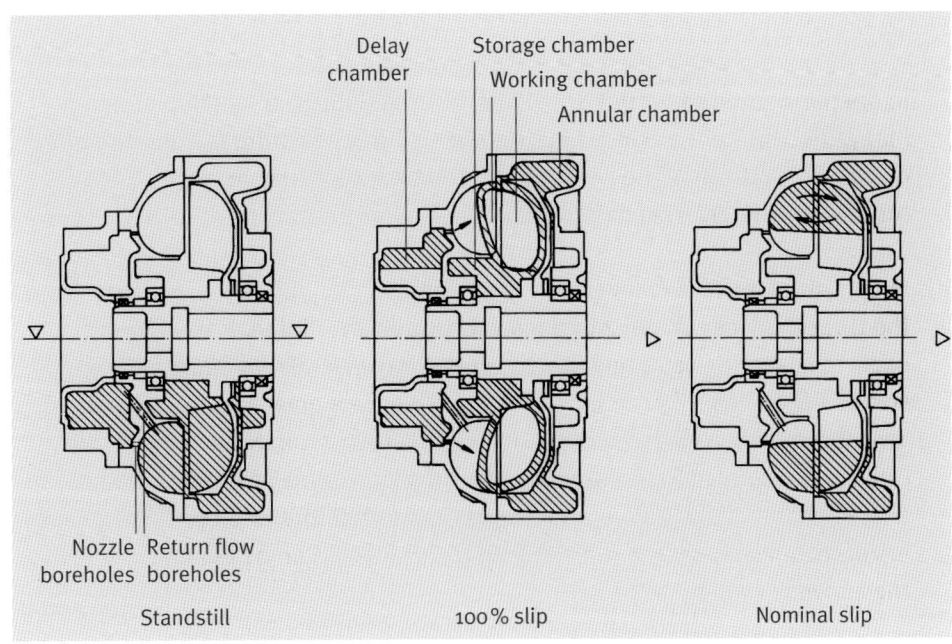

Delay chamber
Storage chamber
Working chamber
Annular chamber

Nozzle boreholes | Return flow boreholes

Standstill   100% slip   Nominal slip

Flow conditions of the hydro-dynamic coupling type TVVS, developed for starting up belt conveyors. Example with coupling filled to 50%.

Distribution of the oil volume
– at standstill
– after motor run-up while belt conveyor is still at a halt
– after run-up of the belt conveyor

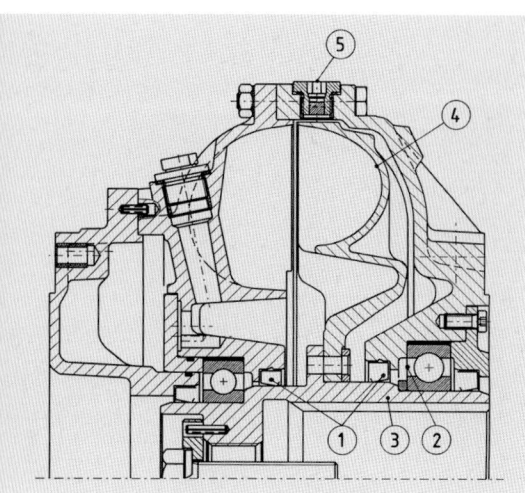

Series-produced water coupling

1 Bearing with double seal
2 Liquid grease filling
3 Hubs made from anti-corrosive steel
4 Low-copper alloy with hard anodic layer
5 Operating temperature up to 100 °C (fuse 110 °C).

2. New operating media. The primary task of the operating fluid is to transport its inherent kinetic energy in the closed circuit between pump and turbine. From a hydrodynamic point of view, water is highly suited for this purpose. In earlier couplings, it was therefore the preferred choice. Yet its corrosive effects soon made it unpopular and it was replaced by lightweight oils. These oils were, however, easily flammable; their use in mining environments was therefore out of the question. Although the first synthetic fluids were flame-resistant, they had toxic effects and had to be taken off the market. After all these detours, the Voith engineers made another attempt at water.

Prior to producing the coupling on an industrial scale, several questions had to be answered regarding:

– Corrosion resistance and chemical compatibility of the chosen materials
– Mechanical resistance against cavitation and abrasion
– Lubrication of the bearings
– Reaction on the low boiling point of the water
– Operating behaviour at temperatures below freezing during transport.

Unexpectedly, it was the corrosive behaviour of the conventionally used AL casting alloys towards the raw water available underground that proved to be the biggest development challenge. In a sub-zero test with operating fluid in temperatures as low as -40 °C none of the components showed any damage.

3. Start-up couplings with lock-up function. The special characteristics of the turbo coupling are indispensable for heavy or load starts of individual driven machines. In continuous operation, the systems-inherent slip is, however, often undesirable. For this reason, there have been several attempts in the past to lock up not only flow couplings but also start-up and safety couplings in nominal operation.

The development of such a coupling designated TurboSyn is making excellent progress: a first unit is currently undergoing field tests. The basic design of this type is similar to that of a conventional turbo coupling. The turbine consists of individual segments, some of them with friction coating, that are fixed to the hub and can radially swivel to the outside. The driven machine is started up hydrodynamically. A power component pushes the segment from the torque to the inner

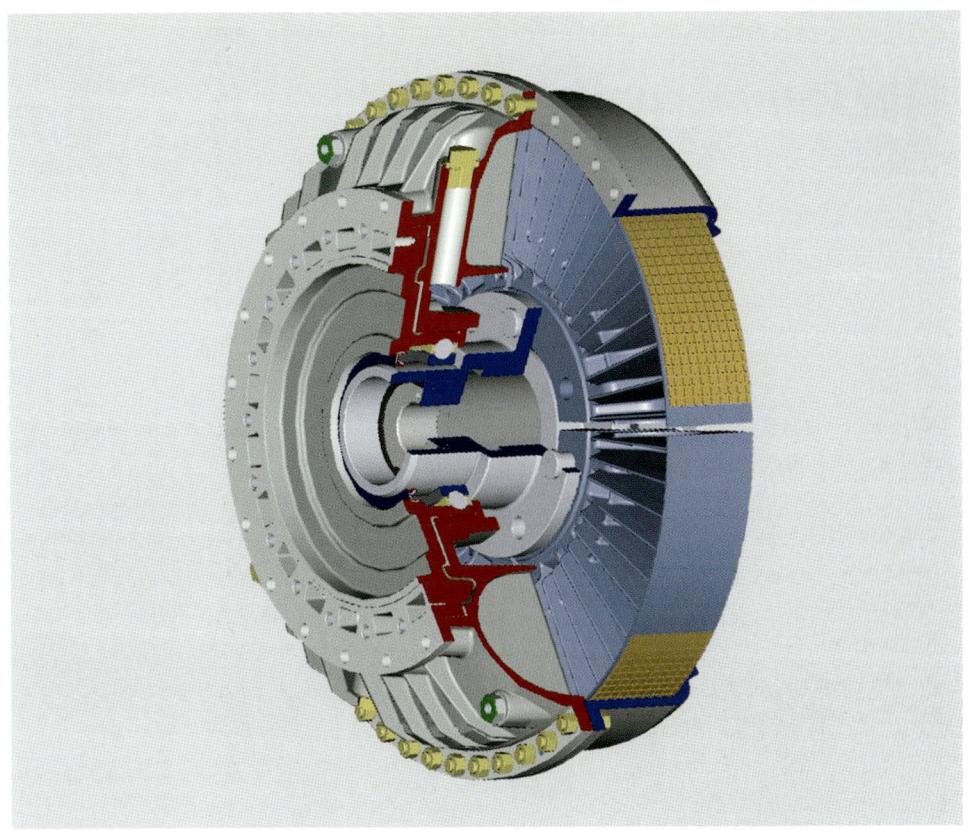

hub. With increasing turbine speed, the centrifugal forces acting upon the segments take over and create a friction connection between inner wheel and coupling shell.

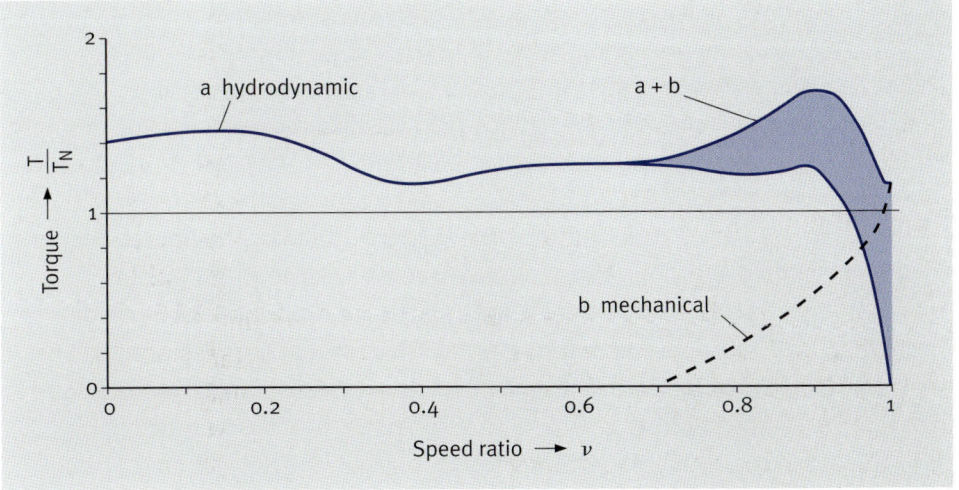

The development history of industrial couplings and converters on the basis of individual characteristic examples described in this section is closely related to their actual applications. It therefore seems appropriate to provide an outlook on the further development of these products, their future markets and utilization areas at the end of this chapter.

Turbo Couplings and Torque Converters in the Control Process

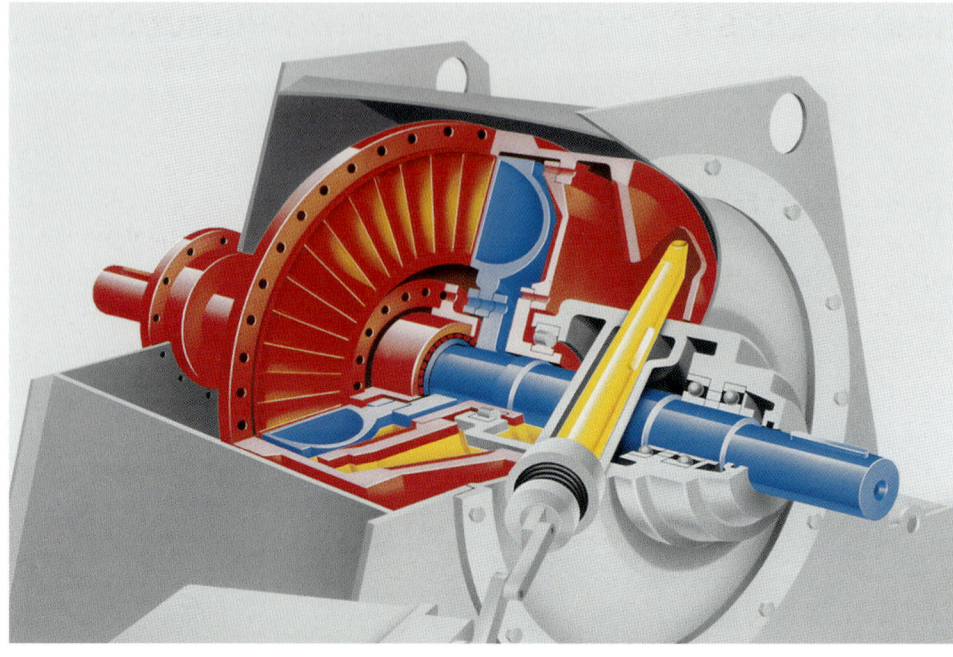

Voith variable-speed coupling of the SV series with adjustable scoop tube.

Voith 682 S-flex turbo coupling in a fan drive, in operation at Hamborn power station since 1953, with approximately 350 000 operating hours.

The license agreement with Sinclair dates back seventy years. Since this time, Voith has been active in the field of variable-speed couplings, not counting a brief spell of experimentation a few years earlier.

The principle "scoop tube", at the time the basic idea for the operation of the Voith Sinclair variable-speed coupling, was such a breakthrough that it is still used in its function as a static head pump. Of course there have been countless mechanical and hydraulic improvements and ideas since then, resulting in an impressive line-up of patents. They also led to a vast extension of applications for variable-speed couplings. The letter "S" standing for "scoop tube" has, however, almost continuously maintained its place in the Voith product designation system. The product catalogue ranges from the S-flex coupling built between 1934 and 1966 up to type R.K, VORECON, and the water-operated type SVTW used today. All couplings excel by their high availability, wear-free operation and excellent operating and control characteristics.

At the Voith plants in Heidenheim, St. Pölten and, from 1960, in Crailsheim, 26 000 couplings of this type have been produced and delivered. They can be found in installations all over the world. Operating times of 350 000 hours are not unusual! It was the S-flex coupling that opened up a major application area for Voith in 1936 – power generation.

Applications in the Power Generation Industry

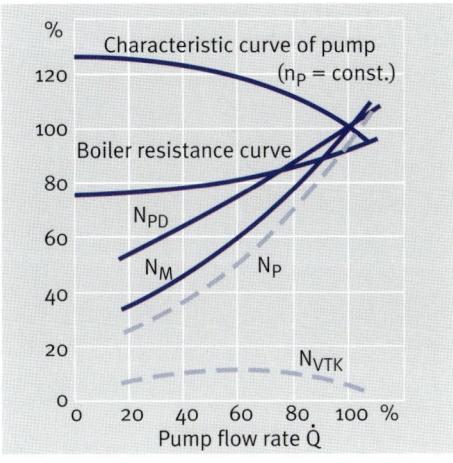

Drives for Boiler Feed Pumps

Steam turbines, electric motors and generators are not part of the Voith product portfolio. The company is nevertheless strongly interested in sophisticated drive and speed control systems in the power station process, where the special characteristics of variable-speed couplings can be employed more advantageously than any other conventional methods.

One of the most important functions in the operation of thermal power stations is the preservation of the water circuit for the supply of the steam generators. In this process, the water flow needs to be continuously adapted to the required power station output. There are two possibilities to do this. The systems resistance is either modified by a throttle valve in the water circuit which provides the required flow adaptation or, as an alternative, the speed of the boiler feed pump is changed.

It was noticed at an early stage that there is no way past speed control. Neither of the two control types mentioned operates without losses, yet these are higher in the case of throttle control applications. If a variable-speed coupling is used, the required motor power is significantly lower than with throttle control, although there are some slip losses. Meanwhile, drive systems with variable-speed couplings have proven themselves all over the world.

First Drives for Boiler Feed Pumps

The first Voith Sinclair coupling for a 100 HP-boiler feed pump drive was delivered by Voith for the Kirchlengern branch of Minden-Ravensberg power station in 1936. It belonged to the S-flex series and had a profile diameter of 316 mm.

In those days, coupling parts had no bearings but were supported by motor and pump. Filling and draining of the coupling system was carried out by a gear pump. Depending on the control process, the oil was either removed from a separate tank or pumped back into it. At the time, the cooling of the operating oil was still separated from the filling/draining process. The stationary scoop tube took the oil from the turbo coupling, led it via a heat exchanger and forced it back into the coupling.

Performance comparison speed control/throttle control for boiler feed pumps.

N_{PD} = *Required motor power for throttle control.*
N_M = *Required motor power for speed control with turbo coupling.*
N_P = *Pump power required with speed control.*
N_{VTK} = *Power loss in the turbo coupling.*

Kirchlengern power station at around 1930.

147

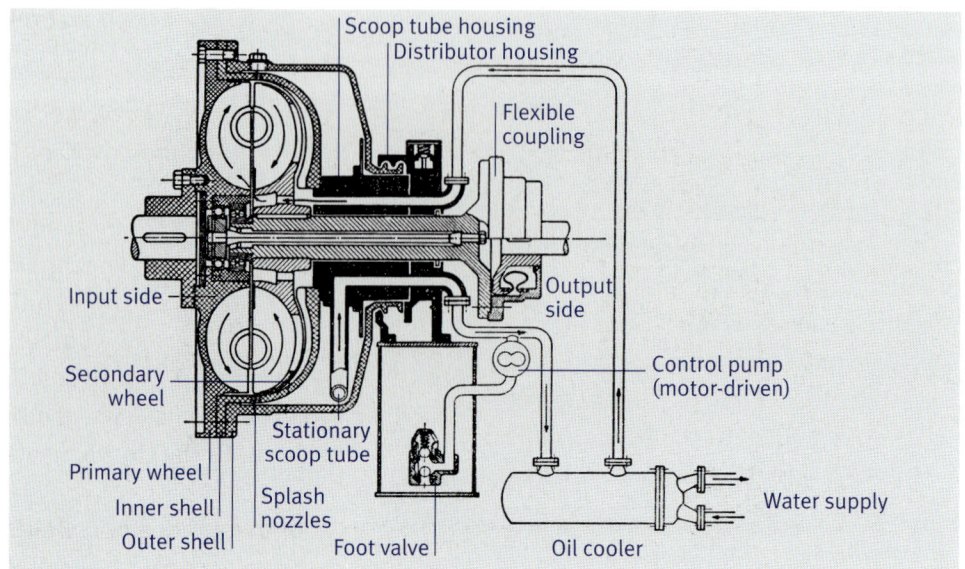

Voith Sinclair coupling size 316 at Kirchlengern power station.

S-flex variable-speed coupling.

The requirements on the control dynamics of the boiler feed pumps with their high volumes were fairly low at the time, and this coupling type proved to be perfectly suited for this purpose.

Applications with Boiler Feed Pumps for Once-Through Boilers

Rapidly increasing power station outputs with ever more complex control dynamics made it necessary that not only boiler, turbine and generator manufacturers but also the suppliers of peripheral systems and machines adapted their products to these rising demands. Fritz Kugel had anticipated this development

Mannheim super power station.

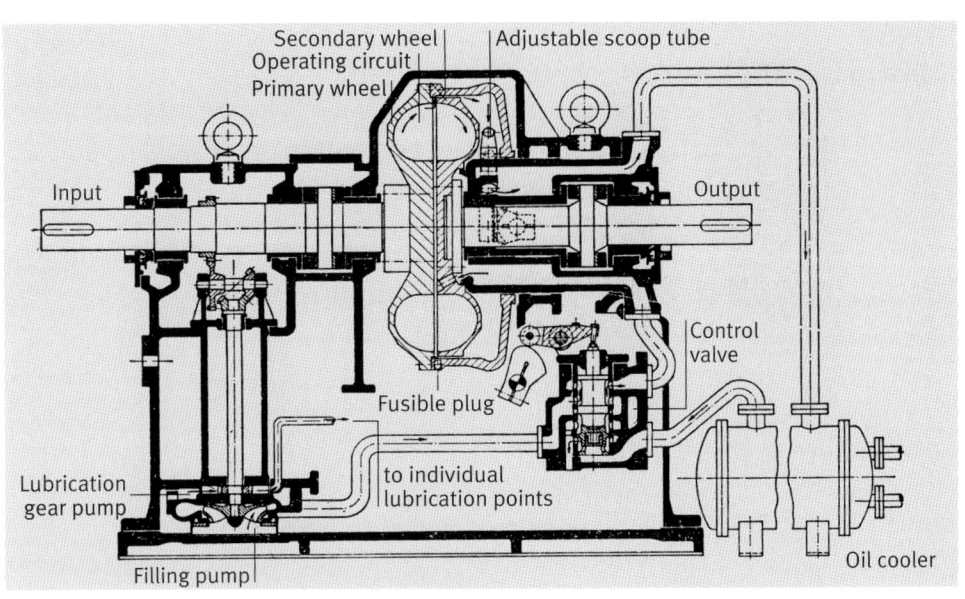

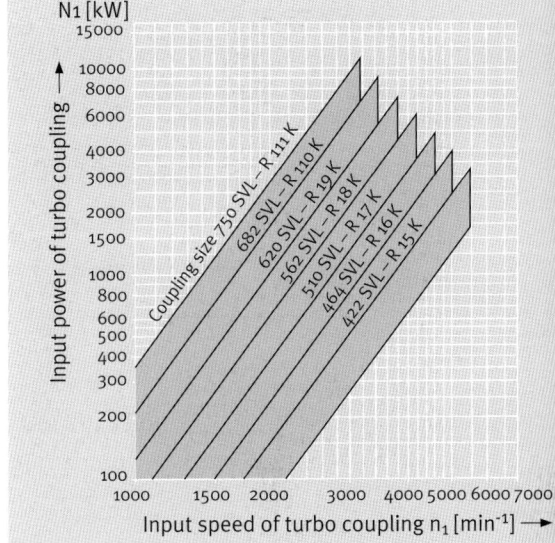

Voith variable-speed coupling series SVL.

Performance diagram (replaced by P – power) of SVL and R1.K for selection of coupling size.

at an early stage and recognized that the filling process of the coupling had to be adapted at a much faster rate. The scoop tube became an active element.

This gave rise to the development of the variable-speed coupling type SVL with adjustable scoop tube and plain bearings in the early fifties. A control valve allowed a volume flow of the operating oil that could be adapted in the control flow range depending on the slip loss. This allowed a reduction of the required oil. The oil supply system was dimensioned to provide sufficient lubrication oil for the bearings of the electric motor and the boiler feed pump. In 1953, the first SVL couplings (size 620 with a power transmission rate of 1 700 kW) were delivered. They were installed in Goldenbergwerk power station.

Geared Variable-Speed Couplings

Ever sharper pencils were used to calculate the efficiency of power stations. The investment and operating costs of all components were subject to ongoing scrutiny by investors and operators alike. The need to reduce costs extended equally to boiler feed pumps. Their economy had to be maintained or preferably improved. When faced by such demands, manufacturers normally look first at the size of a machine, which can be reduced if the speed is increased. The standard values of electric motors of 3 000 or 3 600 min^{-1} were, however, no longer sufficient for this, necessitating the addition of separate helical gear stages both before or after the variable-speed couplings.

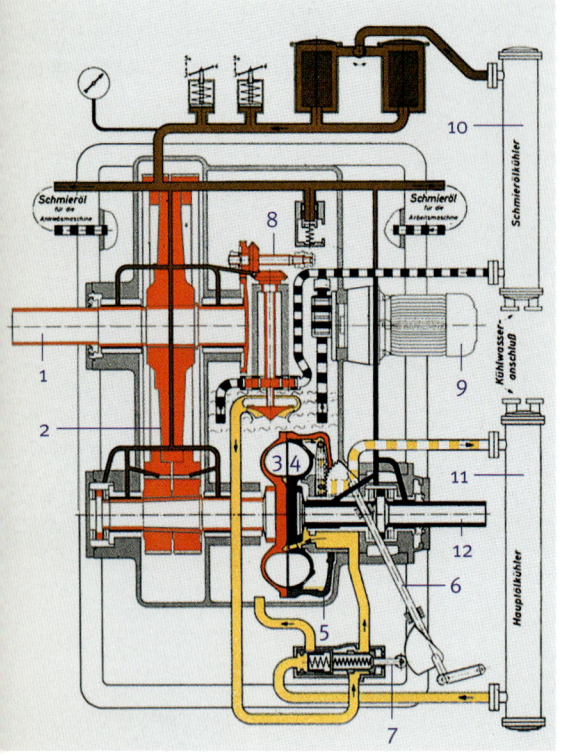

The diagram labels within the image read:
Schmieröl für die Anwendemaschine — Schmieröl für die Arbeitsmaschine — Schmierölkühler — Kühlwasseranschluß — Hauptölkühler

R 1.K geared variable-speed coupling
 1 Input
 2 Gear stage
 3 Primary wheel
 4 Secondary wheel
 5 Shell
 6 Adjusting shaft for scoop tube
 7 Control valve
 8 Pump combination
 9 Starting pump
10 Heat exchanger for lube oil
11 Heat exchanger for operating oil
12 Output

While unavoidable, this measure logically resulted in larger dimensions and therefore higher plant cost. The Voith design engineers who, like all others, paid special attention to the economy of their product, found a solution for this dilemma. They put their proven SVL coupling system with an added double helical gear stage into one common housing. As required, the higher speed also led to a smaller diameter of the coupling, since – as with all fluid machines – the transmittable power increases with the third power of the speed and with the fifth power of the reference diameter. The geared variable-speed transmission as a compact, space-saving and economical transmission unit was born. It was given the type designation "R1.K".

Meanwhile, in the seventies, the demand for ever higher outputs and speeds took on extraordinary dimensions. By then, "Voith Turbo KG" in Crailsheim had become responsible for the production of geared variable-speed couplings on a company-wide level. Now the senior engineers in Crailsheim were expected to implement further developments in order to enable the transmission of even higher outputs. The performance chart gives an impression of the values that have been achieved since then. This coupling series, extended by an additional gear, was now able to cover a performance spectrum of up to approximately 20 MW and a speed range exceeding 10 000 min^{-1}.

R16K geared variable-speed coupling in the drive of a boiler feed pump. The first units designated R16K were supplied to the RWE power station in Frimmersdorf in 1956, at an output speed of 4 870 min^{-1}, the transmission power was 2 260 kW.

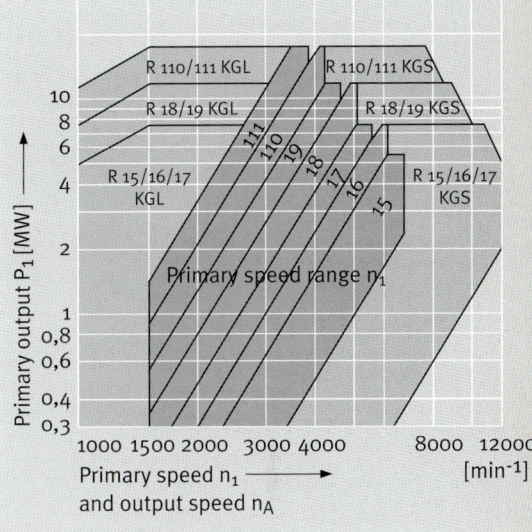

Performance chart of geared
variable-speed couplings
R1.K, R1.KGS and R1.KGL.

Geared variable-speed coupling
R19KGS in Moorburg power
station. The first couplings of
this type with quick-starting
device were delivered to
Moorburg with a transmission
output of 9 860 kW at an output
speed of 4 860 min^{-1}.

But the demands on control dynamics and the handling of emergency situations continued to rise. If a feed pump system fails, the water flow must not be interrupted for more than ten seconds, otherwise there is a risk of severe damage to the sensitive once-through boiler. Within this extremely short time span, a second boiler feed pump on active standby has to be able to take over the water flow at the same level.

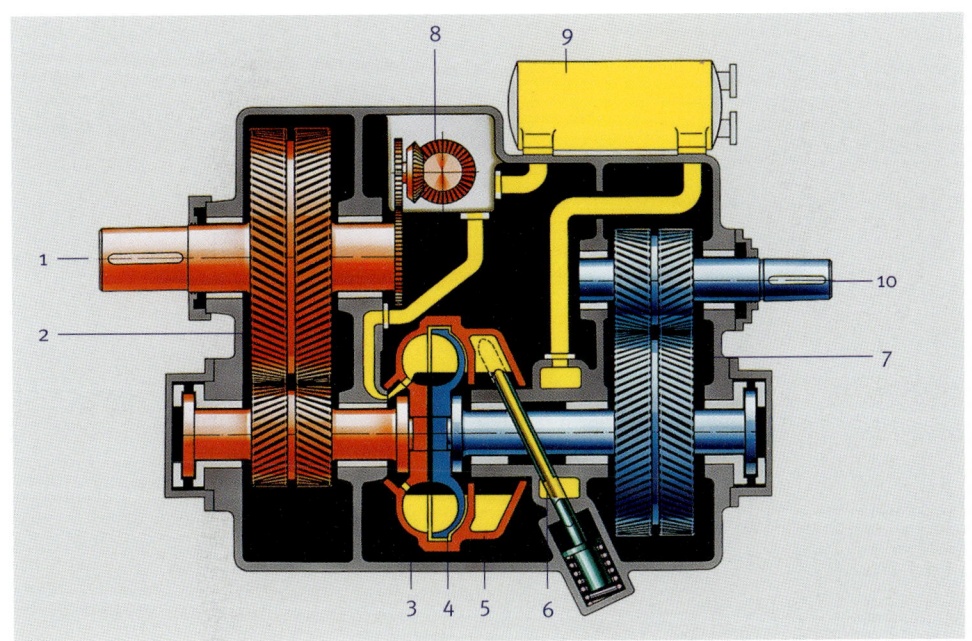

Geared variable-speed coupling
R1.KGS
  1 Input
  2 Gear stage
  3 Primary wheel
  4 Secondary wheel
  5 Housing
  6 Adjustable scoop tube
  7 Gear stage
  8 Pump combination
  9 Heat exchanger for
    operating oil
10 Output

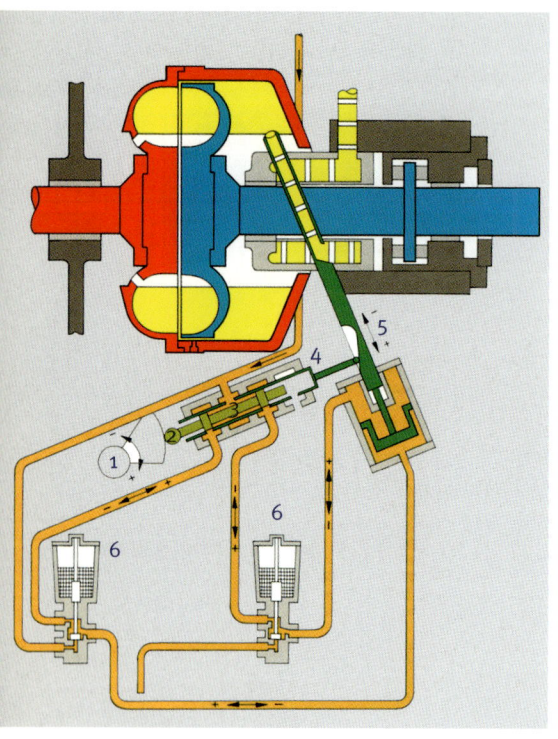

Quick starting device
1 *Adjusting shaft with cam plate*
2 *Guide roller*
3 *Control pin*
4 *Control sleeve*
5 *Scoop tube*
6 *3/2 way-valves*

With its newly developed hydraulic control system, the scoop tube once again turned out to be a flexible and reliable device capable of reacting promptly when required. With the quick starting device, consisting of two selector valves, the scoop tube is kept at "0% filling" stage, allowing the unloaded motor to run up to nominal speed in the shortest of times. The scoop tube then releases the filling at high speed.

Other applications required two 50% boiler feed pumps in the driveline of the main turbine. With this requirement in mind, a special version with only one reducing gear stage for higher speeds and a brake on the secondary side of the coupling was developed. This design allows bringing the boiler feed pump to a standstill, for example for maintenance work. In this case, the scoop tube is blocked in 0% position. The unit is designed for input power of up to 17 000 kW.

For safety reasons, **nuclear power stations** have very high demands on units and materials. Any technical modifications on geared variable-speed couplings have to be reported. In some cases, retrofits are necessary in order to keep the plant in compliance with the latest technical level.

R111GS geared variable-speed couplings at Scholven super power station.

Nuclear power station Isar Block I and II.

Voith variable-speed and geared variable-speed couplings are in service in approximately 70 nuclear power plant blocks throughout the world. Examples are the nuclear plants Brunsbüttel, Philippsburg and Gundremmingen in Germany, Forsmark and Oskarshamn in Sweden, Mühleberg in Switzerland, Olkiluoto in Finland, Tihange in Belgium, Yonggwang in Korea and Tarapur in India.

New Drives for Boiler Feed Pumps

Although Voith engineers are used to adjusting to ever-changing new market requirements or demands for lower manufacturing costs, they were suddenly forced to focus on operating costs at the end of the eighties. This was due to rising energy prices caused by the worldwide oil crisis. The costs for speed-controlled electric motors decreased and the availability and reliability of electric control systems rose, putting competitors at a new advantage. Kilowatt hours used were consequently given a higher rating and thus made a much greater impact on operating cost calculations. Growing competitive pressure finally forced the development of a new drive concept.

The answer to these challenges was a variable-speed superimposed gear with a torque converter in the superimposed section, and later a modular geared variable-speed coupling.

VORECON RW
A Variable-speed coupling
B Clutch
C Torque converter
D Hydrodynamic brake
E Stationary gear
F Rotating gear (planetary gear)
G Multi-circuit control

Illustration of power flow
1   Coupling operation
2.1 Sub-synchronous converter
     operation
2.2 Converter superimposition

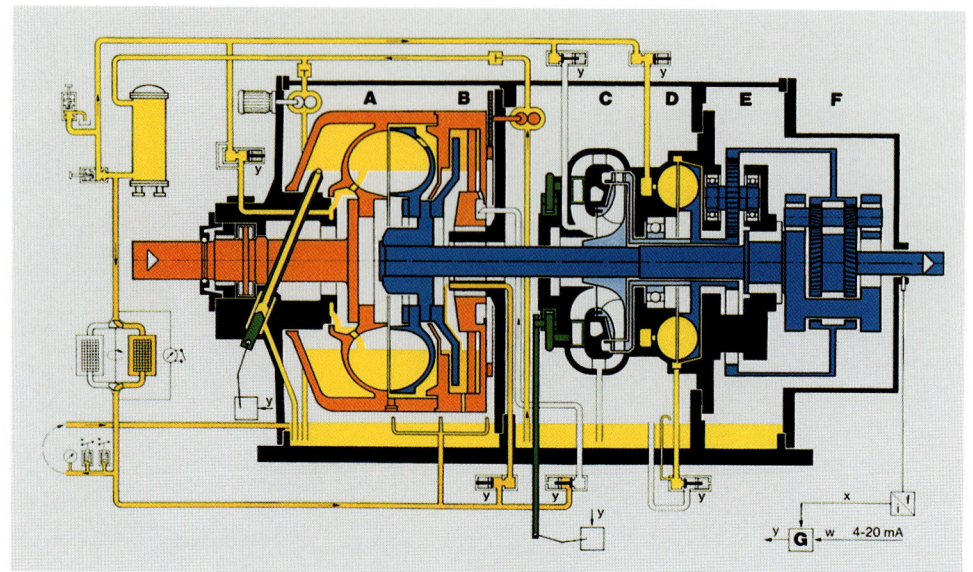

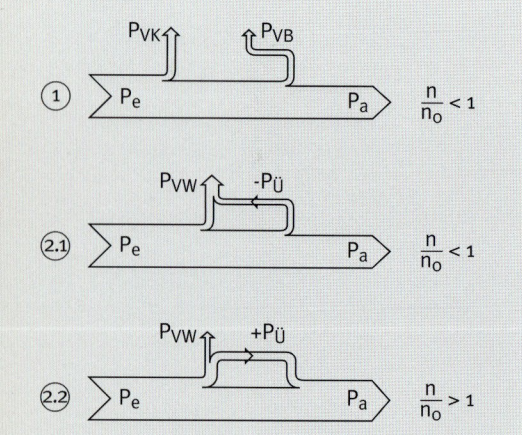

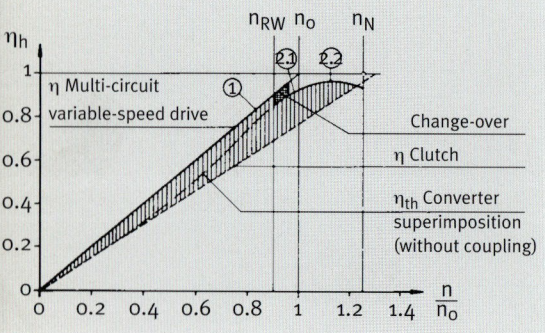

Development of efficiency and
operating collective.

This new transmission called VORECON was an invention by Georg Wahl. It has a coaxial shaft design, made possible by skilfully arranging hydraulic and mechanical components. The superimposition of the power shares generated by different methods results in a control range of 1.5:1, which is significantly higher than that of conventional variable-speed or geared variable-speed couplings. Across this control range the efficiency rate can be as high as 95%.

This high value is explained by the fact that the highest power share is transmitted directly and virtually without any loss via the main shaft and the rotating planetary gear. The smaller power share is transmitted to the rotating planetary gear via the torque converter and the stationary gear, which is more prone to losses. The torque converter is used for adapting torques and speeds.

The VORECON enabled Voith to compete successfully against speed-controlled electric motors. In cases where additional control is required or load-free motor start-ups are essential, the VORECON can be adapted by a variable-speed coupling with a lock-up function or a hydrodynamic brake with variable blades. The type designations RW, RWE, RWS and RWC have been introduced to distinguish between such design variations.

The VORECON was certainly an innovation at Voith, and most definitely a novum in the market. Although it consisted of Voith components that had proven

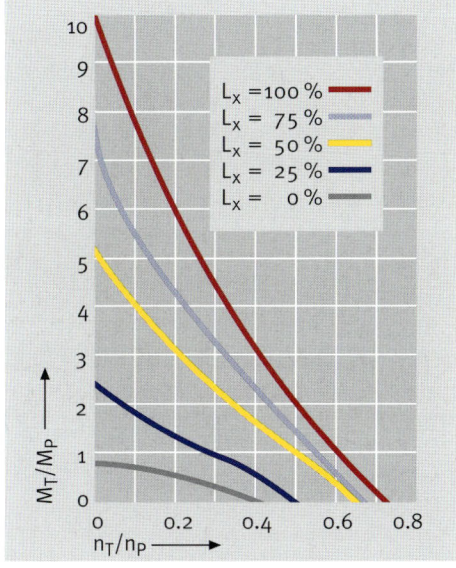

Torque/speed characteristic curve of gas turbine starting device.

In 1970, the first example of a gas turbine starting device was delivered to the Altbach power station. It transmitted 3 200 kW at 3 000 min^{-1} and received the type designation RL111YFG.

With this order, Voith had opened up a completely new field of applications for hydrodynamic torque converters.

Left: Gas turbine starting device: E-motor and converter type EL9 G 6.0. Converter housing in new Voith design.

Changing requirements to outputs, availability and maintenance that originated primarily from the US market, resulted in a **new converter type** in modular design with plain bearings. An integrated turning device can be added as an option. The shape of the housing followed the new Voith design. The first converters of this type were delivered to General Electric in 1996.

Generator Synchronization (Phase Alternator)

Maximum use of an electrical three-phase or alternating current network is only possible if the current and voltage values change simultaneously and in the same sense, i.e. if they are "in phase". In practical operation, their phases are, however, always displaced against each other due to inductive and capacitative loads with different magnitudes. This unfavorable situation results in efficiency losses, but can be improved by "Phase Alternators". This designation is used for generators that run at nominal speed but are connected with the network without creating power of their own. To fulfill their task, they have to be accelerated and synchronized from standstill at high break-away torque, i. e. they must be connected to the network at precisely the same frequency and phase range.

750 SVNL variable-speed coupling in the drive of a district heating water pump in Herrenhausen power station.

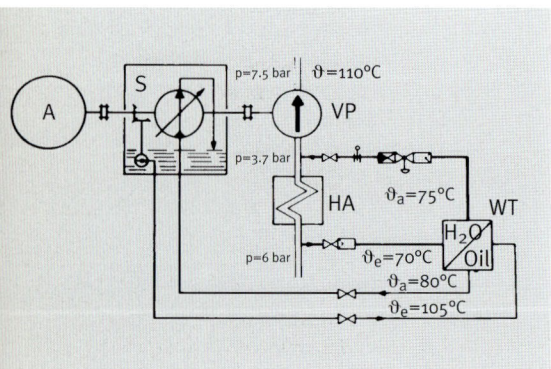

Heat recovery in a heating water system
A   Drive motor
S   Variable-speed coupling
VP Flow pump
HA Water heater
WT Heat exchanger

As it turned out, starting devices similar to those used for starting gas turbines proved to be suitable for this difficult undertaking. Thanks to their special starting conversion and their ability to adjust over-synchronous output speeds, Voith torque converters were ideal for such applications.

District Heating Plants

In a district heating network, the consumer demand for heat changes with the seasons and the time of day. District heating plants have to react to this and specially equip their heating water pumps that are used as flow and return pumps so that they can cope with these fluctuations. In such cases, variable-speed couplings type SVNL and SVTL are used to adapt the speed to the respective demand. As a result of the increased temperatures in which they operate, hydraulic oils with higher viscosity and correspondingly designed oil supply systems are required.

Heating power stations work after the so-called cogeneration principle. Part of the steam of the power generating process is extracted and forwarded to heating condensers. The overall efficiency of such plants reaches up to 80%. A welcome side-benefit that is only possible in a heating power plant increases the total efficiency of the supply system: the slip heat created inside the couplings is utilized for pre-heating the return water.

Drinking Water Supply, Seawater Desalination

The principle tasks of public supply systems are very similar to those of district heating plants. While the operators of such systems do not have access to slip heat, they can still utilize the Voith coupling type SVTW. This new development that can be operated with water is particularly suited for applications in waterworks. Allowing easy installation, this compact drive unit combines the advantages of hydrodynamics, unrivalled environmental compatibility and increased economy due to its natural operating medium: water.

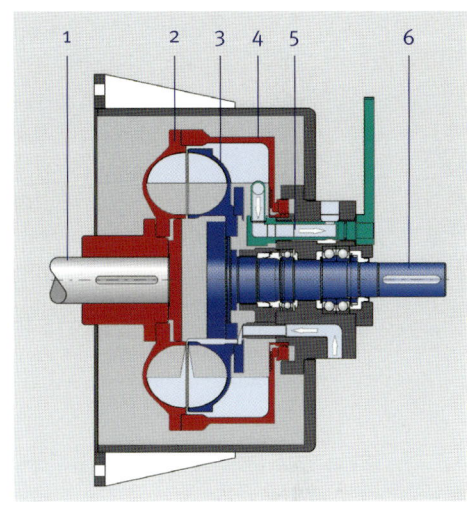

SVTW variable-speed coupling in
a water processing plant.
1 Input
2 Primary wheel
3 Secondary wheel
4 Shell
5 Swivel-mounted scoop tube
6 Output

SVNL vertical variable-speed
coupling in the drive of
water pumps in a seawater
desalination plant.

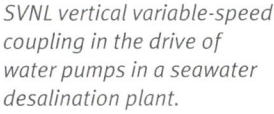

163

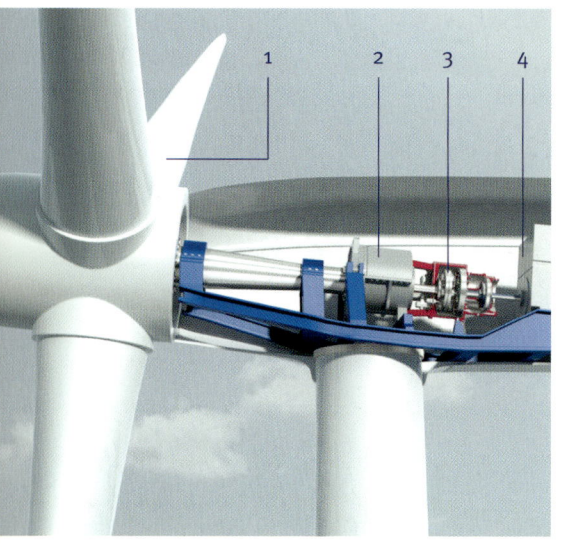

WinDrive superimposed gear.
1 *Wind rotor*
2 *Main gear*
3 *WinDrive*
4 *Synchronous generator*

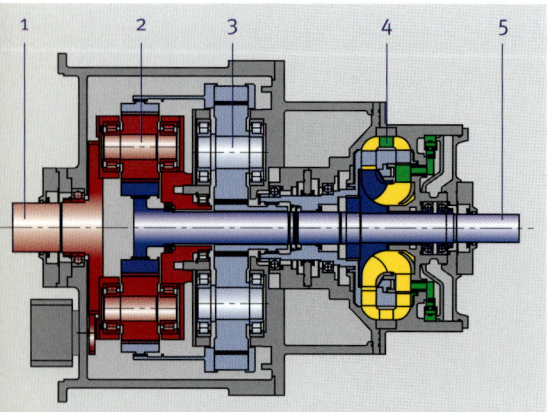

Longitudinal cross section
of WinDrive.
1 *Input shaft*
2 *Planetary gear*
3 *Stationary gear*
4 *Torque converter*
5 *Output shaft*

Drives for Wind Turbines

Limited resources of fossil fuels and an increasing awareness of the need for climate protection have resulted in a more intensive utilization of renewable energies in recent years. In 2004, some 1 200 wind energy plants with a combined output of over 2 000 MW were commissioned. The total value installed has grown to more than 16 000 MW. Today, approximately 5% of Germany's energy production comes from wind turbines. With a projected annual growth of more than 10%, wind energy is likely to become more and more interesting! Voith is still discovering further potential applications for fluid couplings and transmissions in this field, because machine control plays an important role for wind energy projects. Deriving from the VORECON, Voith engineers developed a special gear for wind power systems, called WinDrive.

The hydrodynamically controlled superimposed WinDrive connects the variable-speed wind rotor with a synchronous generator running at constant speed. As a result, it is possible to produce a constant frequency in the generator without the need for an inverter. The WinDrive adjusts the speed of the wind rotor in such a way that the system always runs at optimum aerodynamic efficiency, whatever the wind speed.

The driveline of a wind energy plant with a WinDrive consists of a two-stage main gear with a permanent ratio of 20 – 30, as well as the hydrodynamic gear as a "third stage" with a variable ratio. Through the controllability of the intermediate hydrodynamic components, the variable rotor speed can be controlled to a constant output speed. A synchronous generator that is directly connected to the grid is hence operated without any power electronics.

With its adjustable guide blades, the WinDrive can almost perfectly control and regulate the four areas of required output characteristics of a wind turbine
- parabolic operation
- operation at constant speed
- operation with constant maximum torque and
- temporary energy storage at fluctuating wind speeds
due to the parabolic torque curve of the torque converter and the torque setting. By 2006, the construction of two prototype systems will be realized with two partner companies.

Applications in the Oil-, Gas- and Chemical Industry

Applications in these industries make the highest demands on drives across the entire process chain – from raw material extraction over transport to processing. These plants must function reliably and with maximum availability in extremely tough conditions, for example in open-air installations, highly diverse climatic zones, on offshore platforms, in the desert or in arctic environments. Country-specific regulations and user-related specifications, both for the mechanical and the electrical design and equipment, must be observed. Frequently, various explosion protection protocols apply to such plants. Voith hydrodynamic products meet these protocols for hazardous areas (explosion protection).

All over the world, Voith variable-speed and geared variable-speed couplings are used in these industries as efficient and reliable elements for transmission powers of up to 30 000 kW and speeds of up to 17 000 min^{-1}.

Such applications follow a long tradition. As early as 1934, a variable-speed Voith Sinclair coupling type 316S was delivered to Amöneburg Chemical Works for a fan drive with 15 HP at 1 500 min^{-1}.

Drives for Crude Oil Pumps

In most cases, centrifugal pumps are used to push the crude oil through a pipeline from the place of extraction to offshore loading stations or directly to the refinery. Along these pipelines are booster stations (booster pumps) for compensating pressure losses and overcoming geographical altitudes. Their speed must be variable in order to adapt to the prevailing flow, pressure and density.

The environmental and operating conditions under which the Voith variable-speed couplings have to work are extremely rough and difficult. Oil is often extracted in desert zones. There, temperature fluctuations from up to 50° C in the shade during the day and below 0° at night are normal. The open-air plants have to withstand these extremes. They also have to endure sandstorms without being damaged. Voith variable-speed couplings have been designed for such applications and are widely used due to their reliability. In many cases, couplings have been in operation for 150000 hours.

R17KGL geared variable-speed coupling in the drive of a booster pump in an oil field in Saudi Arabia.

VORECON in the drive of a pipeline compressor.

Oil Production

Oil well derricks present numerous application opportunities for Voith variable-speed couplings and torque converters. Casing pipes, flushing pumps and lifting gear all need drives. Drilling speeds must be steplessly adjusted to the prevailing underground conditions. Scavenger pumps with varying pressures and flows flush out drilling waste. Lifting gears operate with very high accelerations and drastic retardations when casing pipes are pulled from large depths and have to be quickly dismantled to save time. The torque converters have fixed guide blade positions that are adapted to the engine characteristics and the process requirements. They significantly reduce material shocks and fluctuations of the drilling torques. The service life of the diesel engines is thus considerably increased.

Drives for Compressors

Gases are either transported through pipelines across long distances to faraway processing plants, or they are used immediately on site, for example for processes in the chemical industry. In both cases, compressors ensure the required pressures. They have to be controllable, as the high gas pressures in the chemical plant always depend on the process cycle. Additionally, flows have to be adapted to seasonally fluctuating requirements. All over the world, this is ensured by hydrodynamic variable-speed drives. They control the speed and the daily operation of the compressors so reliably, that operators can forgo replacement or standby units confidently. To the operator, high availability of all components is therefore a matter of course, and Voith components have proven to be very reliable. They always meet international explosion protection regulations, as well as the standards of the American Petroleum Institute for oil supply systems and monitoring devices.

Drives on Offshore Platforms

Offshore platforms are production or processing plants for crude oil and natural gas. They are either afloat or anchored at the bottom of the sea. They are used in the world's shelf regions, up to 200 meters deep and are surrounded by waves. With their help, coveted, rich oil and gas reserves can be extracted, although under extreme conditions. Platforms are often hundreds of kilometers away from the coast.

Oil well derrick with power packs including torque converters for reducing engine loads.

R..KGS geared variable-speed coupling in the drive of a gas lift compressor.

SVNL variable-speed coupling in the drive of an oil pump on an oil platform.

VORECON RWE
1 Input
2 Mechanically driven pump
3 Torque converter
4 Stationary gear
5 Planetary gear
6 Output

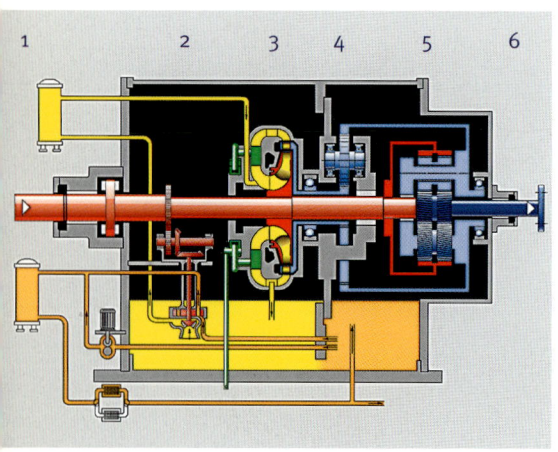

The entire power requirements of these self-supporting "islands" are covered by on-board gas turbine power stations. They supply electricity for centrifugal pumps used for extracting crude oil, separators for separating oil and gas, compressors for transporting gas, as well as various auxiliary systems. These platforms do not offer much space and have to accommodate all the people working on them. Oil and gas are highly flammable media. Technical systems are subject to strict regulations regarding explosion protection.

Variable-speed couplings, geared variable-speed couplings and VORECONS are ideally suited for these confined places. They require little space, and their weight is comparatively low. Both factors have a cost-saving effect on the construction of the system as a whole. Especially during the planning of floating platforms, these characteristics serve as convincing arguments. In addition, Voith components have clear advantages over competing electrically driven alternatives when it comes to explosion protection.

Over the years, Voith engineers have got used to handling the extraordinarily complex specifications that come with the tenders for such platforms. It began in 1976 with variable-speed couplings for the Ninian offshore platform near the Shetland Isles. Customer requests that can normally be accommodated on an A 4-page, turned into piles of paper several centimeters thick. The specifications normally fill several files. The regulations for deliveries and services are primarily based on US standards such as ANSI, ASTM, AISI, SAE, ASME and API. Voith specialists have to study numerous meticulous rules regarding the selection of raw materials, piping, appliances and electrical installations. Customer-specific requests in terms of instrumentation, piping and surface coating have to be taken into account as well. The design of oil supply systems presents extra challenges, because they have to be operable even when seas are rough.

As a result of added instruments with pipes and armatures made from stainless steel, oil supply systems with screw pumps, pressure control valves, stainless steel oil tanks and extensive electric installations, the coupling often takes on a new appearance that even the Voith engineers find strange at first sight. The couplings have become part of a large and complicated system in which everything must operate perfectly.

Voith Complete Solutions

Thousands of Voith variable-speed couplings and torque converters for the drives of crude oil pumps, process, pipeline, gas transport and gas lift compressors in virtually any application imaginable constitute a wealth of experience. In addition to their wide range of engineering skills, especially when it comes to torsional vibration and FEM calculations, Voith engineers are able to act as consultants at an early stage of an industrial application when it comes to the

Complete system. Drive motor, VORECON and compressor on a common base frame.

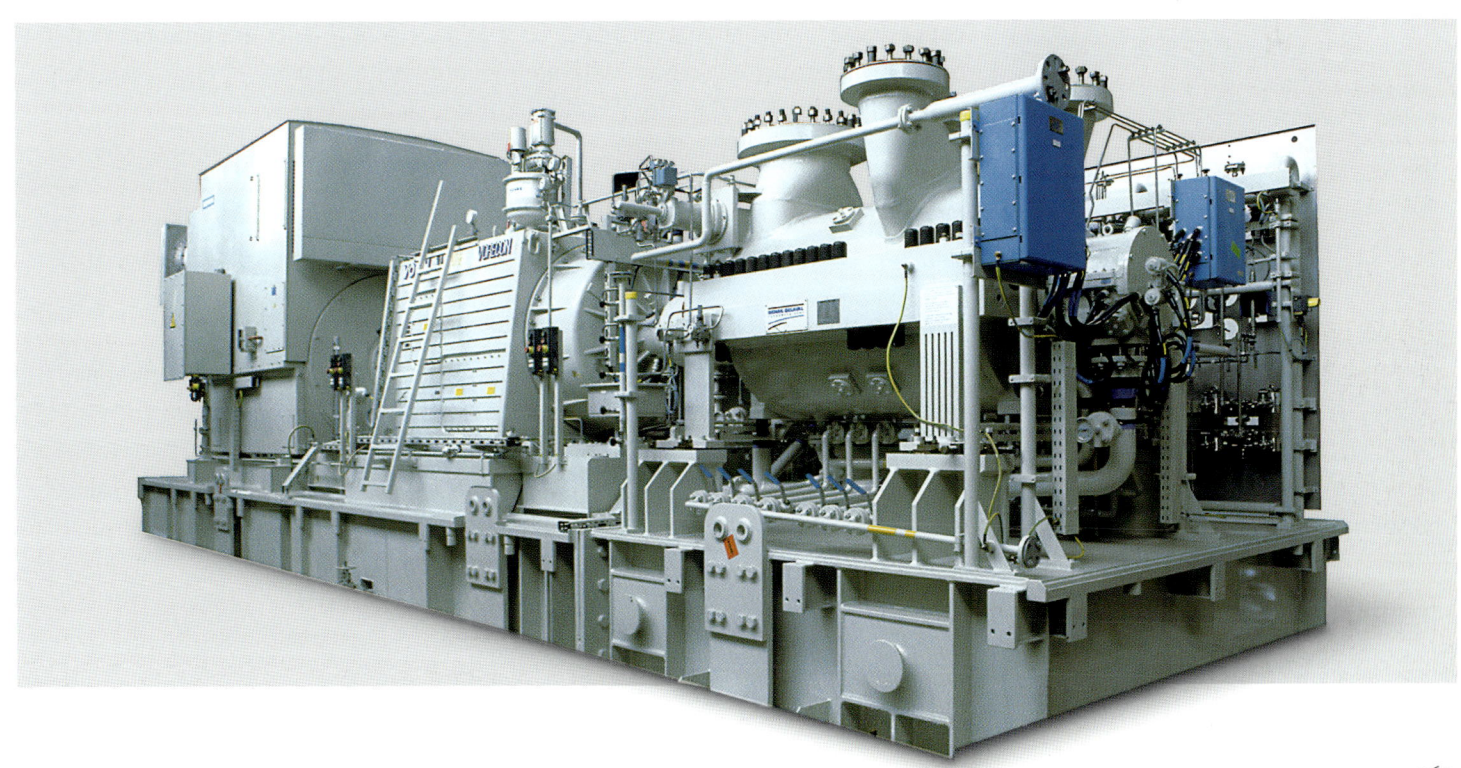

*Subsea variable-speed drive
with e-motor and pump.*

selection, the design and the construction of the drive system. In this way, individual solutions can be developed in cooperation with the end users.

During such talks, the participants are always unanimous that the delivery of a complete system with a reduced number of interfaces results in significantly reduced costs and labor regarding transport, assembly and commissioning. This has led to the systems concept as a preferred alternative to individual component deliveries. Complete drive units, consisting of a motor and a suitable variable-speed drive are developed to suit specific application requirements. As a result, all elements, from a sophisticated oil supply system up to an integrated, stainless-steel oil tank, are mounted on a base frame, ready for transport and installation.

Subsea Variable-Speed Drives

The exploration of new offshore and onshore oil reserves has revealed that deposits are located at ever deeper layers. It is realistic to expect depths of up to 3 000 meters below the sea surface. The gas pressure across the oil fields can therefore not provide sufficient natural lift. This disadvantage is particularly apparent among the numerous small oil fields in the North Sea. The application of delivery pumps in large depths will therefore become the norm.

Voith engineers were then faced with new challenges. While they had been concentrating exclusively on drives on a geodetic height of zero or above, they now had to ensure that their variable-speed drives also work at large depths.

A well-known method in the oil industry for extracting crude oil from depths between 600 to 3 000 meters is the installation of pump systems on the ocean's floor that ensure a static delivery head. In view of anticipated pressure fluctuations at the borehole and the fact that the water and gas contents of the crude oil can vary widely, the pump speed has to be continuously adapted.

For Voith design engineers, this task was a challenge with many unknown quantities. They nevertheless succeeded in designing a nearly perfect drive solution on the basis of a torque converter with adjustable blades. The torque converter itself, the monitoring devices and the drive for the guide vane adjustment, all have to function under ambient pressures of up to 300 bar. A trial on the test stand was to supply ultimate proof. A prototype of the new variable-speed drive with an output of 1 800 kW at 3 600 min⁻¹ was designed, constructed and successfully tested at a simulated water depth of 1 000 meters. The oil pressure in the housing was 120 bar.

The converter is installed directly between a constant-speed asynchronous motor and the pump. The projected output range is 10 000 kW. Once again, the hydrodynamic drive has revealed itself as the more economical solution due to its simple construction, wear-free power transmission and considerable savings in terms of investment and maintenance cost.

As a next step, plans are to install and test a prototype in realistic conditions in deep seas.

Applications in Gas Liquifaction Plants

Natural gas is playing an increasingly important role as an alternative to oil, also for political reasons. It can be transported to the end user via pipelines, causing little commercial or technical problems. Other possibilities, such as the transport by ships, can however, only be realized economically, if the gas is liquified.

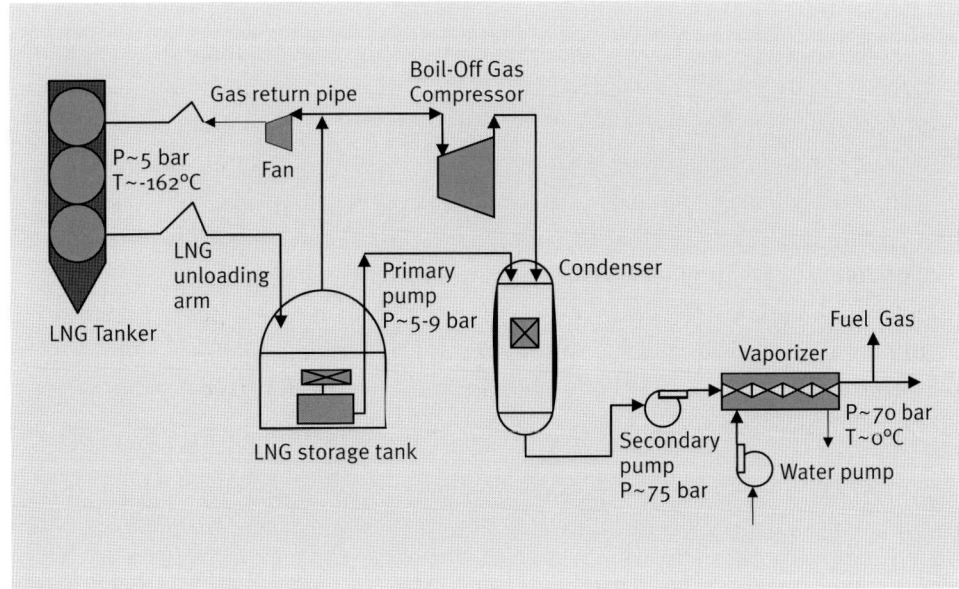

Process scheme of an LNG receiver terminal.

The liquifaction of the gas takes place at very low temperatures. The compressors of the cooling system require powers of up to 70 MW. During transport, the temperature must be kept at approximately -160° C and at 1 bar excess pressure. Evaporating gas is used as fuel for the drive motors of the ship. Compressors are also needed while the liquid gas is loaded and unloaded. The compressor speed must be variable, so that it can adapt to changing conditions during operation, including changing molar weights. If very large AC motors are used, they, too, require hydrodynamic starting devices in order to avoid excessive starting currents.

New Drive Element VOSYCON

In recent years, Voith engineers have developed torque converters with extremely compact designs for the starting devices of large gas turbines with outputs of over 40 MW and large AC motors. An integrated gear stage enables the turbine runner to turn slowly during the shut down phase. The first unit entered service in the Altbach works of Stuttgarter Neckar AG as early as 1970. Systems such as these can accelerate the turbine up to self-sustaining speed.

VOSYCON CSTC
1 Input shaft
2 Torque converter
3 Guide blades
4 Synchronizing device
5 Output shaft

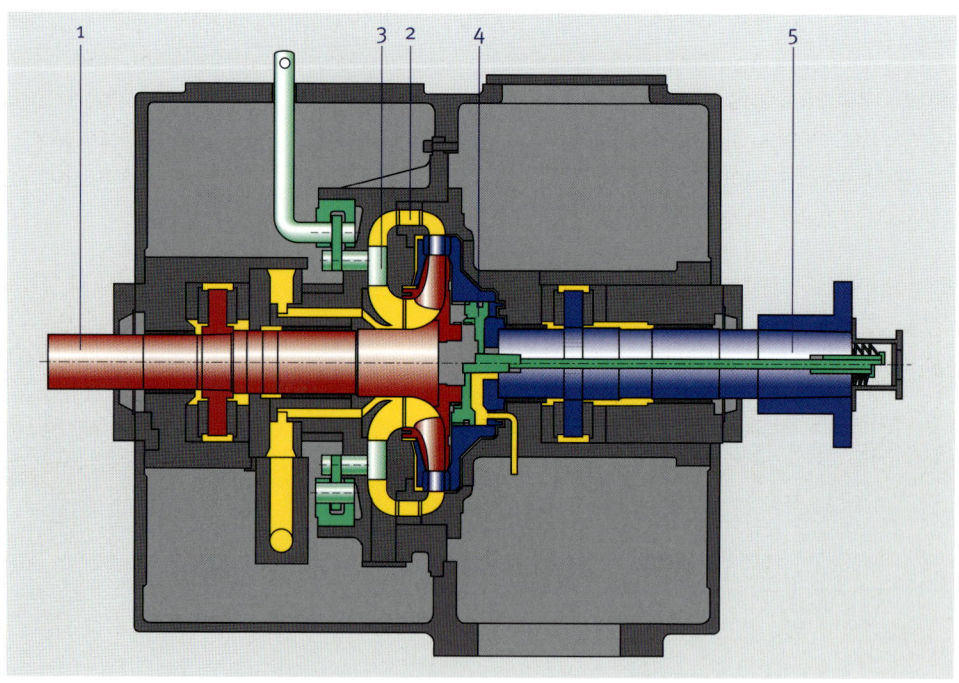

The newly developed VOSYCON, consisting of a torque converter with a lock-up coupling, enables large machines to run up load free to nominal speed. It can be used, for example, as a separating and coupling element between driving and driven machine (compressor). While the driving machine (gas turbine or electric motor) runs up, the torque converter remains inactive until nominal speed has been reached. The driver can then run at rated speed and accelerate the compressor via the torque converter that have meanwhile been activated. As soon as the impeller and the turbine wheel run synchronously, the two wheels are mechanically connected via a gear coupling. Now, driver and compressor are connected mechanically and the torque converter can be drained to eliminate ventilation losses and achieve 99.5% overall efficiency.

Torque converter in the drive of an agitator.

Drives in the Chemical Industry

The range of applications for hydrodynamic couplings and torque converters in the chemical industry is extremely wide and diverse. The following chapter provides a brief list of characteristic installations.

Agitators and mixers benefit from the hyperbolic torque development of the converter, which allows excellent control of chemical processes. By adjusting the guide vane positions, it is also possible to set the desired degree of "performance output" of the mixing material. For **piston pumps, piston and screw compressors** operating in the fertilizer industry against constant feed pressures, the favorable characteristic curve development of the torque converter is particularly beneficial. An additional advantage are the torsional vibration damping capability of the converter.

Torque converter in the drive of a piston pump (from right: E-motor, torque-converter, gearbox, pump).

Centrifuges often have a very high mass inertia and cannot be started with conventional AC motors, unless a torque converter is used. Additionally, the torque converter also takes the filling medium into account. If it is sensitive, it starts the centrifuge slowly at constant torque up to its nominal speed. In other cases, it can accelerate very rapidly due to its inherent high torque conversion capacities. It can also adapt the speed to suit certain process stages of different lengths and intensities. It can even be used for braking the centrifuge at the end of its operating cycle. It generates a braking torque, when the squirrel cage motor initially used for driving is reversed and rotates into the opposite direction.

Very high starting torques and hot, viscous materials under high pressure are characteristic operating conditions of **extruders in the plastics industry**. In order to cope with different raw materials or during a change of the processing method, the speed of the worm drive must be variable. The unusual physical conditions combined with frequent requests for explosion protection make Voith torque converters the optimum drive unit for such applications.

General Industrial Applications

The spectrum of applications in general industrial installations is very wide. Variable-speed drives with variable-speed couplings or torque converters are particularly sought after, when the demands on reliability and availability are high, even if the climatic conditions are unfavorable. Due to their inherent capacity of starting smoothly and isolating vibrations, they provide these characteristics effortlessly in addition to their actual tasks, or concentrate on these features if they have been installed for just this purpose. The following examples demonstrate how versatile these products are.

Conveyor Systems
These systems are installed in all climatic zones of the world for transporting bulk materials across short or long distances, and on upward or downward gradients. Depending on their applications and locations, their drives present the design engineers with a challenging task. Electric motors have to run up without load, the torque development and the drive torques must not exceed defined values, and belt oscillations must be avoided. In most cases, specially optimized Voith constant-fill or fill-controlled turbo couplings, as described in the following chapter, are used for such installations.

The conveyor system of Phosboukraa runs from the Sahara desert up to the loading terminal at El Ayun.

For specific applications, for example if speed control is required, variable-speed couplings can be equipped with hydrodynamic or mechanical brakes.

The 100-kilometer conveyor system of Phosboukraa built by Krupp Anlagen-bau in Essen at the beginning of the seventies is an excellent example in this context. It is used for transporting phosphates from the place of extraction in the Sahara desert to the Atlantic harbor of El Ayun. At the time of its construction, the special functional requirements of this system could only be met by variable-speed turbo couplings.

The system consists of ten individual belt conveyors, each with a length of ten kilometers. A total of 51 drives operate at the main, intermediate and end stations. They have to be started up in a controlled and synchronized way. Belt vibrations must not occur at any time. Maximum permissible belt loads have to be taken into consideration both on upward and downward gradients. Naturally, the speeds of 4.5 m/s of all conveyor belts must be adapted to each other, not just during start-up but throughout the entire transport operation of 2 000 t/h. No-load inspection runs at belt speeds below 1 m/s must also be possible.

Already faced unusual technical conditions, Voith engineers had to ensure that the extreme daily temperature differences, flying sands and even occasional sand storms did not affect the equipment. It was finally decided to use 650 SVNL variable-speed couplings with an output of 380 kW at $1 485 \text{ min}^{-1}$. The couplings of each drive station are connected to a common oil supply system. In 2000/ 2001, the plant underwent a complete overhaul, after which it re-entered service.

Steel Mills

Using short, sharp water jets, descaling plants in rolling mills remove scale from the red-hot material surface prior to the next rolling cycle, in order to achieve the desired surface quality. Variable-speed and geared variable-speed couplings are used for the continuous switching on and off of the pumps.

Fast reaction and absolute reliability were also the characteristics of a variable-speed coupling installed to prevent production disasters in a Turkish steel mill. Within just six seconds, it activates an emergency fan if one of the **furnace fans** breaks down.

Rolling train in a steel mill.

The most powerful Schlüter tractor (320 HP) with turbo coupling.

Below left:
Interaction of the primary coupling characteristics in connection with a naturally aspirated diesel engine.

Below right:
Interaction of the primary coupling characteristic with a turbo-charged diesel engine. The coupling with the start-up curve (Ak) enables reduced transmission of torque during the starting procedure (see page 135).

Compared to naturally aspirated engines, the drive torque of turbo chargers is less homogenous. It was therefore necessary to suppress the coupling characteristics (torque transmission) during the starting and run-up procedure of the engine. This led to a series of examinations aimed at reducing the "Start-up parabola".

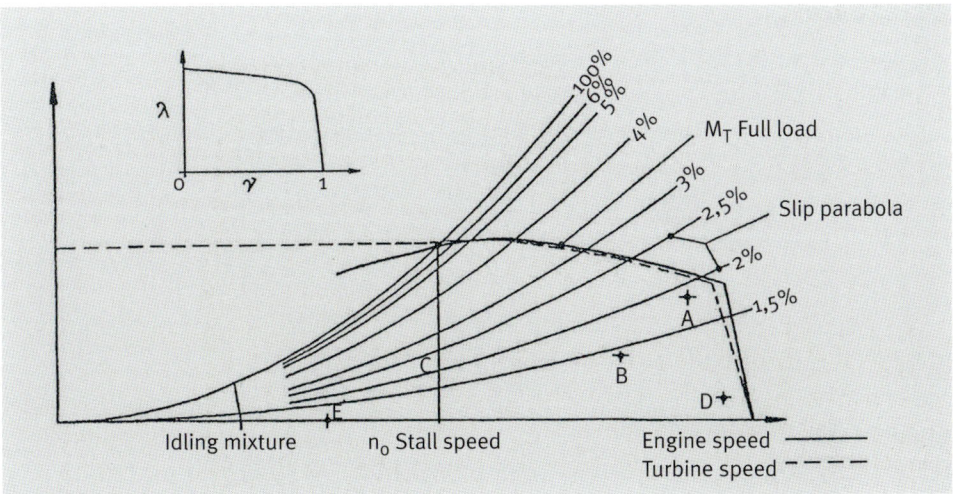

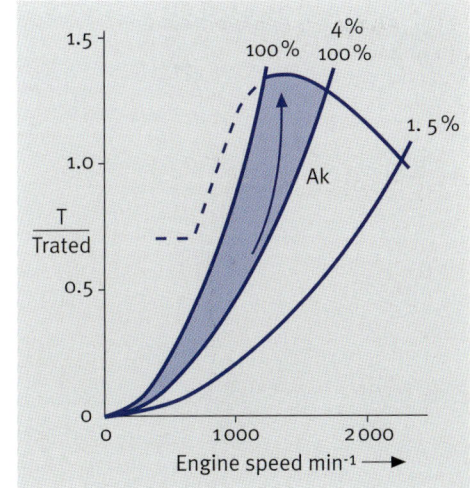

From approximately 1960, the coupling business for agricultural tractors gathered momentum due to the growing interest from other tractor manufacturers. At the time, Schlüter produced the most powerful vehicles. In view of outputs that would soon reach up to 320 HP, the conventional installation of the coupling in a housing with cooling air supply was no longer sufficient for dissipating the slip heat. For this reason, constant-fill couplings, until then traditionally used for this type of application were replaced by fill-controlled units.

In this coupling type, the oil circuit that normally transmits the torque is directed via an external heat exchanger and hence cooled for a longer period. Higher tractor engine outputs could then be transmitted without having to use a larger coupling type.

The introduction of this concept to the international market was made possible by a cooperation with the companies Bobar and Valmet. As a manufacturer of special vehicles such as vineyard tractors, Bobar attaches high importance to smooth start-ups, even with less experienced drivers. During service on extremely steep slopes is must be guaranteed that the vehicle does not tip over, even if it is driven inappropriately. The coupling was therefore given a filling that ensured that even jerky gearshifting movements would not present any hazard. The Valmet drive concept was similar to that of its competitors Fendt and Schlüter. Other positive product features were high driving comfort and easy handling of the mechanical gears.

Due to the development of power-shift or hydrostatic transmissions, hydrodynamic couplings are now less frequently used for tractors. But manufacturers of specific power classes or customers in emerging markets still prefer to utilize the inherent function characteristics of the couplings.

At the same time the first applications of turbo couplings in mobile agricultural machines were developed, the first generation of automatic transmissions for **passenger cars** began to emerge. When these transmissions were launched in the market, Daimler-Benz decided in favor of a drive concept with turbo couplings, being fully aware of their advantages over torque converters.

This decision led to a cooperation between Daimler-Benz and Voith, resulting in the first mass production of hydrodynamic Voith couplings.

Voith also developed new business contacts with a number of **bus manufacturers,** starting in the 1960s in Switzerland (FBW), in 1968 in France (PAM), in 1975 in the Netherlands (DAF) and in 1988 in England (London Buses). All of them had used semi-automatics in their city buses in the past. These semi-automatics operated with a mechanical clutch between engine and transmission, whose thermal capacity was extremely limited and led to frequent failures.

Bobar vineyard tractor.

Between 1961 and 1974 over 200 000 couplings were delivered for Mercedes cars.

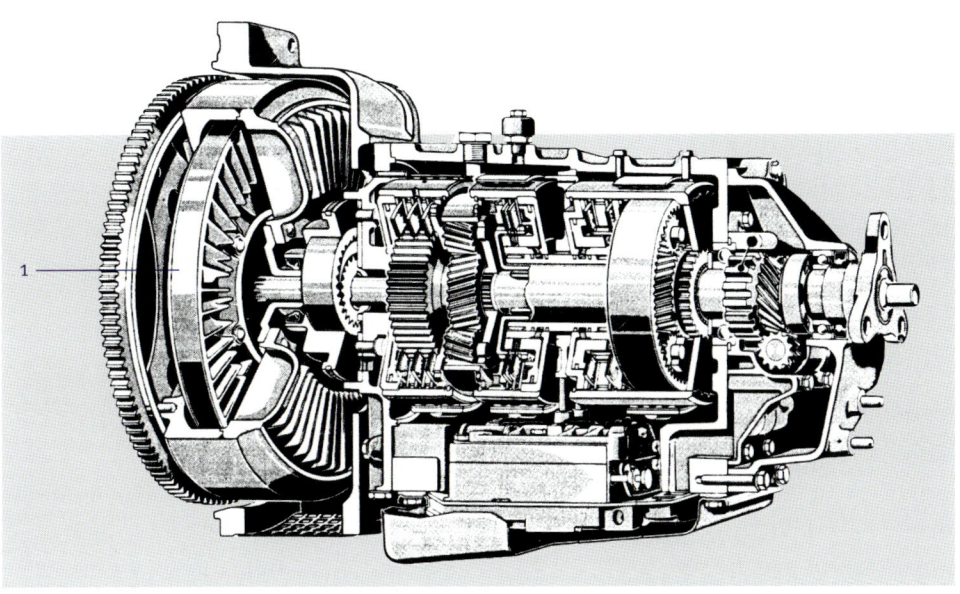

In order to protect this component, a hydrodynamic coupling was installed for starting and accelerating that also prevented gear jolts and jerky start-ups. Thanks to this measure, the driving comfort rose significantly.

After a while, a few drivers reported that for reasons of convenience, the transmissions were often not put into neutral as directed. The coupling therefore entered into 100% slip operation, generating intense heat in the process. But this shortcoming was quickly turned into an advantage. The coupling received an external cooling circuit and was hence ready for this initially not intended operating mode. The drivers of PAM now enjoyed maximum driving comfort.

Over the years, automatic transmissions with converter units were gradually given preference over turbo couplings in automotive engineering – around 1975 in the passenger car sector and at London Buses shortly after 2000. Newly developed turbo couplings with diagonal blades were able to meet the extreme demands on noise reduction and vibration damping in the lower speed range of turbo-charged engines, and led to successful prototype tests at BMW and Porsche. But eventually, car manufacturers could no longer provide the required installation space, ending the utilization of turbo couplings in mass-produced vehicles.

Yet the engine industry will, at varying degrees, always be of interest to Voith as far as ancillary drives are concerned. A turbo coupling was successfully used in a cooling fan of KHD. The unit in question was the first flow-regulated, fill-controlled coupling that had been specially designed for this application. Over a long period, features such as power savings, speedy warm-ups after cold starts, less wear due to constant engine temperatures and reduced noise during

Left: Longitudinal cross-section through an automatic transmission by Daimler-Benz with directly flanged turbo coupling (1).

1984: Convincing prototype test of the diagonally bladed coupling type 250 TD-P in the Porsche 944.

Frequent starting and stopping of London buses make high demands on the reliable operation of turbo couplings.

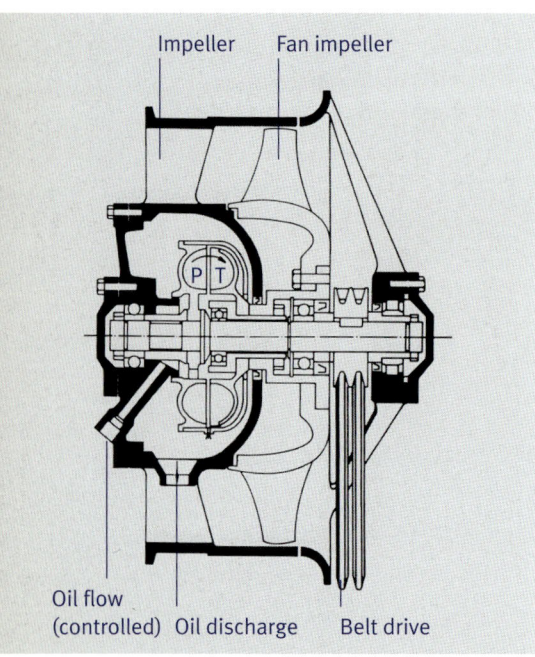

Impeller   Fan impeller

P T

Oil flow
(controlled) Oil discharge   Belt drive

*Above: First fill-controlled cou-
pling with flow control for fans in
air-cooled KHD vehicle engines.*

*Magirus fire engine with speed-
controlled cooling fan.*

part-load operation were some of the best features of this drive concept. In rail traffic, the turbo coupling is used in the drive of the heating generator of the DB locomotive type 218. This generator must be suitable for frequent on and off switching. As a rule, turbo couplings are always installed when vibrations in a driveline must be absorbed or blocked off completely.

Marine Drives

The Voith Schneider Propeller (VSP) is an application area for turbo couplings in combustion engines that is highly important for Voith and has been growing continuously over the years. The most relevant reasons for this application are the damping characteristics of the hydrodynamic coupling used for physical-ly decoupling rotating mass inertias between the slow running diesel engine with its inherent irregularities and the ship's propeller as a driven machine. The turbo coupling also separates the engine from the propeller during starting and stopping.

The market-driven further development of the Voith Schneider Propeller in terms of size (output, thrust) and new fields of application (floating cranes, drilling vessels) resulted in higher power requirements and a need for different drive concepts between diesel engine and propeller. The standard drive of the Voith Water Tractor that is so maneuverable and flexible in harbor zones, is a constant-fill coupling type TM1 with an output range from 150 HP to 2 500 HP. Some thousand units are already in service.

In order to fulfill special functions such as idling mode of the diesel engine with a decoupled propeller, high outputs and corresponding design require-ments, self-supporting couplings with external cooling circuits were developed. Specially adapted hydraulic couplings are also used in ships with conventional axial propellers.

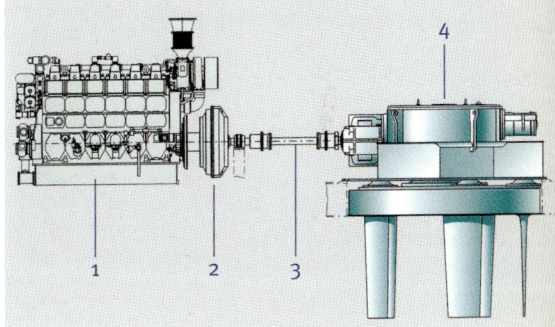

Countless extra functions have been and still are being developed for special tasks. Examples are the ships of Russ shipyards that were designed in the mid-fifties for applications in icy conditions on the Baltic Sea. While these ships were expected to work in rather rough surroundings, numerous ice-going polar vessels had to operate in really harsh environments. From the 1980s onward, they were fitted with variable-speed and fill-controlled turbo couplings of the TR series.

Hydrodynamic couplings in the VSP drive of police boats in 1949. Output 73 kW.

Scheme of the drive concept of a Voith Schneider Propeller that can be steered into any direction.
1 Engine
2 Voith turbo coupling
3 Cardan shaft
4 Voith Schneider Propeller

Above left:
Voith Water Tractor maneuvering a car transporter.

Roll-on-roll-off vessel of the 15 000 BRT class of the Russian Merchant navy for use in the Arctic.

Ship's main drive with central output shaft and four input shafts, on which the Voith turbo couplings are mounted (Renk). Special feature of the coupling: quick draining valves allow variable-speed operation.

Applications such as these make high demands on the drive system. The engines must not stall if the propeller is blocked. The coupling needs to adapt itself automatically, so that it continues to transmit the required torque against increasing resistance, while, at the same time, the overload protection has to be activated in the event of a sudden blockage. These requirements were met by the design of a hydrodynamic operating circuit, where the utilization of various flow conditions at differing slip rates generated the "load-adapted torque increase" and "overload protection" functions.

Drive configurations depended largely on the size of the ship and the number of engines. An interesting example is a four-engine marine drive delivering power to one single propeller shaft. If necessary, the individual engines can be switched on or off very quickly. This is achieved by a special quick-draining valve, which can adapt the operating circuit of a coupling within the shortest of times. Again and again, the Voith engineers are faced with such special requests. Newfoundland Ferry, a Canadian ferry operator, for example, wanted to have the coupling efficiency optimized for summer and winter service. In summer, with no ice in the water, there was no need for the Voith turbo coupling to protect the engine against overload. The coupling could therefore operate with a higher filling volume and consequently reduced nominal slip. Voith was able to fulfill this request. Operation in winter was, of course, a different matter.

Frigate Horizon, CODOG drive allowing diesel engines to be switched off.

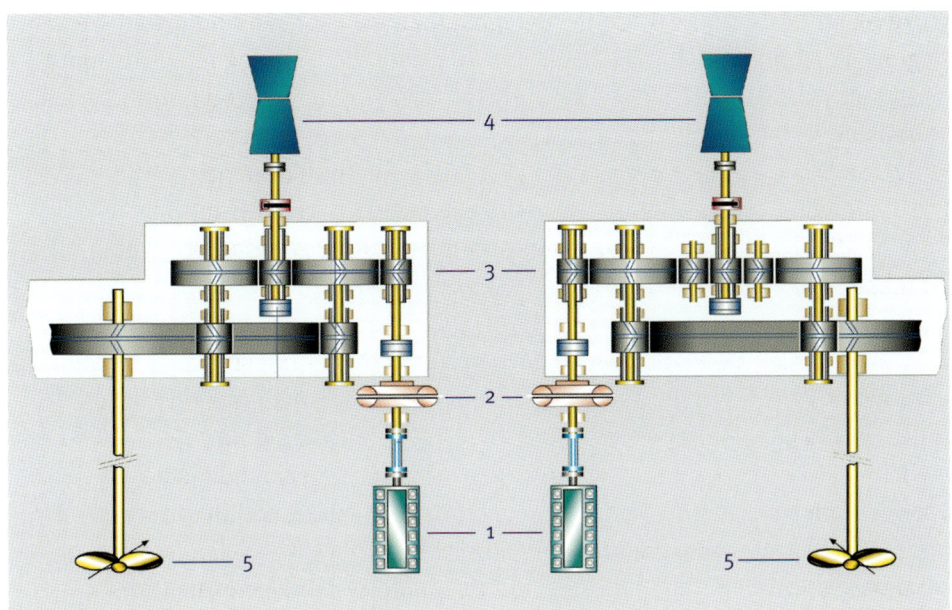

CODOG scheme (**CO**mbined
Diesel **O**r **G**as Turbine) in
a two-propeller ship's drive.
1   Diesel engine
2   Voith turbo coupling
3   Ship's main drive
4   Gas turbine
5   Propeller

Today, ships that need different sailing profiles due to their service routines often have combined drive units. These can be an arrangement of gas turbines and/or diesel engines that are switched on or off depending on load conditions. Voith turbo couplings are used to ensure varied use of these drive units and establish relevant harmonization between the systems.

Apart from being installed in main marine drives, turbo couplings, similar to land applications, are also used in **auxiliary marine drives**, for example as PTOs (power take-off) for on-board power supply, or as PTIs (power take-in) in the drivelines of exhaust gas turbo chargers.

Stationary Systems

As early as 1934, shortly after the signing of the licensing agreement with Harold Sinclair, MAN in Augsburg ordered two turbo couplings type 2 605 T as torsional vibration dampers between their diesel engine type G6V84 ($P = 2\,000$ HP/214 min^{-1}) and their transmission to the generator drive. A remarkable feature of this application is that Voith, although still in its early days as a hydrodynamics supplier (coupling no. 14) produced a unit whose size – outer diameter 2800 mm – was quite remarkable even at today's standards.

Main components of a coupling with a profile diameter of 2 605 mm compared with the smallest coupling size 316 from the early product range.

MAN diesel generator system from the thirties.

But Voith turbo couplings were not just used in rigidly installed systems with combustion engines for stationary industrial applications. Very soon, operators discovered the advantages of the turbo coupling for **semi-mobile plants**.
The first references involve sports gliders. In the fifties long before high-powered automatic car transmissions were available in Germany, hundreds of **towing winches for gliders** were fitted with Voith turbo couplings. These winches are self-propelling. Their engines sit on the chassis and drive the cable drum.
The special task of the intermediate coupling is to tighten the tow rope smoothly and to accelerate the glider swiftly under optimum utilization of engine output and characteristics. Additionally, it is also capable of recognizing temporarily occurring "kite-flying effects" and counteract them by a slip condition of > 100%, in order to continue the towing process safely.

The following application is another example of a semi-mobile installation. Processing and reusing materials of all kinds, **recycling**, is of high importance in modern industries. The trend is moving increasingly toward processing on site. The machines for grinding wood, construction rubble, metal, plants or waste are taken to where they are needed. For this purpose they are designed as semi-mobile, self-driving machines that can travel on public roads. Used as driving units for these applications, diesel engines have to work in extreme conditions. They are protected by the turbo coupling that does not just dampen jolts and shocks, but also allows frequent starting under load and ensures that the engine is not overexerted.

Another example of a virtually unrivalled application of a fill-controlled turbo coupling is the drive of a **high-pressure piston pump** which has operating pressures that can sometimes exceed 1 000 bar. Due to its operation, this driven machine is frequently started against full nominal pressure. A mechanical friction coupling with its often rather low thermal reserves and limited operating cycles can perform this task only to a certain degree.

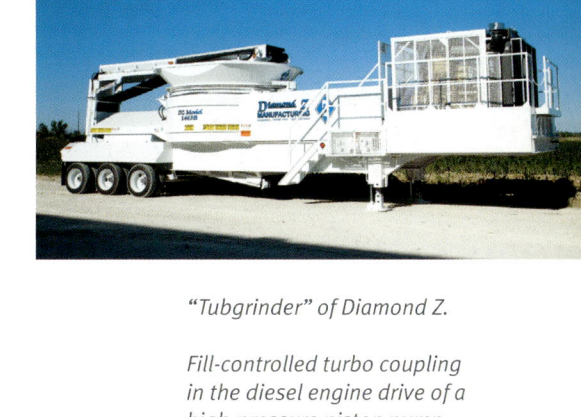

"Tubgrinder" of Diamond Z.

Fill-controlled turbo coupling in the diesel engine drive of a high-pressure piston pump.

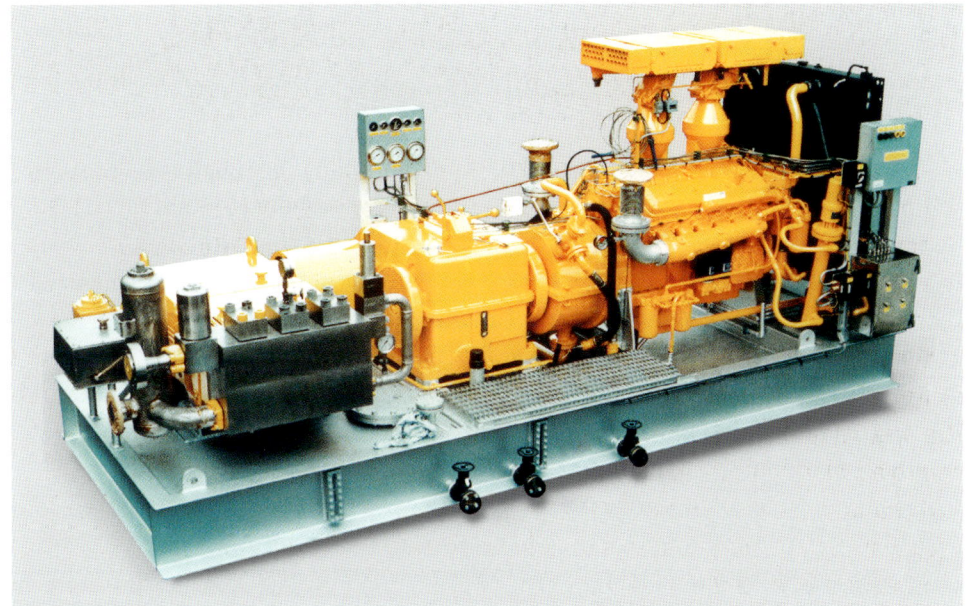

Drives With Electric Motors

In stationary industrial drive technology, asynchronous or squirrel-cage motors are by far the most frequently used power source. Over the years, these machines have proven themselves as virtually indestructible, robust and low wear drives for all kinds of industrial installations. But their torque curves during the instationary phase of the run-up from standstill to nominal speed, i. e. their starting characteristics, are often of limited use or even completely unsuitable for powering operating processes or driven machines directly. The wide spread

First industrial application of a turbo coupling in the pump drive of a resin press rated at 2 HP and 1420 min⁻¹. It was delivered in January 1933.

between minimum and maximum torque (pull-up torque and pull-out torque) and their extremely high electricity uptake during the run- up that can be up to seven times the nominal current are presenting serious problems. Another negative aspect is that the speed cannot be controlled during the run-up period and is predetermined by the net frequency and the number of pole pairs. But it was soon discovered that the installation of a hydrodynamic coupling into the driveline eliminates these disadvantages and establishes the desired start-up behavior of a driven machine or an operating process.

Couplings in Mining and Materials Handling

In the following years, mining and materials handling developed into the key industries for these products. Both sectors have had considerable influence on the wide range of Voith products and the development of their sizes. The first business connections to German mining suppliers were formed as early as in the forties, especially with the union of Eisenhütte Westfalia GmbH in Lünen. The continuous pursuit of the mining companies of finding more efficient and economical production methods led to the development of driven machines that allowed uninterrupted extraction, and also resulted in a machine for the further transport of the mined coal that became widely known as an "Armored Face Conveyor". It is used in the earliest stage of coal production, at the coal face.

Armored face conveyor of Eisenhütte Westfalia with Voith turbo coupling in a housing to prevent direct contact during operation.

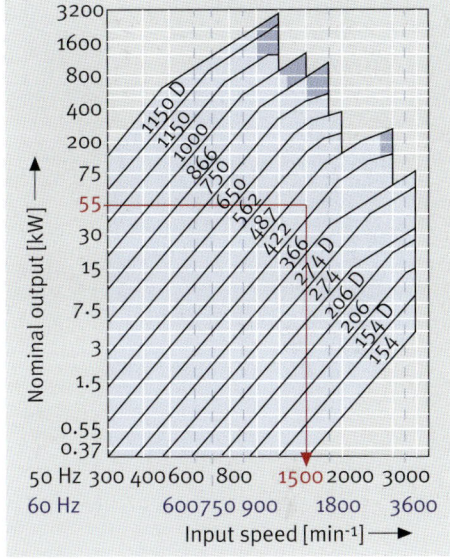

Performace diagram for selecting coupling sizes from the constant-fill range.

Left: Voith turbo couplings are in service worldwide – in the drive systems of bucket wheels to protect against overloads, in the drive stations of belt conveyors for smooth starting and accelerating.

There, it operates under extreme and harsh conditions. It has to be able to withstand stochastic shocks and cope with unpredictable and changing loads caused by falling materials.

When being asked to design a drive concept for these ambient conditions, the Voith engineers were aware that several different aspects had to be taken into account.

The robust squirrel cage motor was the clear favorite right from the beginning. It was widely recognized that its pull-out torque made it more suitable for breaking away overloaded armored face conveyors than any other motor type. But, when starting from standstill, the required torque could not be achieved with a direct, stationary coupling. Another aggravating fact was that the torque characteristics of the squirrel cage motor depend on the amount of grid voltage, which is, of course, prone to interruptions due to the limitations of underground electric supply. These circumstances suggested a separation of the motor from the armored face conveyor in the start-up phase. Installed between these two components, a hydrodynamic coupling with its parabolic torque curve enables the motor to run up to its nominal speed even if drive speeds are increasing. Afterwards, the motor is able to generate the required drive torque for the armored face conveyor.

This application marked the successful entry of the Voith turbo coupling into the mining industry.

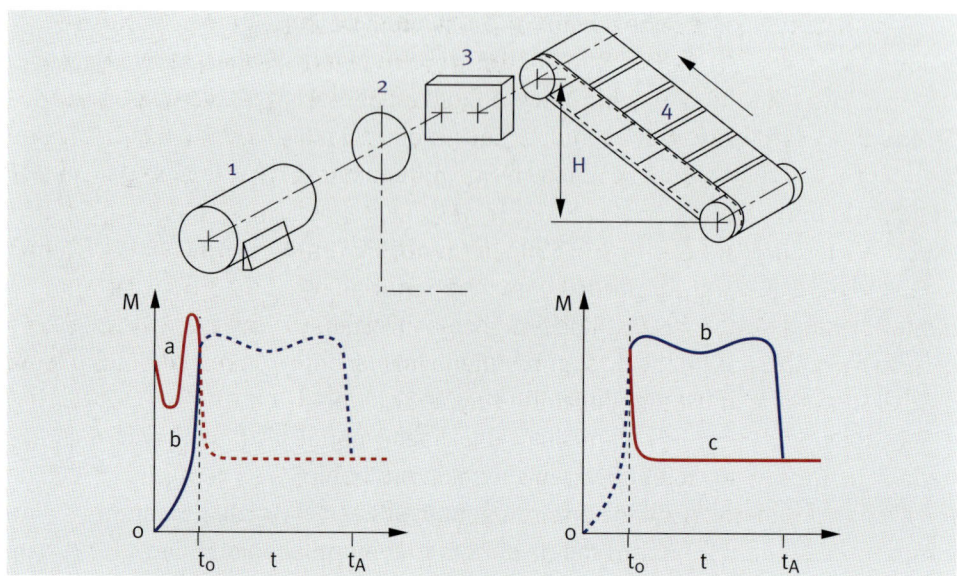

Schematic illustration of the separation of functions by the Voith turbo coupling on the input and output side.

1 Motor
2 Turbo coupling
3 Transmission
4 Driven machine armored face conveyor

a Motor characteristic curve
b Turbo coupling characteristic curve
c Load curve

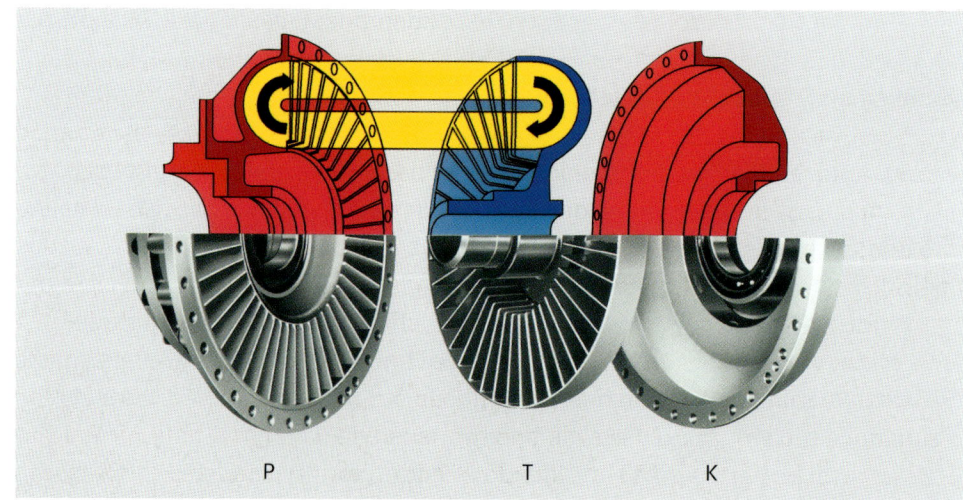

In its basic, constant-fill version, the hydrodynamic coupling only needs three major components to carry out its functions:

P Pump wheel
T Turbine wheel
K Coupling shell.

This version is designated "T-coupling".

The increasing use of technology in underground mines also meant a rise in the demand for electricity, so that existing networks were often overloaded. In the event of high, short-term energy requirements, for example when several drives are started simultaneously, voltage dropped quickly. As a result, the motor torque declined to such a degree that it could no longer accelerate to its nominal speed, despite its load being physically decoupled via the turbo coupling. The Voith engineers therefore looked for ways of optimizing the system to ensure that full motor run-up was guaranteed even under unfavorable network conditions. This meant that even more load had to be taken off the motor during its run-up phase.

These considerations resulted in the development of a turbo coupling with an antechamber, the so-called delay chamber that fulfills the function of an oil reservoir. This coupling version provides high protection for the motor during its run-up phase. One part of the oil filling is redirected into the oil tank during standstill. Now the coupling operates with a lower, yet effective oil volume in the bladed operating circuit, as a result of which the start-up parabola is reduced, and extra load is taken off the motor. After reaching the nominal speed, the delay chamber is gradually drained, and the retained oil is directed back to the operating circuit. After this procedure, the coupling operates on the basis of the conventional T-functions. The German term "Verzögerung" (delay) was used for the new designation TV coupling.

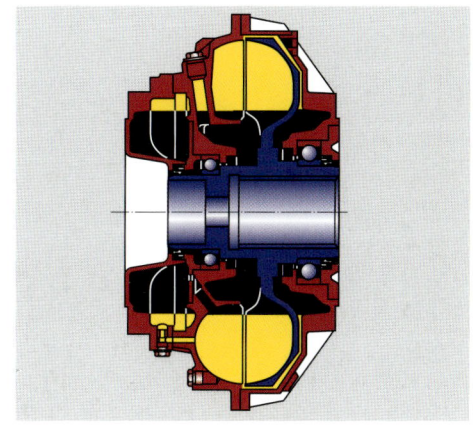

Turbo coupling type TV.

This technical development laid the foundation for turbo couplings with active antechambers in 1947. The findings gained by the installation of the coupling in armored face conveyors were now also applied to couplings in belt conveyors that were increasingly used for the continuous transport of mined coal. Here, too, the motor must be protected during the start-up phase. But there are other crucial points. While the drives of armored face conveyors require high break-away torques, the major concern with belt conveyors is the protection of the belt itself, usually the most expensive component of the entire system.

Belt conveyor in a German coal mine. Often, it is also used for taking miners to their workplace.

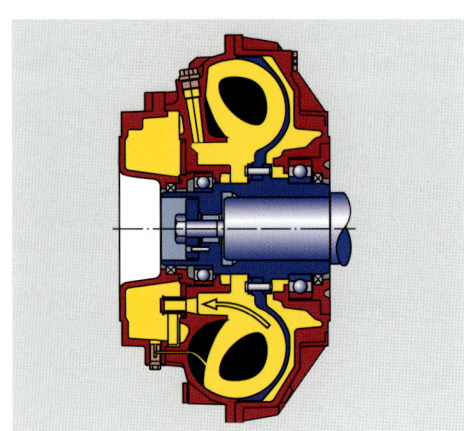

Turbo coupling type TVF with centrifugally-controlled valves for refilling the delay chamber with operation medium during the motor run-up.

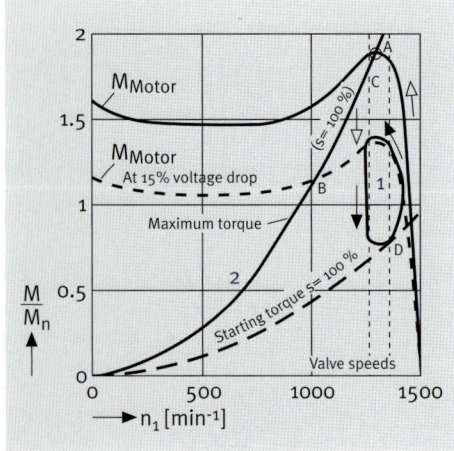

Reduced torque transmission of the coupling during motor run-up owing to centrifugally controlled valves, with subsequent utilization of motor pull-out torque in dependence on the prevailing grid voltage.
1 *Start-up parabola with effective valve function*
2 *Torque curve of the coupling while valves are inactive*

These application-specific differences also led to different product developments. The coalface section of the mining industry increasingly asked for a better utilization of the motor pull-out torque as the maximum start-up torque for breaking away full or overloaded armored face conveyors. In combination with proven squirrel cage motors, this torque needed to be available at short notice, especially for repeated start-up procedures, despite unfavorable conditions caused by overloads within the electric supply system.

This demand was met by Voith in 1964 by Willibald Meyer who clearly identified the series of tasks and largely inspired the development of the new coupling type TVF based on the TV concept. With the help of function "F" (**F**liehkraft [centrifugally]-controlled valves), i. e. with its special valve technology, the active antechamber – the delay chamber – can be used more consistently. Depending on the load condition of the armored face conveyor, the oil exchange between the delay chamber and the bladed operating chamber could now take place much faster. As a result, the required high break-away torque is available immediately upon reaching the nominal speed and/or the motor pull-out torque. While the power efforts of the motor were ineffective during the first run-up, the new valve technology allows further start-ups in quick succession, so that the motor pull-out torque can be used to maximum effect.

A demand that specifically applied to belt conveyors was to ensure a start-up during which the belt is protected as much as possible. This meant that the torque introduced during the start-up phase had to be limited to a low level. This led to a number of theoretical observations of the start-up behavior of conveyor systems, carried out in collaboration with universities and scientific institutes. These theories provided quantitative criteria that served as a basis for determining the functions of the coupling. They were also used as a starting point for possible designs and dimensions.

The main task was now to find design solutions that optimally fulfilled the demands for a limited starting torque, but simultaneously allowed economical operation at low slip. Meeting these new requests on constant-fill couplings without external interference in the control system turned out to be a huge challenge.

But the Voith engineers succeeded. Extensive internal examinations and test runs of the newly designed delay chamber were performed until usable results for solving the problem were finally available. The first delivery of a VTC type TVVY to Kali and Salz GmbH in 1968, marked the introduction to the market of a characteristic that had been especially optimized for belt conveyors and allowed a 150% limitation of the starting torque related to the nominal torque.

At the same time, coal production methods became ever more efficient with ever higher extraction quantities. Belt conveyors had to be adapted to this development and offer improved transport performances. In order to do this they needed more powerful drives, and it appeared as if the dimensions and outputs of the couplings had to be increased accordingly. But it soon turned out that the hydrodynamic similarity of the turbo couplings had not been sufficiently researched, so that flow data or experiences with a plant could not simply be converted to other sizes. Empiric methods had to be applied to establish new characteristic curves for the belt conveyor requirements specified by the operators. Expenditure for research and development rose dramatically.

Water as an Operating Fluid

In underground mining, there is always a danger of gas or dust explosions. From an early point in time, mining companies therefore prescribed the use of non-flammable operating media, in order to contain this risk. Soon, mineral oil, which was conventionally used for turbo couplings, was no longer permissible. As an alternative, synthetic oils were developed that had the same lubrication capacities and were equally suited for transmitting power. A disadvantage of these oils was, however, their harmful effect on the environment. In 1973/74, there was even an official ban on their use in mining applications.

Finding an environmentally compatible operating fluid was paramount. Water suggested itself as the obvious choice. A working group that also involved the mining companies put forward some basic considerations directed at the standardization of all turbo coupling sizes used in mining environments. In the course of being converted to operating with water, they were to receive identical transmission characteristics and designs. This affected several 10 000 couplings. The idea of standardization was born. On behalf of Ruhrkohle AG as the head organization, the construction features and characteristics of the couplings were recorded over several years, and they were all rebuilt to a uniform design.

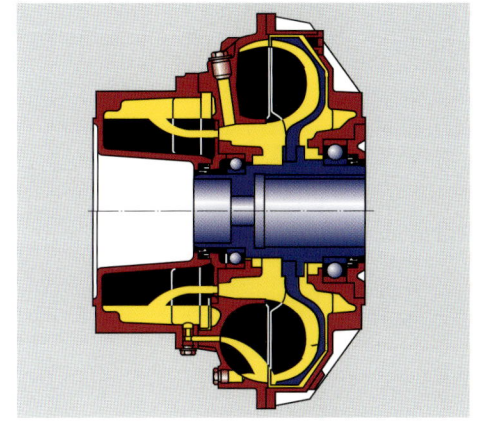

Type TVVY with dynamic refill of the operating fluid into the enlarged delay chamber, in order to reduce the starting torque.

Type TW... with internal bearings seals to allow the use of water as an operating medium.

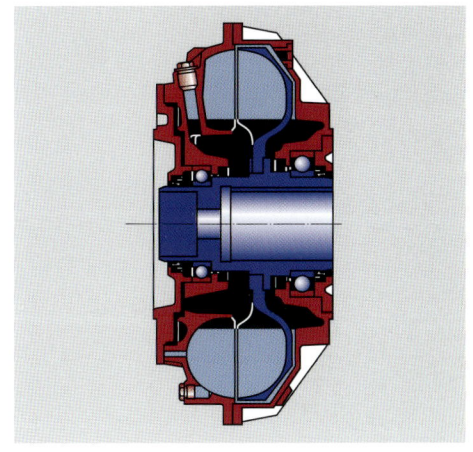

Iron ore harbor system in Saldanha, South Africa, fitted with TVVS couplings.

With water being used as the operating fluid, special attention had to be paid to the corrosion resistance of the coupling materials, especially since the chemical composition and possible aggressiveness of the pit water used for this purpose was not sufficiently known and could only be examined more closely on the basis of individual applications. Practical experience and laboratory tests provided the necessary know-how and supplied reliable data for estimating the service life and hence the economics of the rebuild program.

Thanks to the experiences gained with the German mining industry, exports rose steeply, even if national regulations required repeated product adaptations. In England, for example, the material "Silumin" was not permitted. It was replaced by cast iron that had to be nitrated in order to make it corrosion-resistant. In this adapted form, TVF couplings became particularly popular in England and were soon preferred over the routinely used pole-changing motors in armored face conveyor drives.

The TVVS Belt Conveyor Coupling For Delayed Starts

New findings on the subject of torque introduction to the belt were an incentive for further developments in the field of constant-fill couplings. While the initial demands for tractive effort on the belt were merely used as values for torque limitation, experience increasingly proved that it is not so much the absolute limitation of the torque, but the way in which it is introduced, i. e. the torque build-up across the period between pre-tensioning and braking away the belt, that is really critical.

TVVS couplings in underground belt conveyor drives in a mine of Kali & Salz with special technical demands on corrosion resistance against salt.

Outlook

In industrial applications, hydrodynamic couplings are primarily used for the production of energy carriers such as coal, oil and natural gas, as well as the generation of electric energy. The extraction of these raw materials is tied to their places of origin, while processing plants and consumers usually operate and live elsewhere. This makes it necessary to transform the material accordingly and transport them to the required locations. As this situation occurs all over the globe, relevant technical systems for extraction, transport and utilization are needed worldwide. Hydrodynamic couplings, converters and transmissions from Voith are an essential component of such systems. For this reason, exports exceed 80 percent of the business. The market therefore relies on the continued presence of expertly qualified staff.

Due to their high availability and long service life, hydrodynamic drives will always play a leading role in view of rising global demand for raw materials and ongoing efforts to explore and utilize traditional and new energy forms. Examples are drive developments for land shelf oil production; the application of water technology in mining and ecologically sensitive environments is on the increase. The utilization of wind power is entering a new era, environmental plants optimize their operation by introducing flexible working methods. Hydrodynamic drives will continue to ensure that these futuristic ideas can be put into practice. Outputs of up to 30 MW have already been realized, and projects covering up to 100 MW are underway.

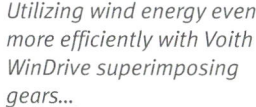

Utilizing wind energy even more efficiently with Voith WinDrive superimposing gears...

Extracting oil and natural gas from sea depths of 3000 meters with Voith subsea variable-speed couplings...

Safely transporting oil, gas and ore across large distances under extreme conditions...

... Voith develops, builds and installs hydrodynamic drive solutions for these and other challenges of the future.

Bernhard Wüst, Klaus Vogelsang

Driving Comfort and Safety

Voith Automatic Transmissions and Hydrodynamic Brakes in Road Traffic

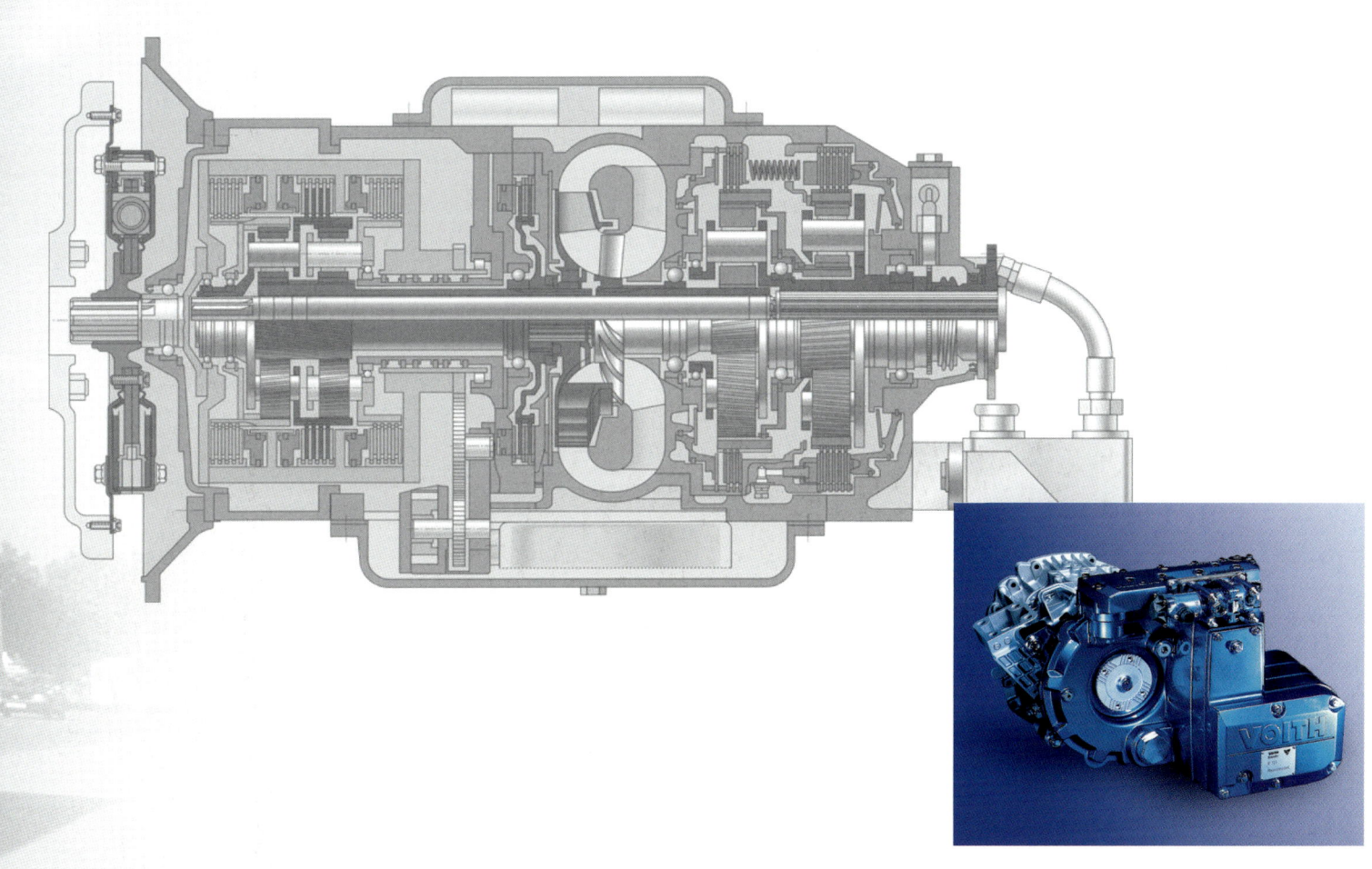

DIWA Transmissions for Road Vehicles

The Beginnings of Hydrodynamic Transmissions for Road Vehicles

No technical innovation was more favorably predestined for use in automatic vehicle transmissions than Föttinger's. But, after the patent had been granted, decades passed until it was utilized for such applications. The reasons for this are obvious. Railways did not need hydraulic converters or couplings as long as drive power was supplied by steam engines, while the automobile was still a new development and had to overcome many other problems.

The first vehicle transmission with a Föttinger converter was presented by Hermann Rieseler in 1925. However, he was not successful. The "Trilok" research team at Karlsruhe was working at approximately the same time, failed equally and never got past the trial stage. It took many years, until the Trilok principle finally achieved a breakthrough. The Lysholm-Smith bus transmissions developed in England and Sweden were far more successful at the time.

At Voith, the positive experiences gained with the first turbo transmission for a rail vehicle – the Austrian Austro-Daimler railcar – encouraged the engineers to carry out a trial installation in a similar design for heavy road vehicles. It was a bus transmission, consisting of a starting converter and two hydrodynamic couplings, fitted into a London Transport double decker bus in 1935. By filling or draining them, the two circuits could be activated or deactivated, depending on the desired driving range. However, these processes took considerable time and therefore were unsuitable for city traffic, especially during downshifts on uphill gradients. In rail vehicle installations, the long gear-shifting times had not really been that apparent. Additionally, the dimensions and the heavy weight of the transmissions were unsuitable for buses. The trials were discontinued.

For passenger cars, hydrodynamic transmissions seemed even more out of place. This did not stop Dr. Walther Voith, at the time Chief Executive of the company, from having a trial transmission installed in his personal car in St. Pölten, Austria. An engineer and inventor at heart, Voith wanted to personally experience this new technology. It is reported that he was absolutely delighted with this car, despite frequent problems with both the cooling and braking system as he drove through the mountains.

Transmission for London Bus in 1935.

In the USA, hydrodynamic bus transmissions with three-stage converters were successfully launched in the market during the war and in the post-war years. Since 1939, there had also been realistic alternatives for passenger cars in the shape of transmissions made by Hydramatic and Dynaflow. All of these transmissions came with well-known disadvantages, such as high power losses, fuel consumption, weight and price. Consequently, there were also no suitable automatic transmissions in Europe. At the same time, the demand for less complicated transmissions, especially for buses, continued to grow. Constant gearshifts and clutch activation became too much for the drivers, and the wear on clutches and other gearshifting elements in congested city traffic presented a huge problem. The first "European" solutions were more driver-friendly, multi-stage gear drives with five or six stages and preselectable gear mechanisms or at least gearshifting support systems. These designs made driving somewhat easier, but their operation was still far too complex. They also turned out to be rather unreliable and prone to wear.

First DIWAbus Transmission With Synchronized Converter

In the light of ever growing urban traffic, many city public transport systems that were dependent on rail or overhead lines, had to make way for far more flexible buses. The one man-operated bus became the norm, especially since personnel costs were also on the rise. Measures to take the strain off the drivers became more necessary than ever.

The introduction of an automatic hydrodynamic transmission could wait no longer. In 1946, Voith began to focus on the development of such a product.

The first in-house produced, fully automatic bus transmission was presented to an expert audience under the name DIWAbus 200 F in 1949. Immediately, it attracted a great deal of attention. This converter transmission with the characteristic hydrodynamic-mechanical power split principle was the first unit to carry the designation **Di**fferenzial**wa**ndler (differential converter), from which its name – DIWA – was derived. A few years later, in 1953, it was introduced to the public at the international automobile exhibition in Frankfurt. The first customers were Büssing and Krauss-Maffei who were very interested in an independently operating transmission without shifting rods for their newly developed rear engine buses. A long period of joint development and testing with both customers ensued.

Despite being launched relatively shortly after the war, the transmission was not the result of rapid developments. Its early foundations had been laid during the war. What the inventor had in mind was to combine planetary gears and Föttinger converters with the then available mechanical components in such a way that they would result in a user-friendly vehicle transmission with improved

Krauss-Maffei rear-engine bus with DIWAbus transmission.

efficiency. The unit was designed to require less shifting of gears. Fritz Kugel and his colleague Wilhelm Gsching jointly developed the operating principle. Both are registered as the inventors.

The idea of improving the efficiency of a turbo transmission by splitting the power flow into a mechanical and a hydrodynamic path was actually not new. Föttinger himself had considered this principle many years earlier. But the credit for developing a completely new transmission based on this operating principle must be given to Kugel and Gsching.

Back in the design department, they still had to find numerous fundamental technical solutions, for which there was very little precedent. Among them were the planetary gears with their toothing and bearings, the overrunning clutches with all their special features and the mechanics of the brake bands. They also looked intensively at vehicles and vehicle drives and studied how different engine types influence the operating behavior or the transmission control of these vehicles.

The Operating Principle of DIWAbus Transmissions

In the differential converter of the DIWAbus transmission, the power of the engine is split into two paths by a distribution gear B (differential gear) with planetary design. One path leads via the hydrodynamic converter (hydraulic path), while the other leads via a gear stage (mechanical path). Depending on the load on one or the other path, higher or lower speed is transmitted, and power is divided, as in a differential. The illustration on page 214 shows the basic operating principle; for better understanding, the differential is represented as a bevel gear differential.

During starting, the planetary gears rotate away from the bevel gear that is stationary with the output shaft and, with increased speed, drive the bevel gear and hence the pump wheel of the converter in accordance with the selected ratio. The ratio of the example shown here is $n_1/n_2 = 2$.

The maximum torque on the output shaft is composed of the support torque of the output bevel gear and the turbine torque increased by the hydraulic conversion. At full throttle, the starting traction of the differential converter is about 4.25 times of its value at nominal speed. The engine speed is lugged down to approximately half the nominal speed, and the engine operates with the higher torque at an economical fuel consumption rate. With increasing traveling speed, the engine speed rises along with the mechanically transmitted power share and

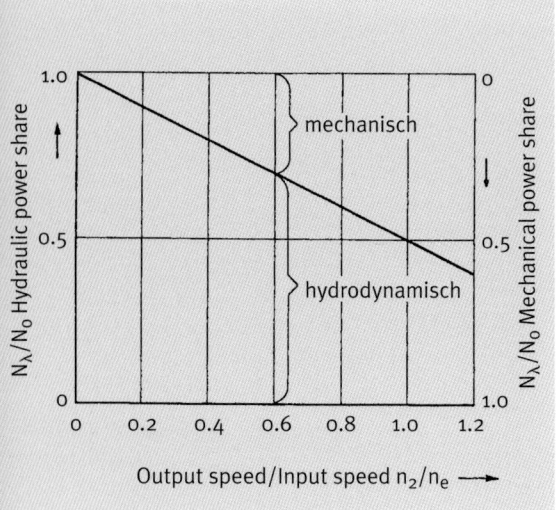

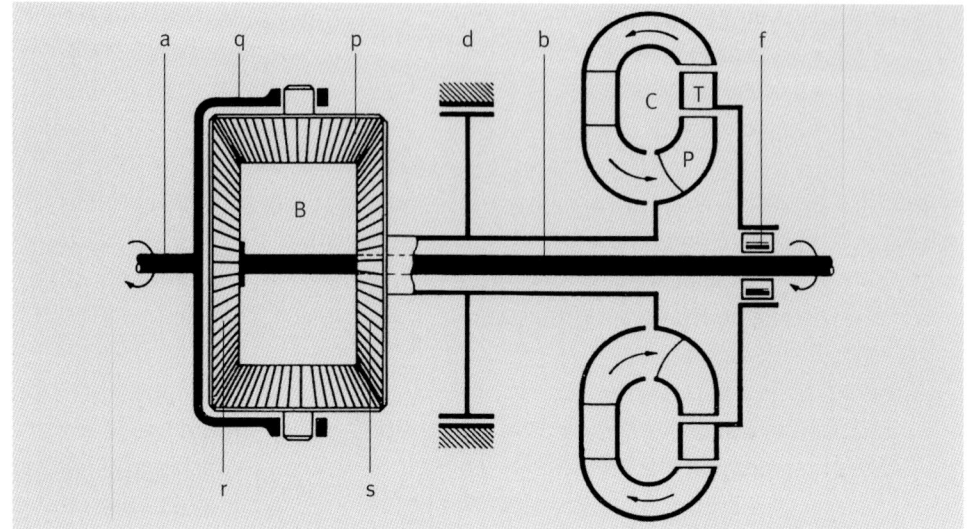

Build-up of hydrodynamic and the mechanical power share.

Right:
Basic principle of a differential converter transmission with synchronized converter.
a Input shaft
b Output shaft
B Transfer gear
C Hydrodynamic converter
d Distribution brake
r Output bevel gear
p Planetary gears
s Pump bevel gear
P Pump wheel
T Turbine wheel
f Overrunning clutch
q Planetary carrier

hence the efficiency, which amounts to 90% shortly before the converter is switched off by the distribution brake.

At the time, the high engine lug-down caused by the power split, was decisive for lower fuel consumption in city traffic, although there was only one automatic shift from the differential converter gear into a mechanical gear. After intensive development work, this process was streamlined.

The gain in efficiency compared to the Dynaflow transmission with its two hydraulic gears representing the latest technology in the fifties and sixties, is significant and did not involve an excessive amount of mechanical gear-shifting elements. The build-up of the hydraulic power share across the ratio output/input speed, that is so typical of DIWAbus transmissions, is shown in the above diagram.

The differential converter application necessitated the development of a special converter with individually optimized blades. Considering that the converter had to be dimensioned only for a part of the engine output, its size and hence its space requirement and weight could be kept at a low level.

The picture below shows the inside of the Voith torque converter with cylindrical blades, a turbine wheel for radial flow and a guide vane ring that is rigidly connected with the housing. Compared to spatial blading, this design had significantly higher efficiencies.

Again it took several years until the development of this converter with its special characteristics was completed. It was an important prerequisite for the realization of the overall concept. With this new development, the foundation stone for the long history of Voith differential converters for vehicle transmissions had been laid.

The DIWAbus transmission with a synchronous converter was concepted in two series that were produced under the auspices of Wilhelm Gsching. These series marked the Europe-wide start of the successful utilization of automatic transmissions in commercial vehicles. Target applications were city buses, fork-lift trucks, earthmovers and off-road vehicles, as the advantages of the high operating comfort for the driver and continuous, jolt-free acceleration far outweighed the higher price of the automatic compared to the manual system.

The bladed wheels of the Voith synchronized converter; guide vane ring with housing, pumps and turbine wheel.

The DIWAbus Series with Synchronous Converter for up to 200 HP

The development of this DIWAbus series, the "Big One", started with pre-trials in 1946. Three years later, in 1949, the first transmissions designated 200 **F/P** were tried out on the test stand and in the vehicle. Another five years later, in 1954, series production of the transmissions, now designated 200 **D**, began. Series 200 **S** followed in 1958.

First leaflet of the 200 D DIWAbus transmission.

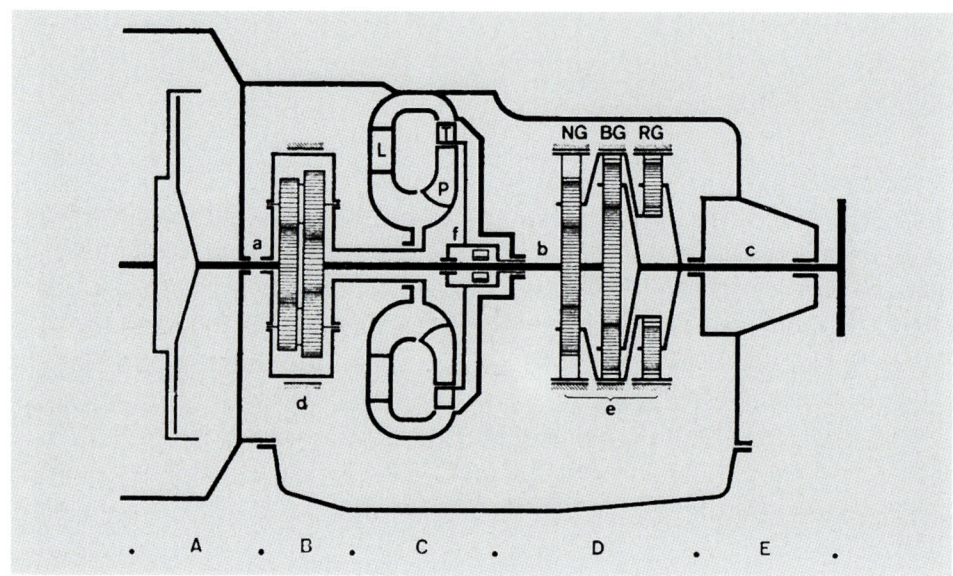

The essential components were the differential converter unit, consisting of a distribution gear and hydrodynamic converter, and the final planetary gear. The units were arranged in separate housings. As a result of this modular design, a wide range of vehicles and applications could be fitted with transmissions from the same construction set.

The design and the operation of a 200D transmission are illustrated on page 217 by a sectional drawing.

This unit is an automatic two-speed transmission with a hydraulic/mechanical DIWA-stage and a purely mechanical gear. The final gear stage that had to be activated at standstill was used to preselect one or two forward gears and the reverse gear.

The transmission was directly flanged to the engine flywheel via the distribution gear housing. In the first few transmissions, the connection to the flywheel was established via a flexible coupling which was soon replaced by a friction

F The multi-disc clutch in the final
   gear stage is activated by a
   "**F**eder" (spring)
P **P**rototype
D The spring actuator is replaced
   by "**D**ruckluft" (compressed air)
S **S**eries production

coupling. The transmittable torque of this friction coupling could be limited by a pressure set. A particularly important component was an additional continuous brake, the converter brake. By manually blocking the overrunning clutch, the turbine wheel was able to rotate in the mechanical gear. The converters therefore acted as retarders. Unlike with modern converter brakes, this feature could only be activated in the converter gear at low speed and utilized after the distri-

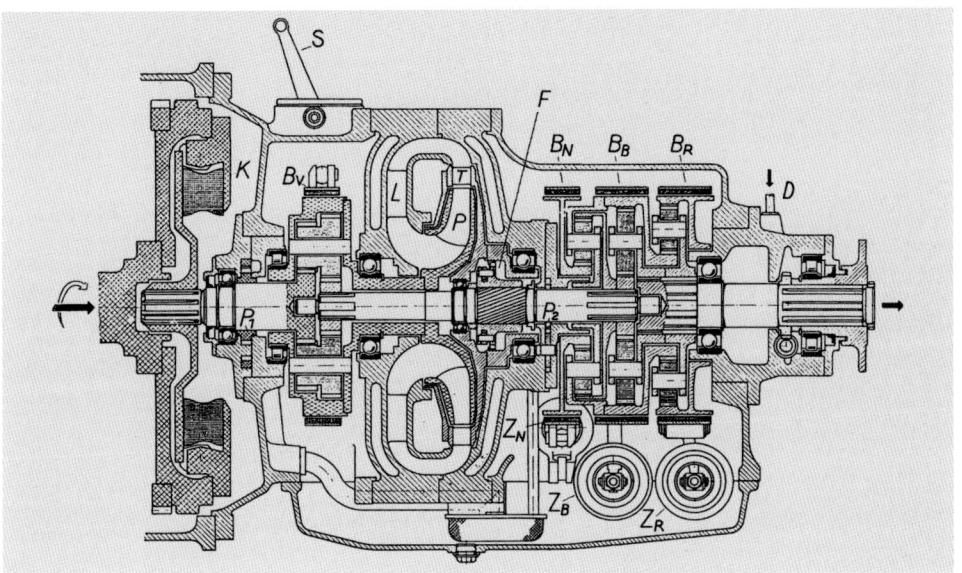

bution brake in the mechanical gear had been shut down. This meant that the driver had to use the mechanical brake in order to activate the converter brake, until the transmission moved into the DIWA gear. When the bus gained speed due to the downward gradient, the transmission moved back into mechanical gear. The driver virtually had to plan this rather lengthy maneuver of activating the converter brake and selecting the ratio of the final drive. Additionally, the converter brake could only be switched off in the hydraulic gear.

The converter housing needed to be supplied with cooling water, especially during braking. For this purpose, the heat exchanger was usually integrated into the cooling water circuit of the engine. The transfer gear was designed as a helical-planetary gear with two planetary stages running in parallel. Depending on the application of the vehicle, the final drive could be designed with different combinations and ratios.

Type/ Application/Weight	Fast gear	Slow gear	Reverse gear	Transmission scheme
JSR/Bus Level road 235 kg	0.83	1.37	0.94	
JS 220 kg	0.83	1.37	–	
SR 215 kg	–	1.37	0.94	
JBR/Bus Mountainous route 235 kg	0.94	1.95	1.90	
JB 220 kg	0.94	1.95	–	
I+BR Bus Mountainous route 235 kg	1.34	1.95	1.90	
J+B 220 kg	1.34	1.95	–	
U+S Rail 220 kg	–	1.44	1.44	

Steering wheel with operating elements (Büssing bus).
1. *Two-spoke steering wheel*
2. *Signal horn push button*
3. *Selector*
4. *Instrument panel*
5. *Selector housing for DIWAbus transmission*
6. *Signal lamp for converter brake*
7. *Push button for converter brake*
8. *Running plate brake valve*
9. *Foot pedal*
10. *Hand brake lever*

The first test drives took place in Heidenheim in 1950 with a Krauss-Maffei chassis. Shortly after, a Krauss-Maffei prototype bus and a Büssing truck were added. The test reports show that the Voith engineers did not just meticulously register technical data but carried out their work with genuine enthusiasm. On the one hand, they were dealing with a product that was truly unique and in whose development they were actively involved, and on the other hand, Voith had been developed hydrodynamic transmissions for a mere 20 years! It must have been this pioneering spirit that helped the initially rather small department TG through most difficulties and gave it the worldwide importance that Voith Turbo has today.

An extract from the 1953 report "With a Voith DIWAbus transmission across Austrian mountain passes" written by the future works archivist Hans Häckert gives an impression of the test drives in those days. The Heidenheim prototype vehicles were used on alpine roads in order to prove the suitability of the transmissions for longer trips through the mountains. For this purpose, a joint trip through the Alps involving the Büssing truck, the test bus of Krauss-Maffei and a prototype bus of the Austrian manufacturer Gräf & Stift was conducted. Häckert writes about it as follows:

"On 9 October, at three in the morning, the two vehicles started in Heidenheim in the first autumn frost and later met the bus from Vienna in Bruck-Fusch in clear, sunny weather. In Bruck (858 m high) the Glocknerstraße begins, a mountain pass leading up to an elevation of 2 428 m and then descending towards the famous village of Heiligenblut. The upward gradient is mostly between 10 – 12%. In Bruck, we were told that there had been rain in the valley the day before, while it was snowing in the higher elevations, and that the Glocknerstraße was impassable and currently being cleared by snow plows. So we could venture to drive up the pass after all. It must be noted that all three vehicles carried iron bars on board to simulate their full service weight. Initially, the road has only a slight incline, nine kilometers later after the Fusch tollhouse (830 m) it ascends steeply. We monitored the stop watch, several thermometers, manometers and speedometers in detail to establish the behavior of the machines during driving, especially on upward gradients." The small vehicle convoy climbed beyond the snowline and the wheels had to be fitted with chains. Adventurous encounters with snow plows on narrow mountain roads and tight curves alternated with stretches of 24% inclines. But the test results were promising. All three transmissions did well, the converter brake proved itself and also ensured optimum temperatures of the engine cooling water.

Double decker buses of Berlin Transport.

One of the first major customers for the 200S transmissions was BVG Berlin Transport with its typical double decker buses.

With the construction of the transmission factories in Heidenheim in 1953 and in Munich-Garching in 1963, another two important events in the company's history of commercial vehicle transmissions took place. In a plant designed for building customized water turbines and paper machines, the production of hydrodynamic transmissions got off to a difficult start, especially as there was a strong increase in orders for these large custom machines.

Voith-Büssing test truck with DIWAbus transmission.

In Heidenheim, the DIWAbus transmissions were largely just assembled. Important components such as gears were manufactured by Maschinen- und Zahnradfabrik Carl Hurth in Munich, as there were no suitable production facilities in Heidenheim. In 1960, the gear production was moved to Albert Hirth AG in Stuttgart, at the time part of the Voith Group, but sold on in 1963, with Voith maintaining some rights in the company.

The opening of the subsidiary in Garching far away from the center of Munich marked the actual start of the company's own DIWA production. The initiative for the construction of the factory was due to the then manager of the department 'Turbo Transmissions for Road Vehicles', Ernst Biefang, who joined Voith in the early sixties from ILO where he had been in charge of the ILOmatic transmission.

Büssing bus in Vienna.

DIWAbus transmission after
25 years of development.

A Friction coupling/spring coupling
B Distribution gear (differential)
C Hydrodynamic torque converter
D Planetary final drive
E Output
P Pump wheel
T Turbine wheel
R Reaction member (guide wheel)
a Drive shaft
b Intermediate shaft
c Output shaft
d Distribution gear (brake)
e Final drive brakes
f Overrunning clutch
g Cooling water jacket
h Drive pump
i Control pump
k Drive pump
l Control lever
m Converter brake solenoid
n Tacho drive
o Oil filler
p Planetary gears
q Planetary carriers
r Spur gear output
s Spur gear input
t Sleeve housing

This was also the time when the DIWAbus transmission acquired huge popularity with municipal transportion operators all over Europe.

After its production had started in Munich and after many detailed improvements, the DIWAbus 200s transmission was renamed type 501 in 1965, and, after further enhancements, designated type 506 in 1968. Soon the transmissions were, for the first time, also simply called DIWA transmissions.

Between 1953 and 1974, Voith produced over 22 000 units of the successful DIWAbus 200-506 series. The vast majority of them, nearly 90 percent, were installed in buses, but they were also used for forklift trucks, earthmovers and off-road vehicles.

The DIWAbus transmission type 506 represented the final development stage of the large 200 HP series with a synchronous differential converter. It incorporated the experiences made with the lamellas and the hydraulic valves of the "smaller" DIWAbus series that had been launched a little later. Due to its special characteristics, especially as far as types U+S were concerned, it was excellently suited for airfield buses with two driver's cabs, and it was still in demand long after being replaced in 1974 by its successor DIWA 851. This new type had the advantage that the bus could drive either forward or backward equally quickly without an additional reversing gear.

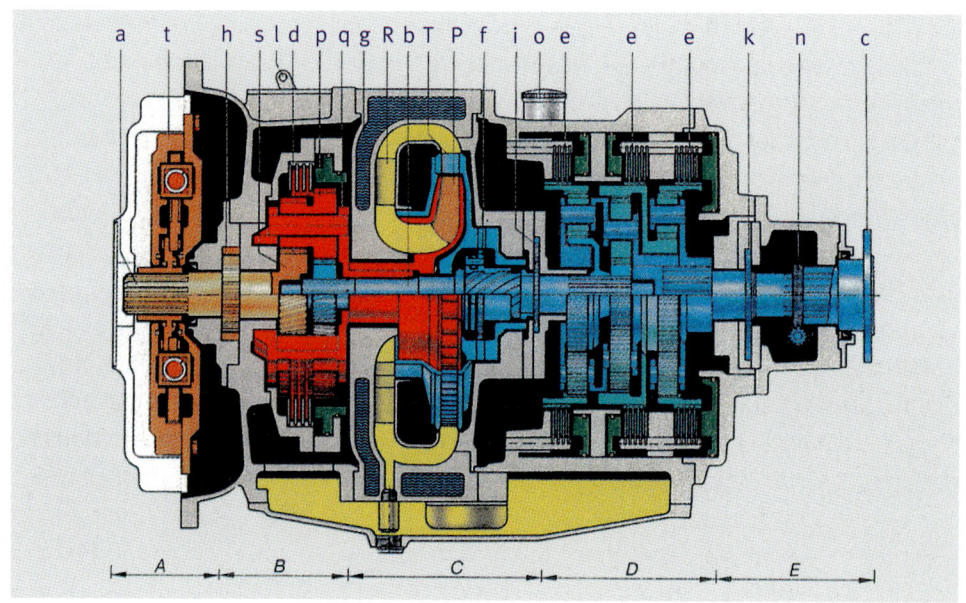

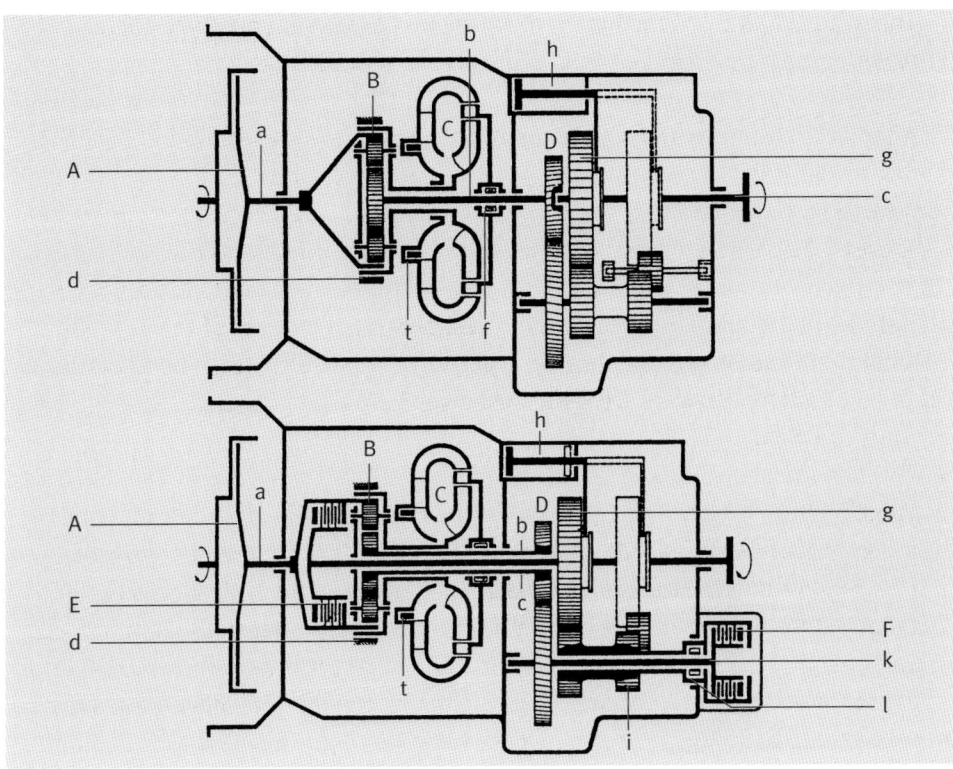

Longitudinal schemes of DIWAbus transmission 145 D2/145 D3.

A Friction coupling
B Distribution gear
C Hydrodynamic converter
D Final reversing gear
E Clutch
f Bridging clutch
a Drive shaft
b Intermediate shaft
c Output shaft
d Distribution brake
f Overrunning clutch
g Sliding gear
h Shift cylinder
l Overrunning clutch
t Sleeve valve

DIWAbus Series with Synchronous Converter up to 145 HP

The forerunners of the one-man buses in city traffic were smaller models with a length of up to 10 meters and fast running engines. But this market was unable to utilize the transmissions of the 200 HP class. They would have been too expensive, as they had to have an additional gear stage for nominal speeds above 2500 revolutions per minute. This would have increased the weight of the already heavy unit even further. What was needed was a smaller, lighter and, above all, more cost-effective transmission offering better comfort and ergonomics for the driver.

This additional market seemed to be attractive, and Voith decided to enter it. The target products were two and three-speed transmissions with outputs of 145 HP. Relevant developments started in 1953, just before series production of the "large" units began. The basic principle of the differential converter, as well as the automatic hydraulic shift from the differential converter to the purely mechanical stage, was retained. The sectional drawing above shows the key characteristics of the new 145 D2/145 D3.

*DIWA transmission
type 145 D2/D3.*

The differential was fitted with reversing planetary gears on just one level. It had a simple final reversing gear D in countershaft design with a forward and a reverse gear. For gearshifting, the vehicle had to be brought to a halt and the flow circuit in the converter had to be interrupted by a sleeve valve.

For the 145 D3, the direct coupling with the input side allowed an additional third gear. It was therefore a "true" three-speed transmission with gears that could be activated by power shifts. But now the converter could no longer be used as a hydrodynamic brake. A multi-disc bridging coupling "F" eventually enabled the transmission of torque for additional utilization of the engine during coasting of the first mechanical gear.

The transmissions had a very compact design, and, weighing 135 and 150 kg respectively, met the demand for a light and also cost-effective unit. In many aspects, for example by its electrically/hydraulically controlled lamella brakes and their lock-up coupling, the "smaller" series was further advanced than the "big" one, thus leading the way to today's DIWA transmission.

Although production continued until 1973, its commercial success was moderate despite favorable initial prognoses. Just over 5 700 transmissions were sold. Contrary to Voith's expectations, developments in the city bus sector moved more and more towards higher outputs and slow running engines. In the end, the 200 HP-version was the more successful product.

In 1960/64, upgrades of the 145 D2/145 D3 into the 145 U2/145 U3 and 150 U models allowed a certain expansion of the transmission functions. It was now possible to shift from forward into reverse gear during driving – a feature that was not required for buses. But it made the transmission suitable for applications in rail vehicles, construction machines as well as forklift trucks. These versions entered series production as type 502-2/-3 in 1966, marking the conclusion of the original 145 series.

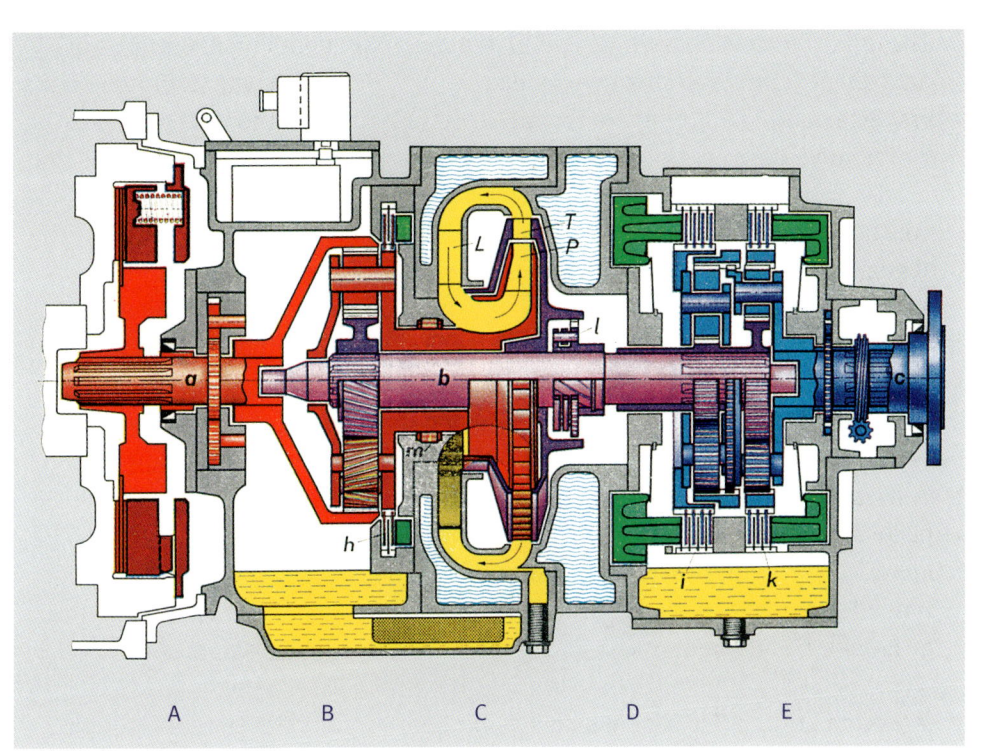

DIWA transmission type 502-2

A   Friction coupling
B   Distribution gear
C   Hydrodynamic converter
D   Planetary gear
P   Pump wheel
T   Turbine wheel
V   Forward gear
R   Reverse gear
E   Output
a   Pinion shaft
b   Intermediate shaft
c   Drive shaft
h   Distribution brake
l   Overrunning clutch
i, k Lamella brake
m   Non-reversing lock
p   Lock-up coupling

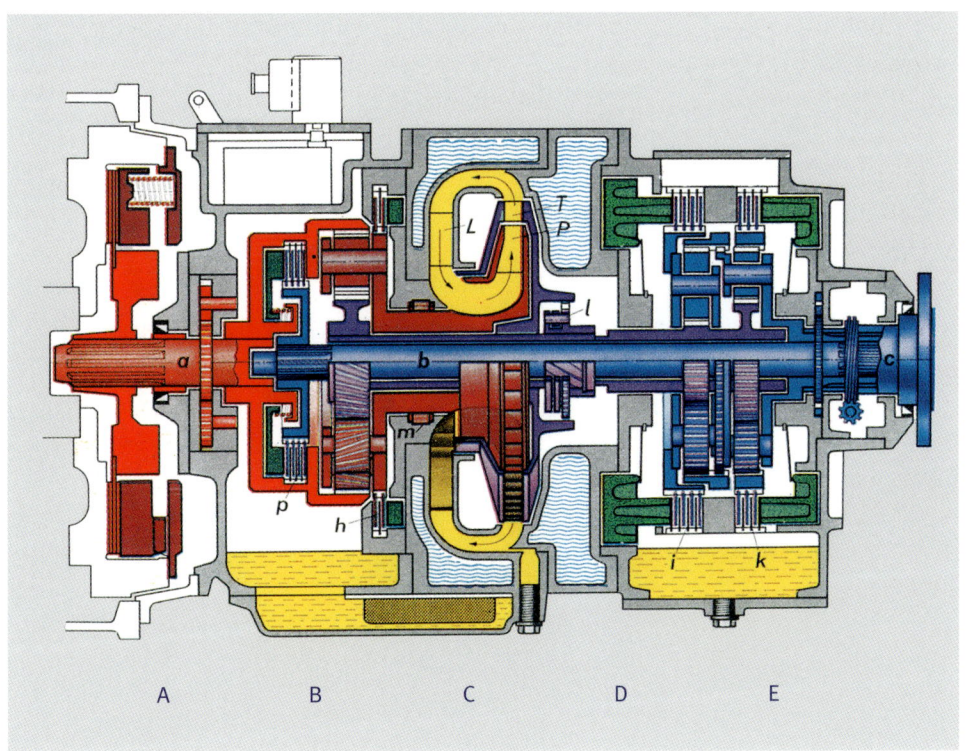

DIWA transmission type 502-3

A   Friction coupling
B   Distribution gear
C   Hydrodynamic converter
D   Planetary gear
P   Pump wheel
T   Turbine wheel
V   Forward gear
R   Reverse gear
E   Output
a   Pinion shaft
b   Intermediate shaft
c   Drive shaft
h   Distribution brake
l   Overrunning clutch
i, k Lamella brake
m   Non-reversing lock
p   Lock-up coupling

DIWAmatic Transmission with Counter-Rotating Converter "System Weinrich"

In the mid fifties, shortly after the market introduction of the DIWAbus series with synchronized converter, Voith decided to produce its bus transmissions on a broader and more economical scale. By acquiring a smaller converter transmission model that could be manufactured on a larger scale, it was intended to supplement the company's own developments and achieve an altogether better business result. By a fortunate coincidence, ILO in Pinneberg wanted to sell its ILOmatic product line – small differential converter transmissions with contra-rotating converters and lower outputs, which ILO had developed in 1955 and sold successfully to forklift truck and smaller construction machine manufacturers. The units were smaller, lighter and more cost-effective than the DIWAbus series. Additionally, the characteristic curve of their converter was more favorable for driving auxiliary vehicle systems. Voith regarded this product line as an ideal supplement of its own range and acquired the license for manufacturing ILOmatic transmissions in 1961.

The design and the operation of the transmission type later designated DIWAmatic largely resembled that of the DIWAbus transmission. The major difference was the converter, which was designed as a "contra-rotating" unit, a version in which pump and turbine rotated in different directions.

Cut-away model and longitudinal scheme of DIWAmatic transmission type 833 k.

a   Input shaft
b   Center shaft
c   Output shaft
e   Gear pump
f   Pump brake
g   Overrunning clutch
i   Fly weight
k   Drain valve
l   Control sleeve
m   Selector fork
n   Adjusting lever
P   Pump wheel
L   Guide wheel
T   Turbine wheel
"V" Forward gear
"R" Reverse gear

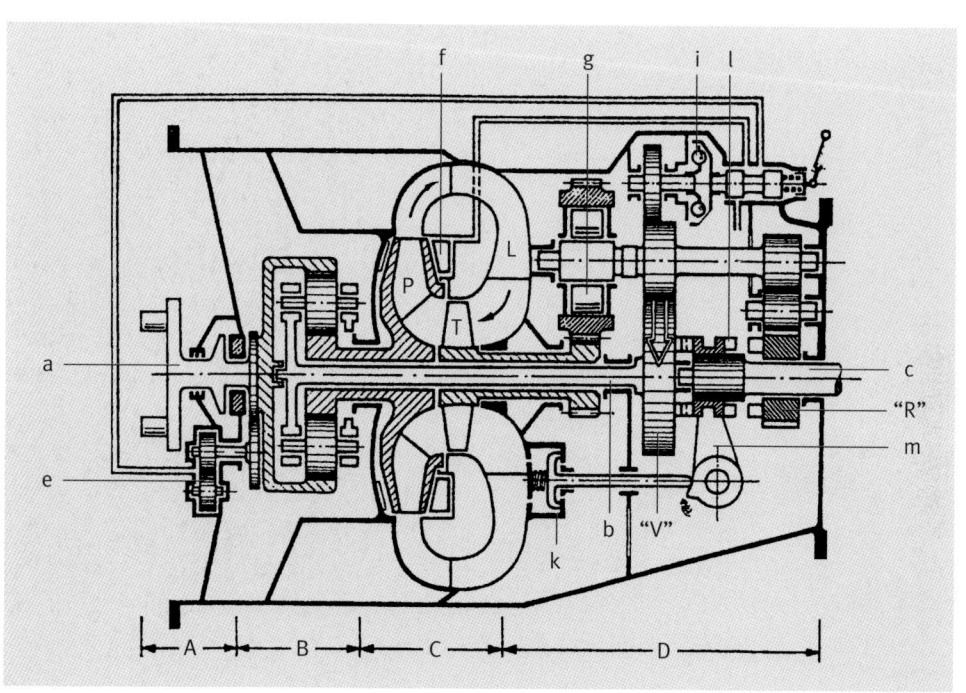

Although it was another differential converter transmission, it differed significantly from Wilhelm Gsching's synchronous converter principle. The power flow was again split in a differential before the converter, but the pump wheel rotated in the opposite direction to the engine. From the outlet side, the oil stream was directed into the axial-flow turbine wheel via the guide wheel in its stationary housing with an axial and radial share. Additionally, the design of the converter differed from that of the DIWAbus transmission and was rather similar to the later DIWA transmission. The small, specifically fast-running turbine wheel was characteristic of the counter-rotating converter principle. The (remote) similarity to a ship's propeller invoked the trademark of the inventor Helmut Weinrich who had developed torpedo drives during the war.

After a number of improvements the ILOmatic, series production began in 1961 at Voith Turbo in Crailsheim, at the Heidenheim transmission plants and also later at the Garching facilities. In 1960 and 1961, expert personnel from ILO were transferred from Pinneberg to Heidenheim, in order to oversee the further technical development of the series now designated DIWAmatic. These employees would later play a key role in the development of today's DIWA transmissions.

First and foremost on the list are Ernst Biefang, the future product manager for DIWA and DIWAmatic, Johannes Peltner who took over the management of the design of the DIWAmatic transmission and became a driving force for the development of the DIWA 851 transmission, as well as Helmut Weinrich, the inventor and owner of the ILOmatic and DIWAmatic patent who worked for Voith on a freelance basis.

The DIWAmatic transmissions were intended for installation in smaller, flexible all-terrain special vehicles with limited input powers. A choice of distribution gear ratios optimized the adaptation to various engines. The picture on page 226 shows a DIWAmatic transmission between the engine and the axle of a BKS lift truck.

As a true series product they were manufactured in relatively high numbers with only a few variations. In 1975, more than 40 000 units had been sold. At the end of the DIWAmatic era, this transmission type, as far as volumes were concerned, over shadowed its big brother DIWAbus. For a long period, the DIWAmatic transmission played a vital role in the commercial success of the company in the road vehicle transmission market.

Assembly of DIWAmatic transmissions.

Above right:
Type 833 K in a BKS all-terrain lift truck.

Aircraft tug with type D 843.

Keen competition with more and more hydrostatic drive systems soon made it necessary to find a better differentiation between drive outputs and hydraulic lifting capacities. The transmissions were therefore fitted with a hydraulic multi-disc input clutch.

An additional operating lever for the clutch allowed a stepless variation of the oil pressure. By "inching" the clutch, it became possible to stop the engine speed from dropping during low driving speeds, even if the hydraulic system was subjected to heavy load. The characteristics of the converter ensured that the relative movement of the clutch plates toward each other was significantly lower than that of mechanical transmissions. This kept power and heat losses at bay.

And yet: at the beginning of the eighties, the successful DIWAmatic era was nearing its end. In the construction machine and forklift truck sector, the hydro-static principle eventually proved to be superior. On the basis of ongoing developments of modern control concepts, this relatively old technology allowed more elegant solutions of splitting drive outputs and hydraulic systems. The drives were also cheaper and more user-friendly.

A further development by Johannes Peltner's design engineering team – the DIWAmatic 843/845 transmission, fitted with planetary gears in the final drive, a variable-speed input clutch and a fill-controlled converter ("Inch Converter") – was not destined for greater success.
In 1986, the production of the DIWAmatic series was discontinued.

These last transmissions with fill-controlled converters for controlling the power uptake of the drive, represented the pinnacle of purely hydraulic control systems in Voith converter transmissions. Improved functionality could only be achieved with electronics that were slowly introduced in the form of analog control systems.

Construction Machine Transmissions, Converter Modules, Special Developments

Below the 70-HP threshold, the market for hydrodynamic automatic transmissions in special vehicles, forklift trucks and construction machines had been well covered by the DIWAmatic transmissions. For higher outputs, various types of Voith converter modules – single units between engine and gearbox – were used.

Forklift truck with Voith C 845 Certomatic transmission.

Driveline of an earthmover with converter module (1964).

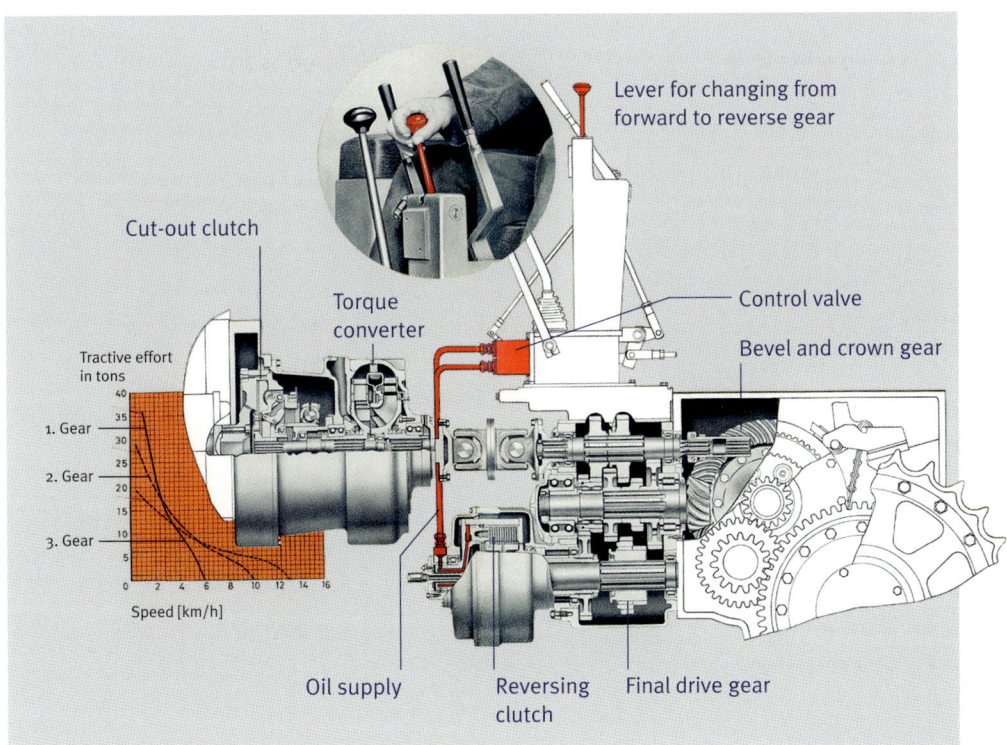

227

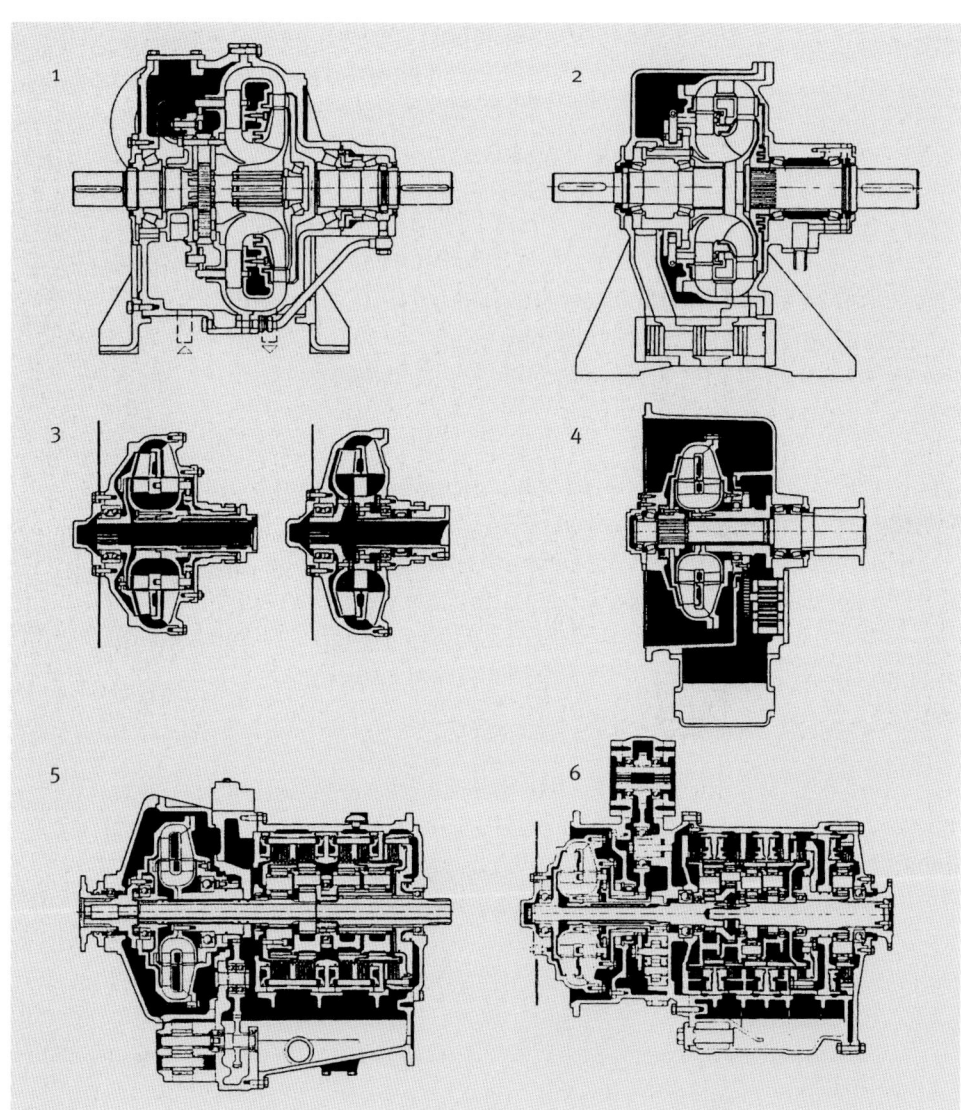

At the time, the individual gear stages had to be selected manually by the driver. Efforts were high and outputs suffered. With increasing urgency, the market cried out for power-shift transmissions, especially for wheel loaders whose operating speeds were significantly higher than that of earthmovers.

In the late sixties, Voith eventually decided to seize a favorable opportunity and acquired the transmission program "Certoplan" of Carl Hurth in Munich. It was intended to cover the output range between DIWA and DIWAmatic.

Voith Certoplan transmissions, as they were called after the takeover, were Trilok converter planetary gears that could be shifted under load. After streamlining the original product program, the transmissions were combined with converters from the Voith portfolio. The designs with converters and power take-offs were now able to split power steplessly between the drive and the hydraulic system. Voith entered the market with this product in the mid-seventies.

At the time, the key customer for converter modules was Hanomag in Hanover. For Certoplan transmissions it was Fiat-Allis who used them for bulldozers, while Faun-Frisch installed them in their wheel loaders. But Voith never managed to gain access to the world market. Large construction machine manufacturers such as Caterpillar, Case, Clark, International Harvester and John Deere had their own transmission designs or installed the well-established Allison series. Voith had to share the European market with other transmission manufacturers. Satisfactory volumes were never reached, and the crisis of the construction machine industry in the early eighties added to the problem. Manufacturers with whom Voith had good business relations had to close or were swallowed up by larger companies.

Voith withdrew from this line of business.

Power flow in differential gear, illustrated at DIWA 4-speed transmission.

1 *Input differential with clutches EK, DK, SK*
2 *Torque converter with counter-rotating pump wheel*
3 *Turbine transmission with brake*
4 *Reverse gear with brake*

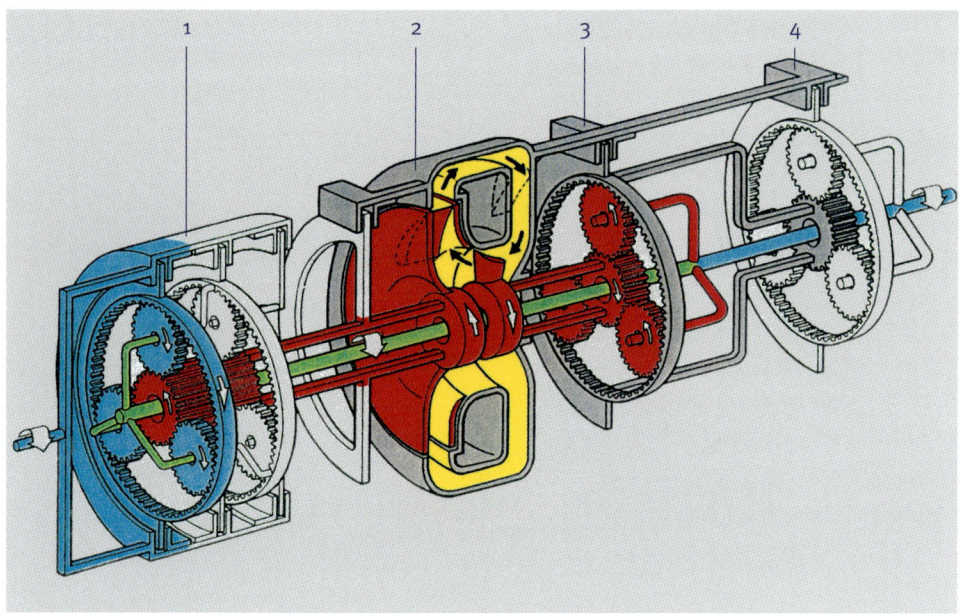

DIWA Transmissions with Contra-Rotating Converter

In the mid-sixties, the market for commercial vehicles began to change. For buses, higher engine outputs were required, especially for the increasingly popular articulated vehicles. The DIWA 506 transmissions were, however, just simple automatic two-speed transmissions with the option of a mountain or a normal gear range that could be selected when the bus was at a halt. There was actually a prototype for gearshifting under load, but its transmittable power was in no way increased. The well-known disadvantages of the converter brake still continued. The series was in danger of eventually becoming technically obsolete. While the smaller type 502 transmissions had meanwhile been developed into three-speed versions with load-shift capacities, they found few customers precisely because of their increased output.

There was definitely a need for further development, especially as the competition was getting stronger. In the USA, multi-speed automatic transmissions with Föttinger converters were launched in the market and installed into earthmovers made by companies like Caterpillar. Buses and commercial vehicles were driving through Allison transmissions. Even in Germany, a transmission with a Trilok converter was introduced for construction machines in 1963.

It was obvious that there would soon be competition for the DIWAbus transmission. A new development was urgently required. It had to meet the increased demands from the industry and replace all previous transmissions.

It was decided to pursue the idea with the counter-rotating differential converter. With its jolt-free acceleration in the differential converter gear, its highly
effective converter brake and a favorable efficiency curve, this concept offered
the optimum solution for applications in buses. It covered a wide range of different engine outputs and characteristic curves without having to change the converter housing. Depending on the design, it would achieve a starting conversion
of 5.8 to 6.5 – significantly above that of a synchronous converter transmission.

The following years must be regarded as a genuine build-phase of DIWA
transmissions, as they would soon be called without any further addition to their
name. The new generation was developed under the direction of Johannes Peltner who was familiar with counter-rotating differential converters from his time
at ILO and also looked after the DIWAmatic at Voith. He would play a vital role in
the future design of the DIWA transmission.

Right from the beginning it was economically feasible to produce the transmission for higher outputs. Initially designed for three speeds, it was configured
for a possible upgrade into a four-speed version. The first drafts of a DIWA 851,
based on DIWAmatic technology, were presented in 1964; a first DIWA 861 prototype with purely hydraulic control was built in 1968. The transmission had a few
special features that were far beyond of what had been available in the market.
Completely new, for example, was the dual function of the integrated component
"Reverse gear brake". Via this brake, the driver shifted into reverse in hydrodynamic-mechanical operation, while the converter brake was actuated in
mechanical gear. The reverse gear was not a full but a "limited" gear that operated only in the hydrodynamic-mechanical range. Another innovative feature was
the introduction of a "Central pressure drop" during gearshifting.

The technology of the new DIWA transmissions was initially viewed rather
critically within the company. It took a great deal of convincing until it was
generally accepted. Helmut Weinrich himself had made a decisive, personal
contribution. In a Ford 17M test vehicle with a converted DIWAmatic transmission, he demonstrated in front of the doubters how to brake with a converter.
It must have been quite something when he suddenly shifted into reverse driving on a downward gradient at high speed. Another vital step towards the development of the new DIWA transmission was the development of an electronic
control system, the very first of its kind for a commercial vehicle transmission.
It represented a significant improvement over the complex hydraulic control
system. Voith even introduced a special series production program for this new
control unit.

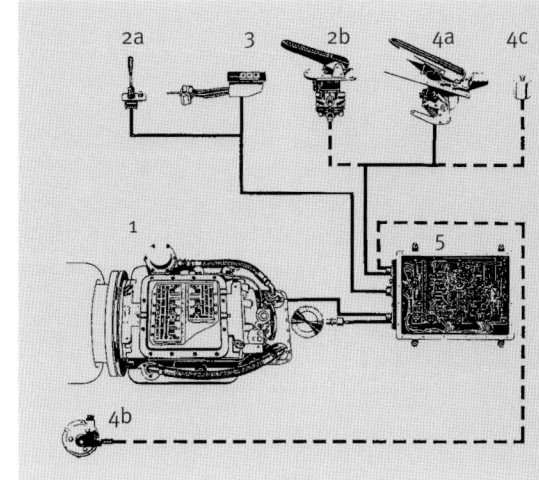

Transmission control
1   Transmission with inductive
     pick-up and control unit
2a   Manual switch for converter
     brake control block
2b   Pedal type clutch valve
3   Push-button switch
4a/b Load transmitted
     accelerator/engine
4c   Transfer switch
5   Electronic control unit

Output range for DIWA trans-
missions 501, 502, 506, 851.

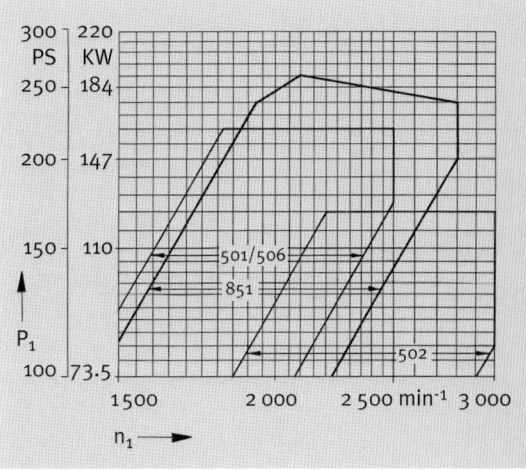

The most important contribution towards this development was made by Heinrich Dick. He joined Voith from the automobile industry in 1968 and was from then on instrumental in the development of electronic control systems for road transmissions for many years.

Voith attracted huge attention when it presented its first DIWA 851 transmission at the IAA in 1973 after several years of testing. Features such as its starting and braking with a hydrodynamic circuit, its high starting conversion and compact design received high praise.

The transmission was connected to the engine flywheel via a torsional vibration damper. This damper replaced the traditional friction coupling of the DIWAbus transmission. It was initially made from rubber elements that later turned out to be unsuitable for this application. By absorbing damping losses they became over heated and failed prematurely. After unsuccessful trials with commercially available vibration dampers, Voith decided to develop its own product, which finally led to the spring coupling in 1979. This component proved to be the ideal solution to the problem. It was later expanded into the "Hydraulic two-mass damper HTVD" (Hydraulic Torsional Vibration Damper). Under the name Hydrodamp it has since become an independent product line of Voith Turbo.

Design scheme of the DIWA 851 three-speed transmission.

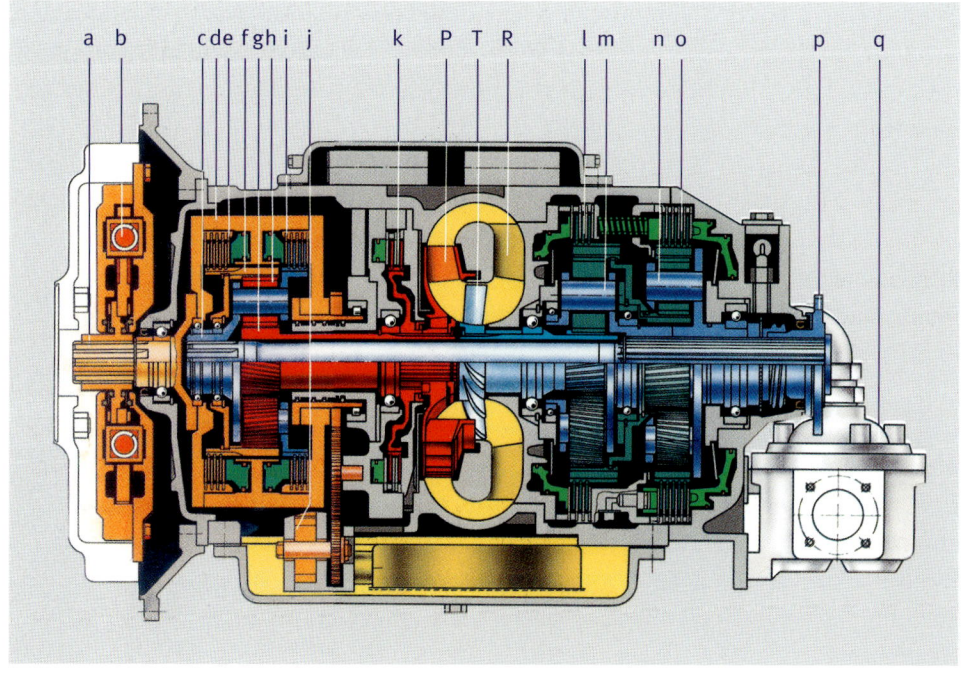

The high braking torque building up in less than a second by the accelerating turbine wheel when the new reverse gear brake is switched on, required especially smooth gearshifts. At the reverse gear brake, the transmission is therefore fitted with a stepped piston that meets this requirement. For the service brakes of buses, this process resulted in significantly reduced wear and became hence a convincing customer benefit for the transmission.

The extremely flexible electronic control unit receives the driver's signal via the push button switch, the driving and the braking pedal, and it is then responsible for all ensuing processes. It ensures that the power uptake of the converters is optimally suited to driving resistance and speed, that the vehicle drives economically and comfortably, and that operating conditions that might be critical for passengers, vehicle or transmission are avoided.

Voith engineers have never ceased to look for further optimizations in this field. The large number of individual transmission applications under different conditions will always be an incentive to find the best possible compromise between operating comfort, component life and retarder efficiency.

The success of the DIWA 851 generation launched a new wave of gear automation among European public transportion operators who wanted to reduce their personnel costs, make life easier for their drivers and protect the brake linings of their buses. For Voith, this meant an expansion of the company's international business. The Italian manufacturer Sicca who built the Siccar 176 L chassis for Inbus was the first one to order DIWA 851 transmissions in 1978.

Siccar delivered seven vehicles to Reggio Emilia, the seat of the Voith subsidiary in Italy. Four of them are still in service. In cooperation with Inbus and IVECO who replaced their in-house produced automatics with Voith transmissions, sales of DIWA transmissions in Italy soon flourished.

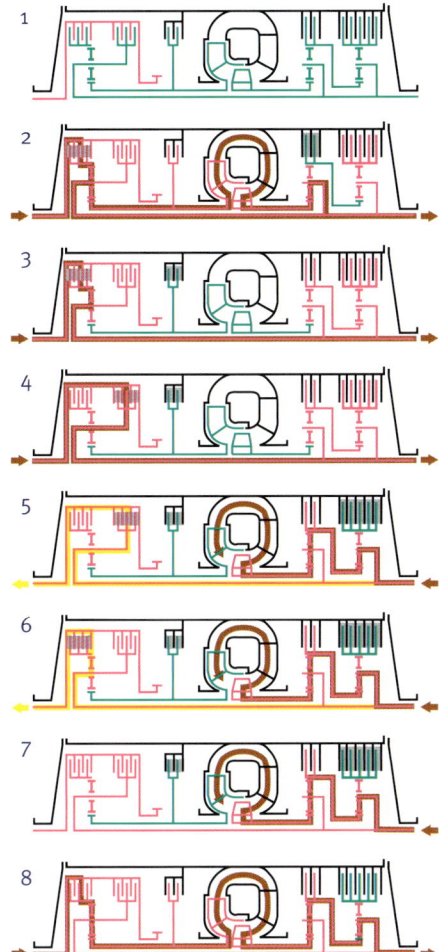

Design scheme and active power flow of DIWA 851 three-speed transmissions.

1. *Neutral position*
2. *1st gear (DIWA gear)*
3. *2nd gear*
4. *3rd gear*
5. *Braking in 3rd gear*
6. *Braking in 2nd gear*
7. *Braking in 1st gear*
8. *Reverse gear*

Brown/Yellow = Active power flow
Red = Rotating parts
Green = Stationary rotating parts
Black = Stationary parts
Grey = Clutch plates closed

MCW double decker.

Sales also took off in England. MCW installed DIWA 851 transmissions in their double decker buses. For these vehicles, Voith developed the first W1 angle drive that allowed the transverse installation of the engine and transmission at the rear of the buses.

At the beginning of the eighties, England and Italy were the most important markets. At one point, over 1 000 DIWA transmissions were delivered to MCW within one year. Other English customers were Dennis and Leyland, where the DIWA transmission had driven out a domestic competitor.

In Ireland, Voith was able to break into the market a few years earlier, through a delivery of DIWAbus 506 transmissions for Dublin Bus. During a visit to Germany, the director of the city's transport operator stopped at Voith in Heidenheim for an unannounced visit, looked at the transmissions and decided to have them installed in his buses.

Soon after its introduction, the DIWA 851 transmission required further consideration. Operating experience had shown that the existing number of mechanical gears was insufficient for vehicles with higher end speeds, especially long-distance buses and articulated buses in cities with ascending routes. Another gear stage was indispensable. As a result, Johannes Peltner's design team set out to develop a four-speed version that had already been considered during the concept phase.

Setra articulated bus type SG 180 for city and long-distance driving.

By adding another planetary gear set on the primary side, the three-speed transmission was upgraded to a four-speed version. By closing the SK coupling, a step-up ratio of 0.7 was achieved.

Although the team had realized its goal with relative ease, there were only limited application opportunities for the new design. As the output speeds after the transmission were significantly higher than before it, they had to be reduced again in the rear axle transmission of the vehicles – a technical process that these units were unable to perform. A differential with the required large ratio was not commercially available. New designs based on the latest technological findings required considerably more space, forcing customers to compromise or not use a Voith transmission.

The market accepted the new four-speed transmission rather hesitantly. Another shortcoming was that high speeds in the turbine wheel could not be fully controlled, making it quite difficult to activate the converter brake. But the product still provided important insights for the next development projects. It was only after the introduction of the electronically controlled DIWA.3 version that the four-speed unit succeeded. The final breakthrough came with the launch of the DIWA.3E.

It was again time for new ideas. An additional planetary gear set after the converter was added and shifted in such a way that it reduced gears 1, 2 and 3. For this purpose, the inner wheel had to be stopped by the group brake, while the output occurred via the planetary carriers. In fourth gear, the group brake was unlocked, and the group coupling between the inner wheel and the planetary carrier was closed, bringing the ratio of the fourth gear back to 1:1.

Designated DIWA 851 G, this model was rather long, complicated, bulky and expensive. Its day-to-day operation was marred by constant technical problems. After 290 units, production was discontinued. The G-transmission entered the DIWA history books as the rather unsuccessful "Gustav". For the sake of chronological accuracy, it nevertheless deserves a mention in this retrospective.

It was therefore almost exclusively the three-speed DIWA.2 transmission that dominated the business of the Voith Market Division "Road". From 1975 to 1985, it was more or less the bus transmission par excellence. During this time, through the market introduction in important European cities and other countries, this product really took off. This resulted in a network of national subsidiaries and service outlets spanning several continents. Through the double

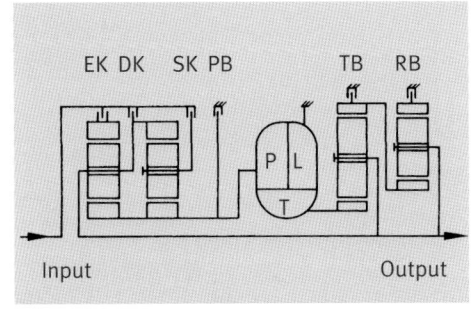

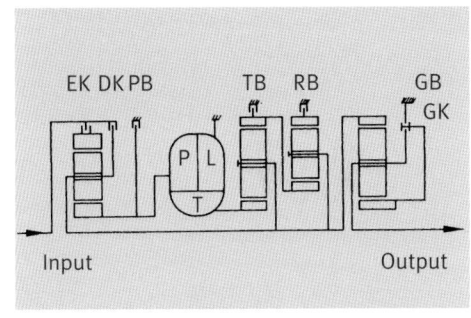

DIWA four-speed transmission with step-up clutch SK.

Slow gear, group brake GB and group coupling GK.

Dennis double decker for Hong Kong.

decker buses made by Dennis in England, Voith also started its first deliveries to KMB in Hong Kong.

The applications in Hong Kong began with a little setback. The first bus with a DIWA transmission had been delivered by Dennis without previous consultation with Voith or sufficient testing on the hilly routes of this city. The transmissions had to undergo several modifications and enhancements before they could finally demonstrate their legendary reliability. This interlude is long forgotten. Today, over 3 000 buses with DIWA transmissions are in service in Hong Kong to the full satisfaction of operators and passengers.

Alongside the technical enhancement and the expansion of international activities, the build-up phase of the DIWA transmission was characterized by the intensive efforts of Voith engineers to achieve a higher service life of the units. The initial goal was a minimum of 300 000 kilometers, but mileages of 500 000 were seriously considered. Voith had achieved service lives of this magnitude before with the technically mature DIWAbus series 502/506. But their production was abruptly stopped in 1974 in order to fully concentrate on the new and commercially more promising DIWA transmissions.

Everything took place under enormous time pressure. After all, the introduction of the new generation of DIWA transmissions had been a great success and should not lag behind other Voith products as far as its overall image was concerned. With this in mind, the Voith service department strengthened its long-established relations with vehicle manufacturers, and also entered into an intensive dialog with large transport operators. For a number of years, these operators had not just serviced their transmissions but also repaired them when possible. The design of the transmission favored this development: mounting the planetary gears before and after the converter was uncomplicated due to the relatively small converter diameter, its central arrangement and its integration into the transmission housing.

Voith systematically incorporated these contacts into its market strategy. They provided immediate technical information and proved to be extremely useful for the further development of the transmissions.

Adding New Functions: The DIWA.2 Transmission

The 1980s were an era of new orientation for DIWA transmissions. Changes in the market had an affect on the product, and the Voith engineers had to face this challenge. The competition had caught up and cost issues became ever more pressing. It was again time to review the existing concept.

Basically, the targets were always the same: higher engine and braking outputs, higher efficiencies, further variations of existing technical possibilities, increased operating comfort, longer service lives for transmissions and brakes and, for the first time and not immediately relevant to Voith, avoiding asbestos-containing materials for clutch and brake linings.

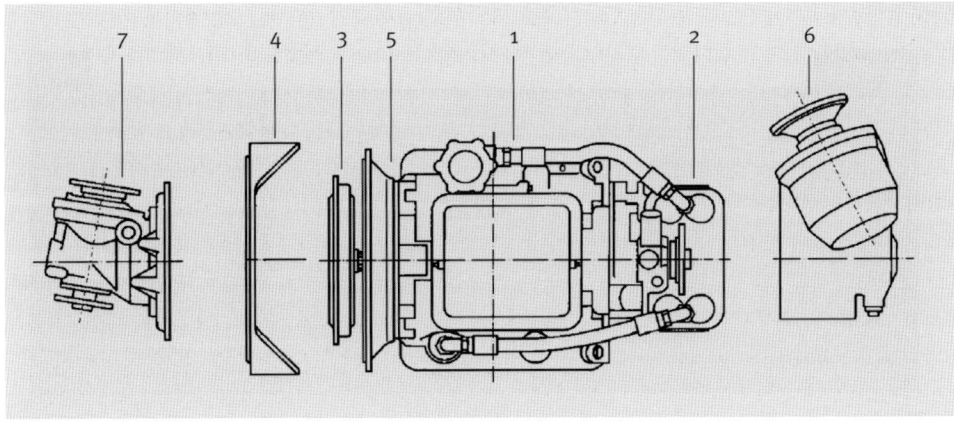

DIWA.2 components

1 Transmission
2 Heat exchanger
3 Torsional vibration damper
   (spring coupling)
4 Suspension flanges
5 Connecting flanges
   engine/transmission
6 Angle drive output side
7 Angle drive input side

The upgraded version of the DIWA transmission that incorporated all of these issues was launched in 1982. Its designation, DIWA.2, documented that it represented the second DIWA generation. The 863/64 series was now able to transmit 245 kW. There were several additional functions, as well as a large selection of angle drives, flanges and numerous suspension variations, even for transverse rear engines. The electronic control systems E100 and E100.2 remained analog, had a modular design and allowed further gearshifting programs. In city traffic, the DIWA.2 transmission yielded fuel savings of 5%.

In the USA, the DIWA.2 automatic transmissions achieved a truly remarkable success: due to the development of a new angle drive on the secondary side, DIWA.2 units were to replace the traditional Allison transmissions in the buses made by Flxible. The ensuing business with commercial vehicle transmissions in the US became and still is an important mainstay of Voith.

City bus made by Flxible, USA.

DIWA Renaissance With Digital Transmission Electronics

But otherwise, the DIWA.2 transmission struggled asserting itself in the market. Despite the upgrading efforts, the improvements did not provide a permanent advantage over the competition. Moreover, there were still a few unsolved problems with some of the additional functions, especially with the four-speed transmissions.

Toward the end of the decade, another aggravating factor was the fundamental structural change of the market. The share of articulated vehicles in bus fleets increased strongly. The reason for this were efforts by the operators to cut down on personnel and overall operating cost. But things became even more complicated: so-called pusher buses, whose engines, transmissions and drive axle were all arranged in the rear, increasingly gained in popularity. Before that, the engines of articulated buses were mostly located in the front of the vehicle, the drive axle was installed before the articulation, and the back was towed like a trailer. This front-drive bus version had been developed after 1951, when traditional bus trailers were no longer permitted for safety reasons. As a result it was possible to install the control units and linkages normally used for solo buses. Pusher buses, in contrast, required that the linkage led from the accelerator to the engine via the articulation, causing endless problems. To improve the situation, E-Gas was introduced, a system in which the linkage was replaced by an electric signal and an actuator.

Low-floor articulated bus Neoplan N 4021 (230/256 HP), 1991.

This first step was followed by additional, increasingly refined measures, resulting in electronic systems and subsystems that were linked with each other via adaptors and plug-in connections. The old analog E100 control could be used on these buses only after the development of expensive additional adaptors.

Voith had meanwhile realized that the competitiveness of the DIWA transmission could only be restored by the systematic development and consistent use of an electronic control system. The pusher buses and the E-Gas systems were used as a basis for the development of a microprocessor control unit. It was time for a thorough revision of the DIWA.2 anyway, because, apart from repeated demands for high engine and brake outputs, reduced fuel consumption and less wear, a new bus type had emerged – the low-floor bus. It left very little space for the driveline and its related components. The first bus of this kind was the Neoplan low-floor articulated bus launched in 1987. It was fitted with a transverse engine, a transmission and a complex angle drive. Voith decided to focus on this new challenge.

A prerequisite for the introduction of a microprocessor control system was the availability of suitable proportional valves instead of the previously used simple solenoid valves. In an automatic transmission, these solenoid valves control the filling and draining processes of the clutches and brakes with pressurized oil, so that gear changes under load take place with a minimum of jolts. With the solenoids in the DIWA.2 transmission that could only be switched on or off. This target could be realized to a small degree by adding further hydraulic elements such as extra reservoir and throttles and introducing other adaptations. In contrast, proportional valves would have ensured accurately controlled gearshifts, but their availability for this purpose was limited. In 1986, Heinrich Dick developed a new, highly responsive control valve that was registered for a patent and successfully tested in a vehicle one year later.

The special solenoid was the key toward the new E 200 control system and future generations of DIWA transmissions. Dick's solenoid valve is still used and currently installed in today's DIWA.5 transmissions.

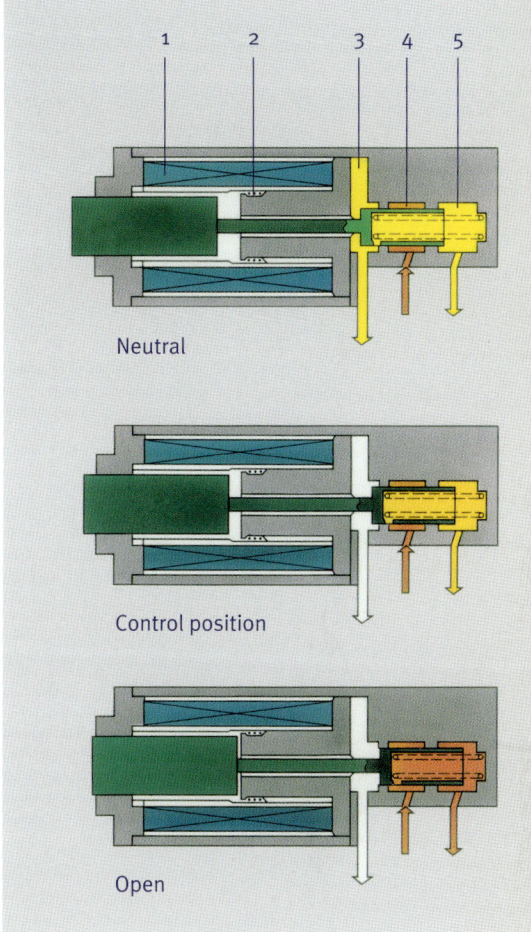

Design of DIWA.3 solenoid valves.

1 Excitation coil
2 Measuring coil
3 Discharge
4 Charge pressure
5 Operating pressure to consumer

*Mercedes-Benz buses
with DIWA.3E.*

The Development of DIWA.3 Transmissions

The development of the new DIWA generation began with the design of a microprocessor control unit, which was designed by in-house engineers. Its purpose was to eliminate the frequent interface problems of electronic systems. Additionally, it was also intended to reduce the complexity of the electronic control system and simplify the hydraulic control in the transmission. The concept for the microprocessor control had been developed as early as 1989. Its main advantage was that its functions were largely performed by systems software, which means that changes could only be carried out after reprogramming. Additionally, digital technology allowed much higher complexity than its analog version. In practical operation, the service teams naturally had to familiarize themselves with the new systems and detect possible failures. Apart from its complex internal control functions, the unit therefore had an incorporated feature that reported such failures and the operating conditions prevailing at the time, and identified them via a PC diagnosis system. These beginnings eventually led to the excellent diagnosis system "Diana", a name deriving from 'DIWA' and 'Analyzer'.

The introduction of the microprocessor control for DIWA transmissions of the third generation was the start of a new life cycle that would turn out to be the most successful within the entire DIWA history.

*Renault articulated bus with
DIWA transmission.*

Voith soon recognized the enormous potential of this technology. Initially, the key mechanics of the transmissions that were now referred to as DIWA.3 had been those of the basic DIWA principle. Only the hydraulic control system had been adapted. But now, a dedicated project team concentrated fully on getting the product ready for series production, which finally began in 1993. Through their new technology, the transmissions set new standards for gear-shifting comfort and fuel consumption in buses, not least as a result of their adaptive gearshifting behavior. During this project phase, the Voith engineers closely cooperated with large vehicle manufacturers such as MAN, Mercedes-Benz, Renault and Volvo. The OEMs were continuously asked for their own experience and were encouraged to identify their expectations on many individual transmission issues.

Volvo city bus with DIWA transmission.

Simultaneously, new production, assembly and testing methods had been introduced in the Voith factory in Munich-Garching. In order to reduce production costs and in view of the projected large volume of new transmissions, sales, service and logistics were improved. In Kazan, Russia, another transmission plant was established as a joint venture.

The Development of DIWA.3E Transmissions

The DIWA.3 transmission had regained a share in the market and held a strong position. Although business was very lively, the Voith engineers were yet again faced with new demands. The transmissions were now also popular for three-axle, long-distance buses with high-performance engines designed for cruising speeds exceeding 100 km/h.

Modules for long-distance buses. Example: Setra.

As a result, further improvements had to be introduced for the four-speed transmissions. The existing transmission program still showed some gaps, so that only some of the long-distance buses offered by the manufacturers could be fitted with DIWA automatic transmissions.

This had a negative effect on sales and the gaps therefore had to be closed. Additionally, engine outputs of up to 295 kW or 400 HP had to be transmitted and the retarder function had to be enhanced. A whole catalogue of modifications was introduced, leading to the new series DIWA.3E launched by Voith at the IAA in 1998. The "E" stands for "Entleeren" (draining) the converter at high output speeds, allowing the utilization of a more efficient converter brake even at higher speeds. It also resulted in a larger oil cooler. Another new feature was the installation of an effective torsional vibration damper instead of the spring coupling. This damper had initially been developed for the DIWA.3 as a damping and isolating system for high loads, but, after a few adaptations, it is now generally used. With the DIWA.3E, a high number of buses entered service with an on board electronic CAN module. This module pools all major data of the key components, such as engine, transmission or brake. In the new Mercedes Citaro, it is efficiently used in a new joint concept between the vehicle service brake and the converter brake of the transmission.

The Development of the DIWA.5 Transmissions

The series DIWA 851-864.3E had been highly successful. During the 30 years since their first presentation at the IAA 1973, some 150 000 transmissions had been built. In 2004, the annual number of units exceeded 10 000 for the first time, approximately 80 percent of these were four-speed transmissions. Over the last few years, their share has strongly increased.

This shift toward the four-speed version was one of the reasons why Voith began with the development of the next DIWA generation. It was launched in 2004 as an exclusive four-speed transmission that was also different in appearance. In the past, the four-speed transmission was created through the additional installation of components from the three-speed module set. As a result, its volume could never be optimized. The transmissions were somewhat too long which occasionally presented a competitive disadvantage when it came to installing them into the vehicles. But the optimization of an input planetary gear set had led to a shorter and lighter transmission. In combination with a new housing without external oil pipes, as well as a new oil filter system, the appearance of the transmission was, for the first time since the beginning of the DIWA history, significantly different.

Some ten years after its development, the E200 electronic control was replaced by the new E300 unit. Once again it was the hardware that provided the incentive for the new software concepts. For example the new ALADIN (Analysis and Diagnostic Network) diagnosis system that would be as pioneering as the DIWAgnosis unit ten years ago with its new features was developed. Now the user is supported by a number of visual aids when looking for failures and trying to eliminate them. This is added by the availability of a data network in which all current repair instructions and spare parts information are stored.

Performance data of
DIWA.5 transmissions
D 854.5 / D 864.4

Input power $P_{1\,max}$ [kW] 220/290
Input torque $M_{1\,max}$ [Nm] 1 100/1 600
Input speed $n_{1\,max}$ [min⁻¹] 2 500
Retarder braking torque M_{BR} [Nm] 2 000
Number of speeds 4
Transmission weight (dry):
Incl. Retarder [kg] approx. 300/305
Max. vehicle weight [t] 28

Software developments have indeed become one of the decisive key competencies in transmission design. The product and operating advantages of the DIWA transmissions have been significantly increased by them, while the mechanical part of the unit has largely remained the same. These product advantages are more and more dependent on electronic control and complex software systems. Since 1993, the number of software developers for this application at Voith has therefore tripled. The first DIWA.5 transmission was installed in a low-floor bus made by MAN. In 2005, Voith and MAN celebrated 50 years of joint DIWA history.

MAN double-decker bus and MAN Lion's city ("Bus of the Year 2005"). Both vehicles are fitted with DIWA.5.

The Voith Retarder

Previous History

It was in 1959 when Helmut Müller, then manager of the special machinery department in Heidenheim, was given an interesting task. The incentive for this had come from the USA.

At the time, the Munich locomotive builder Kraus-Maffei was in sales negotiations with two North American railway companies – Southern Pacific and Denver & Rio Grande Western Railroad. One of the most important jobs of these companies was the regular transport of iron ore, which led right across the Rocky Mountains. As a result of growing demand, iron ore quantities had strongly increased in recent years. The trains had been adapted to this development and by now measured up to five kilometers in length. Total weights of up to ten thousand tons were quite normal. Soon, several high-performance diesel locomotives had to be used as tractors and pushers. Nevertheless, the upward gradients in the mountains did not pose any problems. The increased demand for iron ore could thus be met by either applying conventional methods or, alternatively, by using locomotives with higher outputs.

But the real difficulties began when the trains were rolling downwards. They had to be braked on route sections with downward slopes of up to 30 ‰. For such extreme operating conditions that often lasted for hours, the North American manufacturers fitted the diesel-electric locomotives that used to be the norm in the USA in those days, with electric resistance brakes. By appropriately dimensioning the brake resistors and supplying the required amounts of cooling air, a large share of the braking energy could be taken up wear-free by the locomotives. The mechanical shoe brakes of the goods wagons were thus protected.

At the time, two U.S.-American companies had decided to use a high-performance diesel-hydraulic locomotive that had not yet been proven in the USA. The main contractor, Krauss-Maffei, had asked Voith to supply the hydrodynamic components. Voith therefore had to develop a new three-converter transmission with an output of 2 000 HP, plus – as an indispensable prerequisite and more or less off the cuff – a sufficiently dimensioned hydrodynamic brake.

The basic principle was not unknown to Voith. Some fifty years earlier, in 1912, Wilhelm Spannhake and Walter Kucharski, had built a water eddy brake for inspection measurements at Vulcan in Stettin that was based on the work of their former superior Hermann Föttinger. This brake can be regarded as the forerunner of all subsequent hydrodynamic brakes. Later, Voith considered similar application possibilities, when some of these units were needed for turbine model test stands. Voith was familiar with the design and the operating behavior of these brakes, especially during longer applications where constant braking at low wear was required. Voith's first attempts at solving the problem were carried out on an experimental basis, as soon as the requirement had been identified during the early stages of the negotiations. Basic tests with a turbo coupling soon revealed that diagonally arranged blades provided significantly increased braking torques at maximum filling level.

On the basis of the existing know-how, Voith developed its first hydrodynamic brake under the auspices of Helmut Müller. It had a dual-flow design that virtually equalized the axial forces. The rotor was driven by the secondary shaft at high speed. The mineral oil of the turbo transmission was used as the operating medium.

Voith called its first hydrodynamic brake KB 510, a name that referred to its technical background: coupling brake (German: "Kupplungs-Bremse"), profile diameter 510 mm. It was the first quasi-series development of the "Voith Retarders" department that would soon be highly successful. In 1961, the KB 510 was attached to the Voith L 820 rs and L 830 rU turbo transmissions of the 4 000 HP Krauss-Maffei diesel locomotives that were used for the iron ore transport trains.

By 1980, the term 'Brake' disappeared from the Voith vocabulary for commercial vehicles. It was replaced by the name 'Retarder' to express that the unit was able to effectively slow down a driving process but not suitable for bringing a vehicle to a complete halt like a conventional mechanical braking device.

Hydrodynamic Brakes for Buses and Trucks

Otto Kässbohrer

The B 180 Retarder for Coaches

It was an evening in the early sixties, when Otto Kässbohrer, manufacturer of the renowned and successful Setra coach series from Ulm, stood at his home town's railway station, waiting for the high-speed train from Stuttgart. When the train arrived, Kässbohrer was astonished "how this fast, heavy railcar was brought to a halt over a surprisingly short distance without any braking noises". He still remembered the deafening sound of the screeching and squealing red-hot iron shoe brakes and their flying sparks typical of this situation.

A few days later he met a delegation from German Federal Railways who informed him that Voith in Heidenheim was producing a hydrodynamic brake that provided such characteristics. Voith had developed these brakes for cargo train locomotives used on routes in the Rocky Mountains. The application was such a success that these retarder brakes were now also used by Germany Railways. On the basis of this information, the entrepreneur Otto Kässbohrer came to a decisive and pioneering conclusion: "At Kässbohrer we consequently considered whether this excellent hydrodynamic braking system could also be used for road vehicles".

He thus wrote to the then manager of the 'DIWAbus Transmissions' department of Voith in Heidenheim:

"Dear Dr. Gsching,
I phoned you on 18 July to explain that the coach industry urgently needs a hydraulic continuous brake to be used as a wear-free third braking device. While the wheel brakes that are widely used today are just about sufficient, they are so overloaded during continuous braking and as a result of increasingly faster driving that their service life is definitely affected and safety also suffers. We agreed that the development of a hydrodynamic third brake, i. e. a wear-free continuous brake, would be ideally suited for the production program of Voith, and you, Dr. Gsching, might find it a challenging task to proceed with such a development and see it through to series production. We are thinking of a hydraulic continuous brake whose design characteristics can be summarized as follows:

1) Medium: Water, possibly with anti-freeze, in order to dissipate energy in the engine cooler.

2) Total weight 30 – 40 kg to allow combining the brake with the axle and make it an integral part of the drive axle as an unsprung mass.

I would be grateful if you could inform me, without obligation, whether you actually deem it technically possible to produce such a device, and whether your company would basically be prepared to commence with such a development and provide results in the near future.

Please send me a short note so that our two companies might enter into possible further negotiations. On our side, the problem is a burning issue, while you might find it an interesting challenge."

Kässbohrer's basic ideas were initially more of a strategic nature. The brake, to be arranged on the drive axle as an unsprung mass should not weigh more than 30 to 40 kilograms. Water was suggested as an operating medium, "in order to dissipate energy in the engine cooler".

Voith gladly accepted the inquiry. At that time, the traveling speeds of coaches had significantly increased, and the vehicles themselves had become bigger, heavier and more comfortable. Early indications showed that not only coaches but also commercial vehicle manufacturers were interested in retarders. Additionally, transportation agencies anticipated advantages from the application of hydrodynamic brakes. They were not prone to fading, and the frequent downtimes needed to replace worn friction brakes would be a thing of the past. Voith looked at a prospective new market.

The enthusiasm of both parties led to a lively dialog between Kässbohrer and Voith who were located in two neighboring towns. In autumn 1964, this dialog resulted in the first test drives with retarder prototypes. Afterwards, the new hydrodynamic brake was introduced at the IAA commercial motor show in Frankfurt in 1965, receiving unanimously positive reviews from an expert audience.

While its function exceeded all expectations, the desired weight initially specified by Kässbohrer had not even remotely been achieved. With a profile of DP-275-D, a solid and well functioning prototype unit had been installed into a road vehicle. But it clearly showed traces of its "railway" origins, which Voith did not want to veer away from too much during this early phase. The prototype, very conservatively designed and using plentiful resources, with an overdimensioned heat exchanger and load cylinder, eventually resulted in an extra weight of approximately 400 kg. It must be understood, however, that this solution was exclusively intended for testing the energy dissipation in the vehicle and the driving behavior on the road.

Subsequent design optimizations soon resulted in a first, vehicle-oriented "light" solution. Though manufacturing costs could be reduced, they were still a critical issue with customers. The optimized, double-flow hydrodynamic brake received the name B 180 and had its first public outing at the IAA in 1965.

Based of this model, Voith also developed the B 180 A, which was an air-cooled version used for installation into commercial vehicles (trucks and trailers). But its market was very limited, and therefore, few units were sold. However the development ought to be mentioned here for the record. It demonstrates that Voith can find solutions for niche applications. The unit was based on independent cooling technologies that were available in those days and are still in demand today.

The B 190 Retarder for Coaches and Buses

The advantages offered by hydrodynamic brakes were so obvious that more and more manufacturers wanted to use them. But by 1968, Voith had to admit that the B 180 Retarder that had been such a success in coaches was not suitable for both city and long-distance buses. Its installation was hindered by a lack of space. Coaches had and still have spacious luggage compartments arranged between the axles, a feature that city buses did not need. On the contrary: they were and still are built with low floors, in order to allow passengers easy access. While the installation situation for retarders in coaches was "narrow and high", the message for city and long-distance buses was "wide and low".

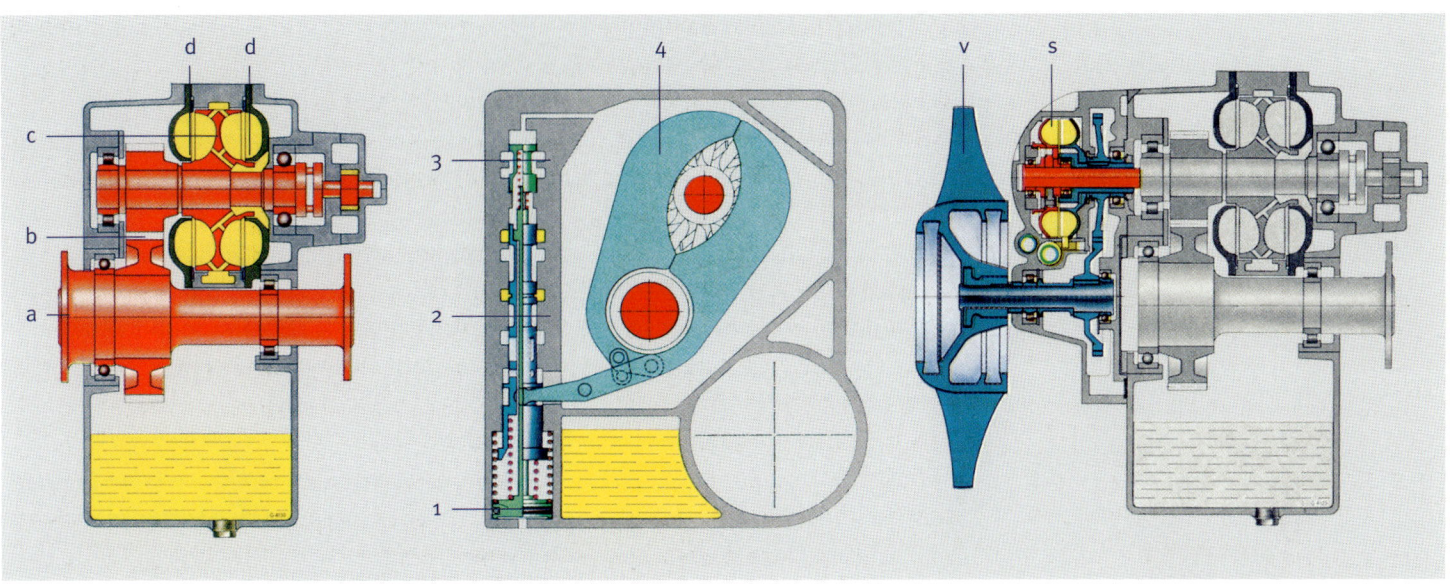

The hydrodynamic brakes had to be adapted to these conditions. In the course of further development programs they were converted to a single-flow design. At the same time the blade geometry and the control variations were revised. The new retarder emerging from the B 180 was named B 190.

This new name led to a brief phase of confusion among users who assumed that the higher model number represented higher outputs. At the time, the Voith sales personnel had its hands full to disperse these doubts and convince the customers that the braking torque was the same for both models.

The B 180 and B 190 retarders were not the ultimate commercial success. They were still too heavy and too expensive. The weight of the retarder far exceeded the expectations of both, Kässbohrer and Voith. Both companies had to accept that there was no way in which the intended "30 to 40 kg" could be realized. The current additional weight, still being 250 kg despite all efforts and hence increasing the axle load, was too high in view of "pressing legal axle load limitations". The number of units sold grew at a moderate pace. Additionally, there were still components that had to be produced individually at high cost. The bladed wheels were a particularly expensive item. These facts stood in the way of the larger-volume production Voith intended to introduce in 1971.

Voith hydrodynamic brakes (Retarders) B 180 M and B 180 A with double-flow circuit.

B 180 M:
a Drive shaft
b Step-up gear
c Braking rotor (impeller)
d Braking stators
1 Control piston
2 Switch on-off valve
3 Overflow valve
4 Ventilation orifice

The longitudinal cross section shows the blade compartments with oil filling. The cross section shows the brake with closed orifices, i. e. without oil filling.

B 180 A:
s Hydrodynamic coupling
v Fan impeller

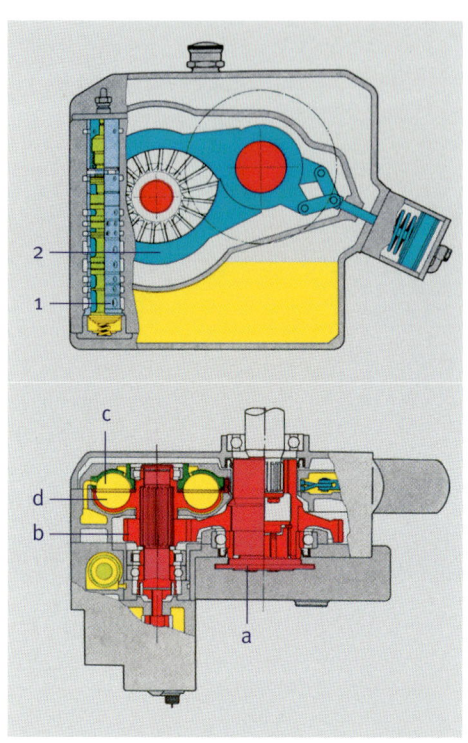

Design scheme of the Voith
hydrodynamic brake type B 190.

a   Drive
b   Step-up gear (gear stage)
c   Braking stator
d   Braking rotor (impeller)
1   Control piston
2   Fan orifice

The Voith design engineers considered starting a full casting series for the bladed wheels. The in-house foundry, specializing in the production of single units weighing several tons, was not suited for this task. Inquiries with other foundries were unsuccessful. All of the companies declared to be unable to meet the requirements. They stated that is was impossible to cast the blade geometry (introduced in 1968) with its thin and simultaneously sharp edges. Consequently, the Voith's in-house foundry demonstrated its technical competence and ingenuity – in an unfamiliar field. Although not accustomed to the small size of the required components, it developed an effective casting and foundry method for these parts and led the ditherers on the right path. It did not take long before a suitable supplier was found.

It did not take long before the company realized that the Heidenheim factory was not the right place for mass production. The custom manufacture of large units such as turbines and paper machines was clearly its specialty. After a short interim period at the company's plant in Munich-Garching, the retarder production was relocated to the then Voith Turbo GmbH facility in Crailsheim in 1971. Crailsheim offered much better conditions for this kind of large-scale manufacturing.

The R 130 Retarder

At Crailsheim, the engineers reexamined the two retarders. They quickly realized that the original thought of converting the KB 510 brake designed for rail applications "mutatis mutandis" for road vehicles, could not be realized. New approaches were made that focused exclusively on coach applications. The everyday experiences gained by Kässbohrer and Voith over the last few years were incorporated into these considerations. Voith consequently moved away from the heavy railway past of the coach retarder after its operating safety had proven itself.

The design engineers also derived from significantly higher unit numbers. Although sales were continuously rising, the volume was not that of a typical series production. The annual production figure of the B 180/190 amounted to an average of 200 units. On the one hand, individual component manufacture and assembly were still too closely oriented on individual or limited series production methods. On the other hand, Voith intended to supply coach and commercial vehicle builders on a large scale. But this industry had much higher production volumes. In order to get a successful foothold, Voith had to adapt to this situation right from the beginning.

The next product to emerge was the VHBK 130 Retarder (Voith Hydrodynamic Brake for Commercial Vehicles), in day-to-day business referred to as R 130. Its weight had shrunk to 85 kg, the systems weight including cooler amounted to approximately 120 kg, and its price had significantly decreased. In 1973 it was presented to a technical audience. The illustration shows the symmetrical and compact design that allowed installation into nearly all vehicles. The solid shape without step-up gear was achieved by the consistent utilization of all parameters relevant to hydrodynamic installations. Type B 180 and B 190 were discontinued from 1973.

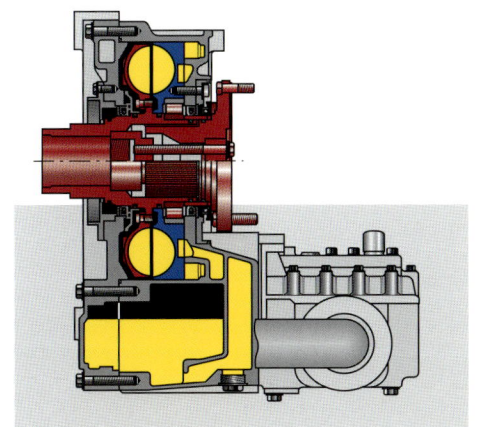

Retarder R 130.

Among the many new employees working in the retarder department at Voith in Crailsheim, the engineering manager, Klaus Vogelsang, who worked at Voith for nearly three decades, stands out through his untiring support and ongoing positive influence on the continued development of this product.

Voith Retarder R 130, single-flow design without step-up gear, consisting of
1 Rotor
2 Stator
3 Housing

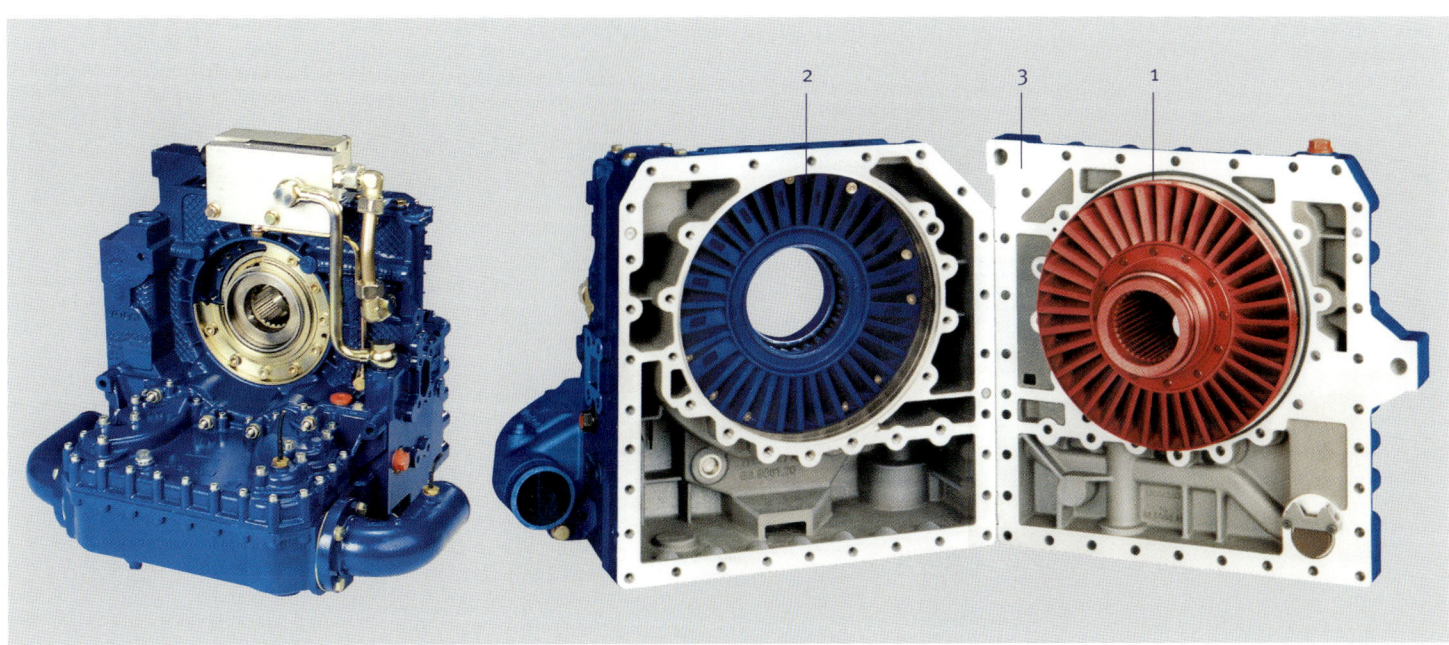

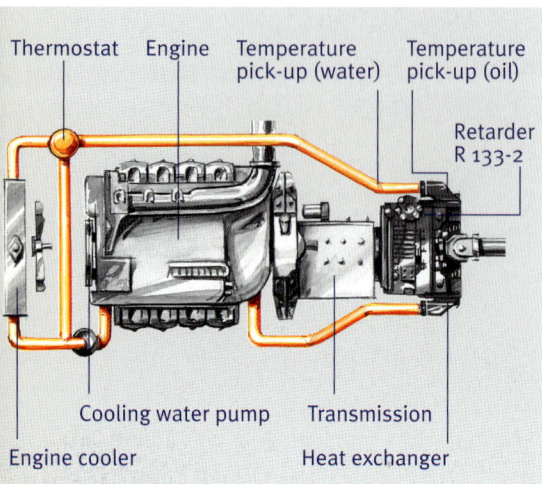

Thermostat   Engine   Temperature   Temperature
                      pick-up (water)   pick-up (oil)

Retarder
R 133-2

Cooling water pump   Transmission

Engine cooler   Heat exchanger

Cooling scheme:
Heat transport from oil to water
to air.

Cooling and Braking

Since retarders are used for both continuous braking and adaptation braking in the high-speed range, the cooling system plays a vital role. Apart from being optimally designed and economically priced, the new models, similar to the disintegrated construction type, had to be built without flexible hose connections in the high temperature range to avoid safety risks.

The cooling scheme shows the typical water circuit of a secondary retarder with the essential power transmission components, optimized and tested in close co-operation with the vehicle manufacturer. The temperature sensors serve to limit maximum permissible temperatures and help to avoid thermal shocks during low temperatures.

Hydrodynamic braking systems (retarders) advantageously utilize the cooling system of a commercial vehicle, especially when it is not fully needed for engine cooling during coasting operation. The "idle" cooling output in this phase corresponds to the constant retarder braking output and usually meets the legal directives for continuous brakes, also in combination with engine brakes.

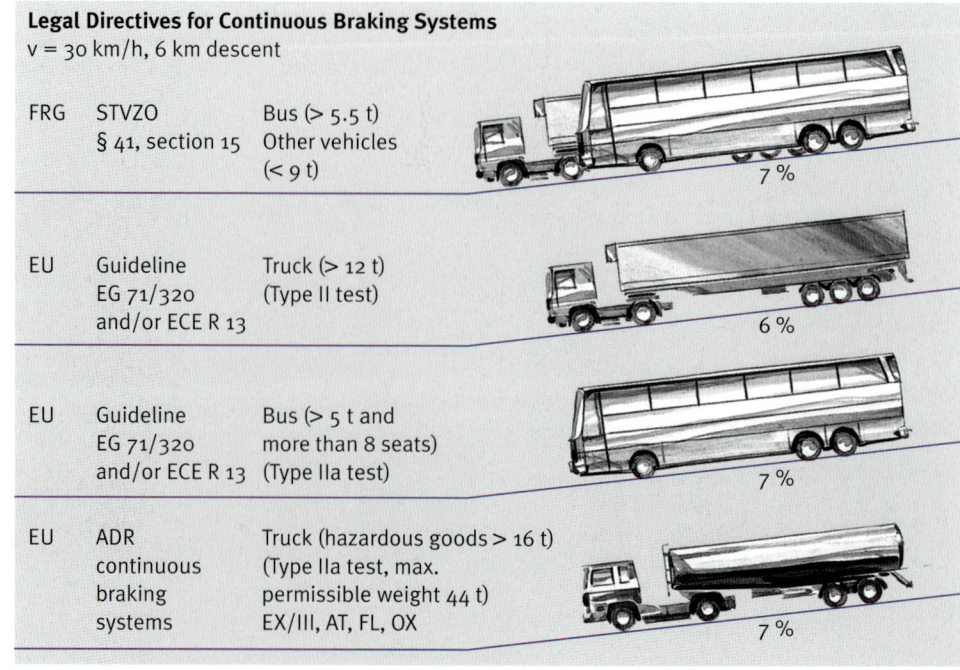

Legal Directives for Continuous Braking Systems
$v = 30$ km/h, 6 km descent

FRG	STVZO § 41, section 15	Bus (> 5.5 t) Other vehicles (< 9 t)	7 %
EU	Guideline EG 71/320 and/or ECE R 13	Truck (> 12 t) (Type II test)	6 %
EU	Guideline EG 71/320 and/or ECE R 13	Bus (> 5 t and more than 8 seats) (Type IIa test)	7 %
EU	ADR continuous braking systems	Truck (hazardous goods > 16 t) (Type IIa test, max. permissible weight 44 t) EX/III, AT, FL, OX	7 %

The internal control processes of the retarder are largely automated and follow the "ON" or "OFF" signals. Previous consecutive control processes used in the preceding models were eliminated.

The braking procedure is controlled electro-pneumatically. The filling level in the blade chambers changes in dependence of the driving speed. Its value incorporates the data regarding water and oil temperature limits collected by an overriding electronic sensor.

Otto Kässbohrer is perfectly justified to point out his company's vital contribution to the successful development of the retarder for coaches. He summarizes it as follows: "Apart from providing the reassuring thought of a permanently available, wear-free braking force, the retarder protects the service brake and its linings, allows higher average traveling speeds, especially on mountainous routes; as a result of the heat dissipated to the cooling water by the retarder, it prevents the otherwise damaging temperature drop of the engine on longer downward stretches. This also maintains the temperature level of cooling water for the heating system. We are very pleased to note that coach operators increasingly specify the new retarder."

Braking torque dependent on retarder speed and oil filling level.

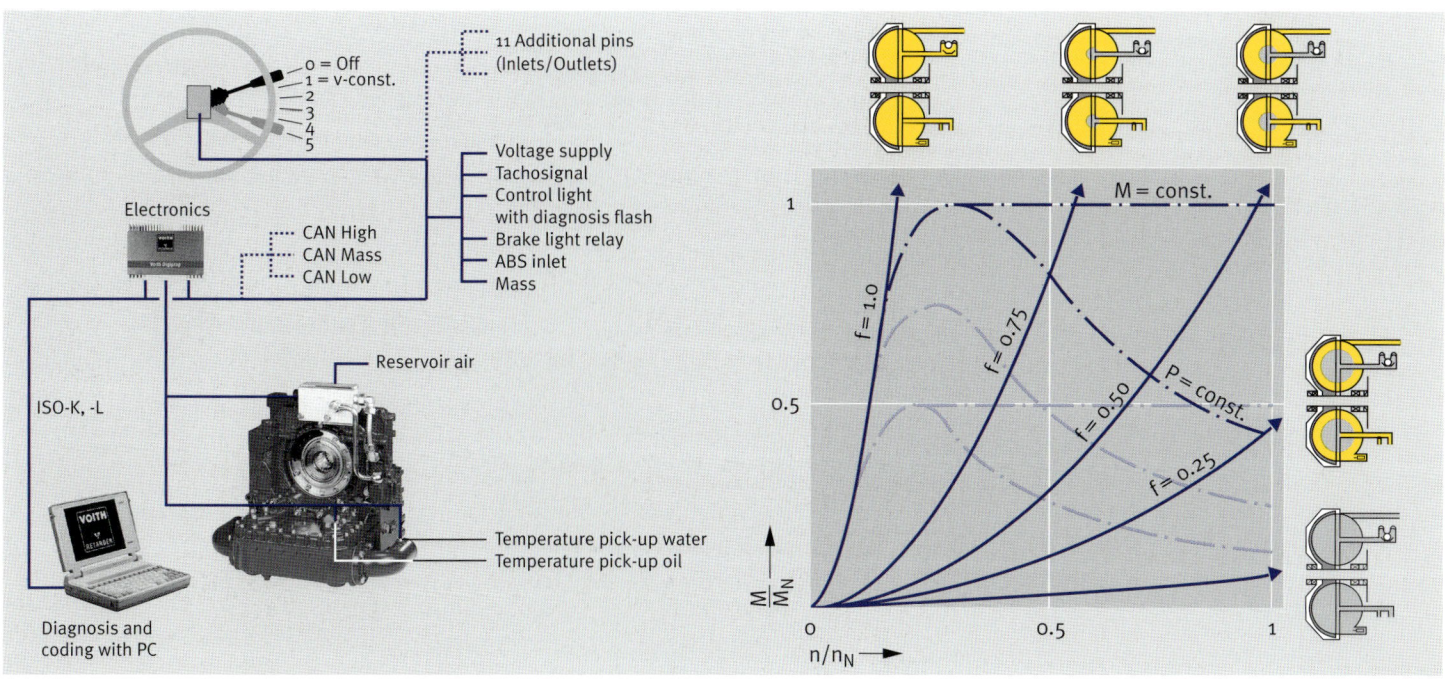

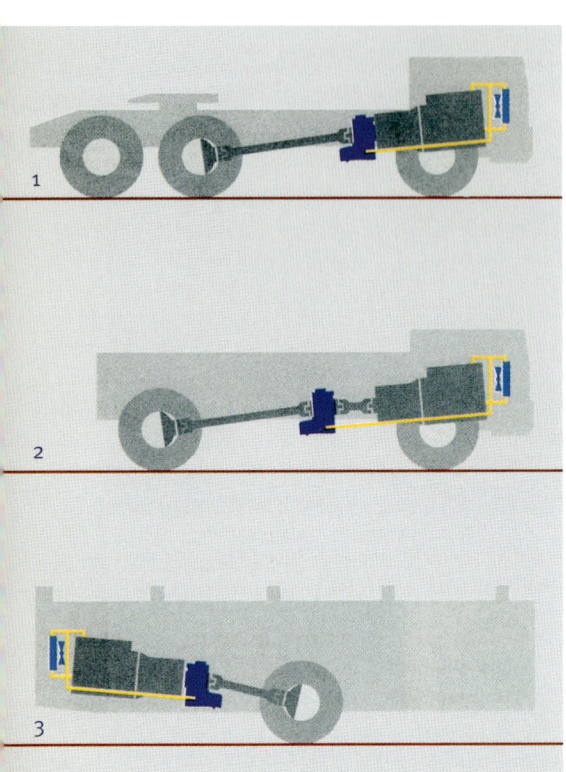

Inline arrangement, i. e. the retarder circuit is arranged coaxially to the propshaft, with several positions to chose from.

1 Retarder installed to the transmission of a trailer tractor.
2 Free-mount installation of a retarder in a truck.
3 Retarder installed to the transmission of a coach.

Inline Retarders

As the first retarder without a step-up gear and new profile geometry, the R 130 marked the start of the "Inline Retarder" series in 1973. In this series, the retarder is either coaxially flanged to the transmission or integrated into of the propshaft:

With its symmetrical outer contours, the retarder met the key prerequisite for universal installation into commercial vehicles. Right from the beginning, its control system was largely designed for this application. Today it offers additional options, such as constant driving speed and thermal power limitation, both of which can easily be integrated into the vehicle electronics.

By 1987, the R 130 had been succeeded by three later versions (R 130-2, 110 and 120), and, in 1994, by the upgraded type R 133 – with 4000 Nm the world's most powerful retarder for road applications.

These retarders had been designed for certain standard load conditions in light, medium and heavy commercial vehicles, as well as all-terrain four-wheel drives. But a few developments for special applications should not be overlooked. Although they naturally never resulted in high unit numbers, they are proof of the never-ending ingenuity of the Voith engineers. Such developments often open up new paths and become the nucleus for innovations.

The "Two-way Application" for rail and road in 1992 is an example of such special development. Another one is the unique utilization of the retarder as a high-performance "Long-term Endurance Brake" in a rack railcar of SLM in Switzerland. This installation required the development of a completely new control system that had to be painstakingly adapted to the individual braking torque ranges. The specified retarder application envisaged almost ideal operating conditions: one hour driving on a descent, braking under thermally balanced energy dissipation, and one hour on an ascent in the opposite direction of rotation. Any additional devices for the dissipation of heat loss were thus unnecessary. The inspection run took place on the Rigi/Kulm Mountain in Switzerland in 1975 and attracted wide attention. A hydrodynamic retarder had once again demonstrated its advantages over the electro-magnetic competition. As a result of their systems design, electro-magnetic retarders require a re-cooling period after emergency stops, which the rail operator could not accept for safety reasons.

Due to their excellent safety aspects and enormous advantages over conventional electro-magnetic brakes, hydrodynamic retarders acquired a respectable and ever-expanding share in the coach market. Gaining a foothold in the much larger truck segment proved to be far more difficult. Manufacturers and operators regarded the retarder as too bulky and too expensive. They were satisfied with friction brakes and were not willing to spend more money for safety. Pioneering customers were told by the manufacturers that the driveline warranties would be voided if retarders were installed.

For the coach segment, Otto Kässbohrer had predicted success at an early stage and prophesied that virtually all Setra coaches would be fitted with a retarder in the foreseeable future. When the 10 000th retarder was manufactured in Garching – now the main production site for retarders – the then Crailsheim sales manager presented Kässbohrer with a pair of binoculars in recognition of his extraordinary vision.

Starting with the third generation of the Setra series S 209 to S 215, Kässbohrer fitted all of its coaches with Voith Retarders. This was the breakthrough for Voith; the advantages of the retarder as a reliable safety reserve were officially recognized in the market.

The safety of hazardous goods transports, be it on highways or steep alpine roads, increased significantly by the installation of retarders. Additionally, traveling times could be reduced.

Slowly, the Retarder Product Group made a name for itself as a supplier to the automobile industry. More and more commercial vehicle manufacturers offered Retarders as ex works fitments. They are now an internationally recognized product. Today, heavy trucks are by far the biggest market segment for Voith Retarders. The key innovation in this context is based upon the original step-up concept and the integration into the transmission. Painstaking individual acquisition efforts have paid off.

Professional and close co-operation can also lead to spontaneous expressions of gratitude. It has been reported that Voith employees were once given a crate of beer after a customer visit. Unfortunately, the precious cargo was destroyed soon after, thanks to the hearty action of the Voith Retarder that had once again powerfully demonstrated its reliability!

Offline Retarders

This series started in 1989/90 with the secondary VR 115 retarder, installed sideways behind the transmission in the tractive flow. It is the result of close collaboration with vehicle manufacturers, driveline specialists and transmission

Offline arrangement, i. e. the retarder is positioned at the transmission housing on the left or right hand side of the propshaft.

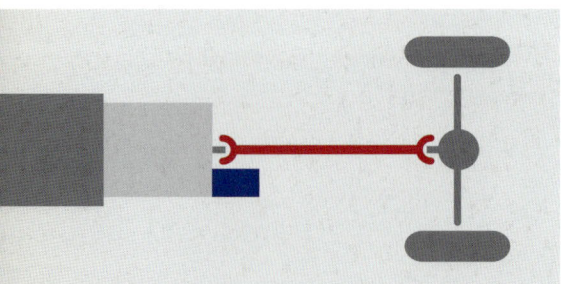

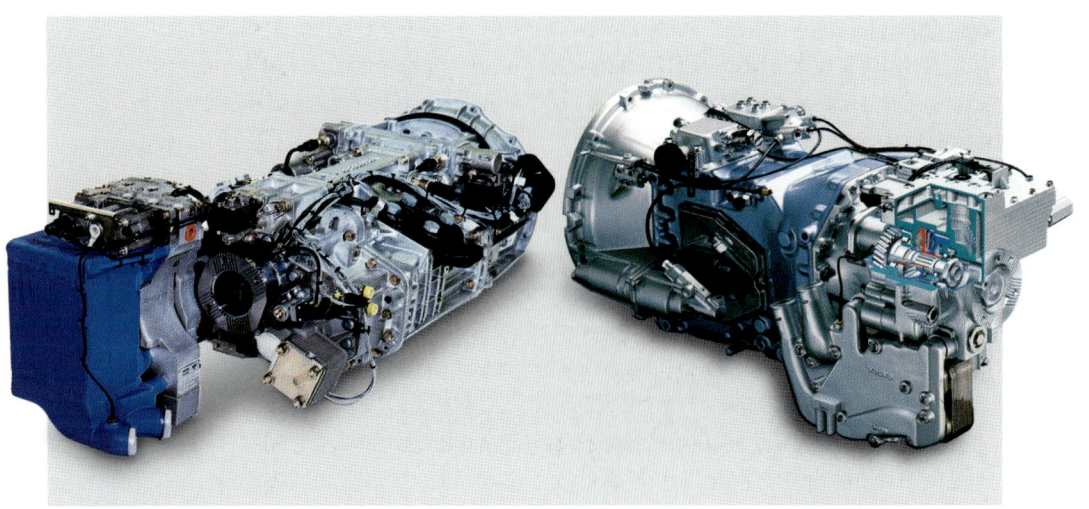

builders. Together with these key customers, specific solutions are developed
and optimized on the basis of highly complex research and delivery contracts.
Here, retarders can be integrated into transmissions, and control units incorporated into vehicle brake management systems in a highly individual manner.
It should be mentioned in this context that, as with all Voith Retarders, the operating fluids of transmission and retarder are strictly separated because of their
different specific loads.

Two further joint developments within the offline series are the VR 115 (HV)
with Mercedes-Benz for heavy commercial vehicles, with a new device for loss
reductions and oil-air separation, and the VR 3 250 (1999) built with Volvo for
trucks and coaches.

With the help of the automated rotor shift, a spring moves the rotor during
non-braking operation into a position with minimum loss. If braking is required,
the rotor is rapidly shifted by the hydrodynamic braking torque and kept
"sharp", i. e. in braking position. Such special new developments are expensive.
Performance, quality and price of the products and their integration into the
driveline are therefore always at the heart of any consideration. Highly special-
ized Voith Product Groups, internal colleagues and partners from commercial
vehicle manufacturers put the ideas and requests of customers of series products into practice. In the 2003/2004 fiscal year the 150 000th VR 115 retarder
was delivered, bringing the total number of all Voith Retarders sold to over
300 000.

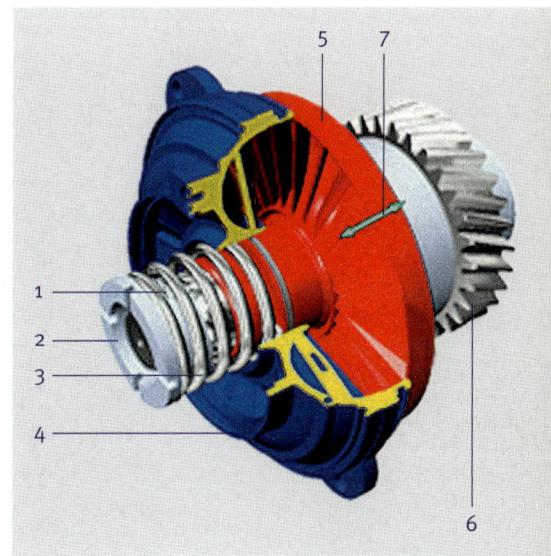

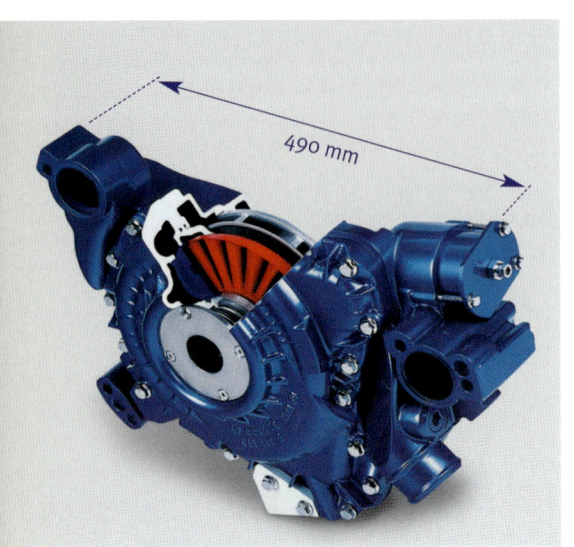

490 mm

Aquatarder

Aquatarder

The primary retarder WR 115 has been designed as a system-integrated solution for installation to the MAN D 28 engine. The engine cooling water is simultaneously used as the operating medium of the Aquatarder. The WR 115, on the market since 2002, enables quick warming of the cooling water, even if the vehicle is at a stop. It also improves cold-start conditions for diesel engines.

Unlike the oil version, the Aquatarder does not require components such as oil-water heat exchanger, oil tank and oil pipes. This allows a very compact design. Its weight is low and its specific performance is consequently relatively high. An enormous advantage is offered by the fact that the high braking torques of the Aquatarder are favorably coupled with matching engine and hence water pump speeds delivering the required cooling agent flow for energy dissipation.

Installation of a retarder (VHBM) into a car dynamometer. The torque reaction rod acting on a pressure measuring cell and hence allowing good torque and power determination can be clearly seen in the foreground.

Industrial Retarders and Special Solutions

Voith hydrodynamic brakes have been used in diesel locomotives since 1961 and in vehicles since 1964. In 1970, Voith decided to take the product to the company's other Market Division – Industry – and look for possible applications.

This market had a completely different structure. It was (and still is) characterized by a wide range of possible applications, but all of them are special cases. The need for industrial retarders is limited. While the Market Division Road is selling large quantities to the vehicle industry, special industrial applications usually require individually designed units. These involve high research and development investments that are afterwards not recouped by profitable series production. But looking into the most diverse applications is indispensable for Voith. Through it, the company remains in the market on all kinds of levels. Voith always anticipates that the effort dedicated to extraordinary problem solutions pays off through synergetic use in other applications. This special market situation is best explained by a few examples.

Retarders for Dynamometers

Dynamometers are primarily used in car repair and service workshops. Routine maintenance and adjustment jobs, as well as output measurements and exhaust analyses of combustion engines, require uniform, precisely adaptable engine loads via the drive wheel. The reliable and wear-free operation of the retarder was simply ideal for tough workshop environments. By introducing increasingly rigorous safety and emission regulations along with regular controls of their observance, the legal authorities had ensured that dynamometers would soon be a standard feature even in small workshops. This business field began to look quite attractive.

Depending on output requirements, retarders type VHBM-216, 260 and 316 are used. VHBM stands for Voith Hydrodynamic Brake Measuring Brake. The end figures represent profile diameters, they are defined on the basis of different torque stages across the filling parabola. The retarder, in this case an Aquatarder, is controlled either in open flow or in the closed operating water circuit. In both cases a simple connection to the water supply network is sufficient. The operating water is cooled by a heat exchanger, which is automatically activated when required. With the help of a very simple control system that is customized to the individual application, it is possible to generate excellently reproducible characteristic curves for output and exhaust measurements on vehicle engines.

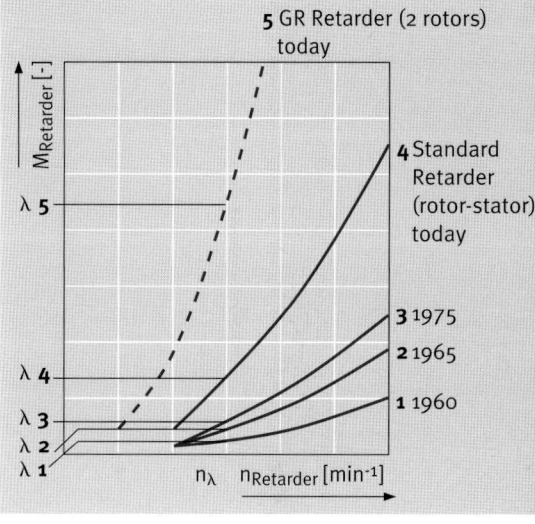

Development of the lambda curve in Voith Retarders.

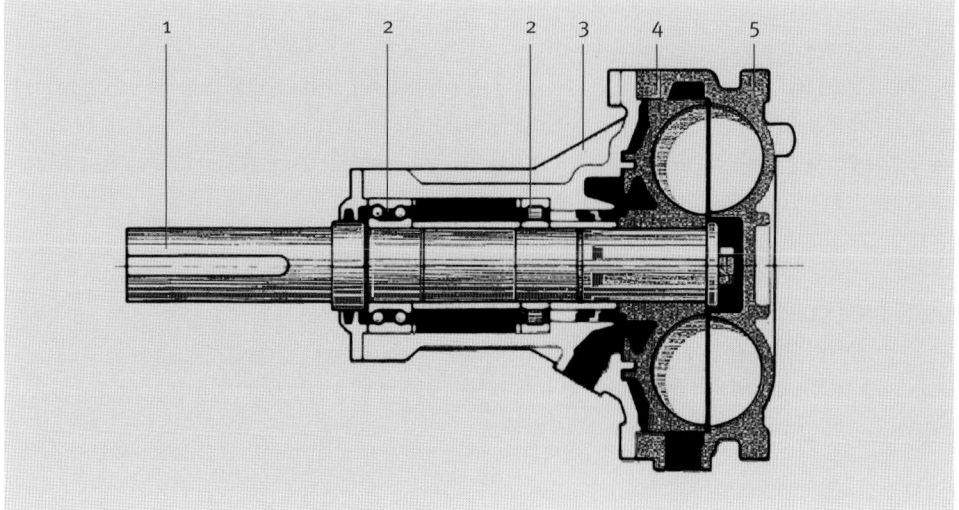

Sectional view of a water-operated Voith Retarder used as measuring brake (VHBM).

1 Brake shaft
2 Bearings
3 Rotor housing
4 Rotor
5 Stator housing with stator

259

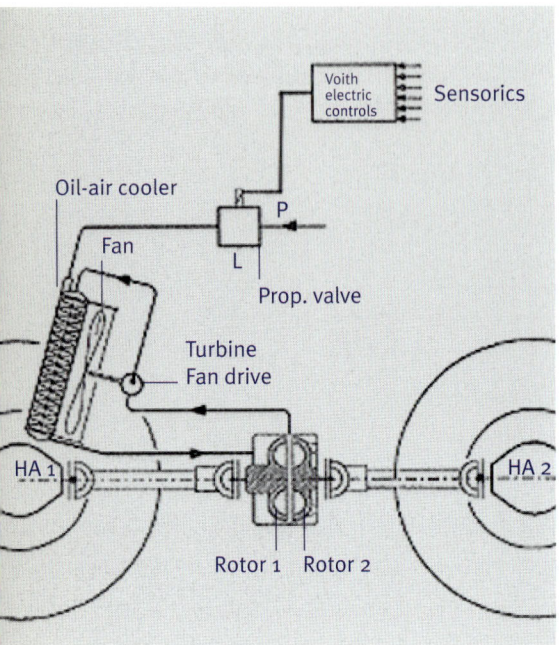

Function scheme of the contra-rotating retarder and installation between two drive axles.

GR 116 Trailer Retarder

An interesting project study was that of a contra-rotating retarder in 1992. Its purpose was to examine the application possibilities of an independent retarder in the chassis of a truck trailer or semi-trailer.

Measuring results at the test stand had already demonstrated that there was an enormous increase of λ values in the braking torque range of a completely filled retarder (see graphic on page 259). If installation space is limited, a small and light contra-rotating retarder is thus able to generate high braking outputs.

At the time, the contra-rotating retarder represented a massive technological leap. Its trouble-free function, self-propulsion and independently operating oil-air cooling system were proven in practical road applications. By their counter-rotation, two rotors of an oil-filled retarder, each coupled to an axle, provide a massive increase in braking torque, especially at low driving speeds.

Industrial Control Brake

A good example of the synergetic use of other projects is the Voith VIR 133-2-S industrial retarder. It has been designed for heavy truck applications and is produced in large quantities.

The German navy ("Marine") uses a number of standardized diesel engine versions from different manufacturers for the propulsion of its ships. At the navy college in Parow near Stralsund, future technicians are familiarized with these engines on test stands under quasi-realistic conditions. For training purposes, engines made by Deutz (P = 1 230 kW, n = 1 000 min^{-1}) and MTU (1 000 kW, n = 1 900 min^{-1}) have been provided. The initial plan was to use single-stage water vortex brakes for subjecting the engines to load. But soon it turned out that these bulky machines could not be accommodated in the unit. The VIR 133-2-S came to the rescue.

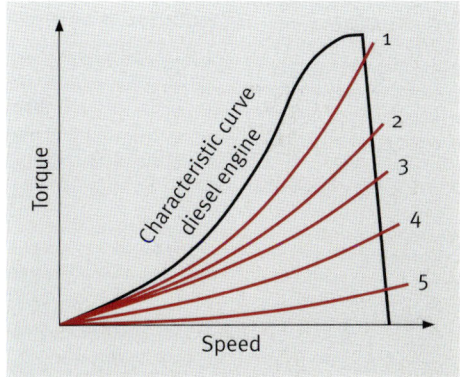

After a few minor amendments, this retarder proved to be the ideal solution for this type of stationary industrial applications. The actual basis for this was, of course, the Voith modular design policy. On its own, the VIR 133-2-S would not have been able to cope with the new task, despite the maximum torque of 4 000 Nm and a maximum cooling output of 400 kW. But combined with two additional retarders of the same type, consecutively connected by torsionally stiff metal disk pack couplings, it was – or rather, the three of them were – in a position to brake the maximum engine output. For training purposes, the control system of the braking unit is preprogrammed for five different propeller characteristic curves.

Heinz Höller

History Has a Future

The Perspectives of Voith Hydrodynamics

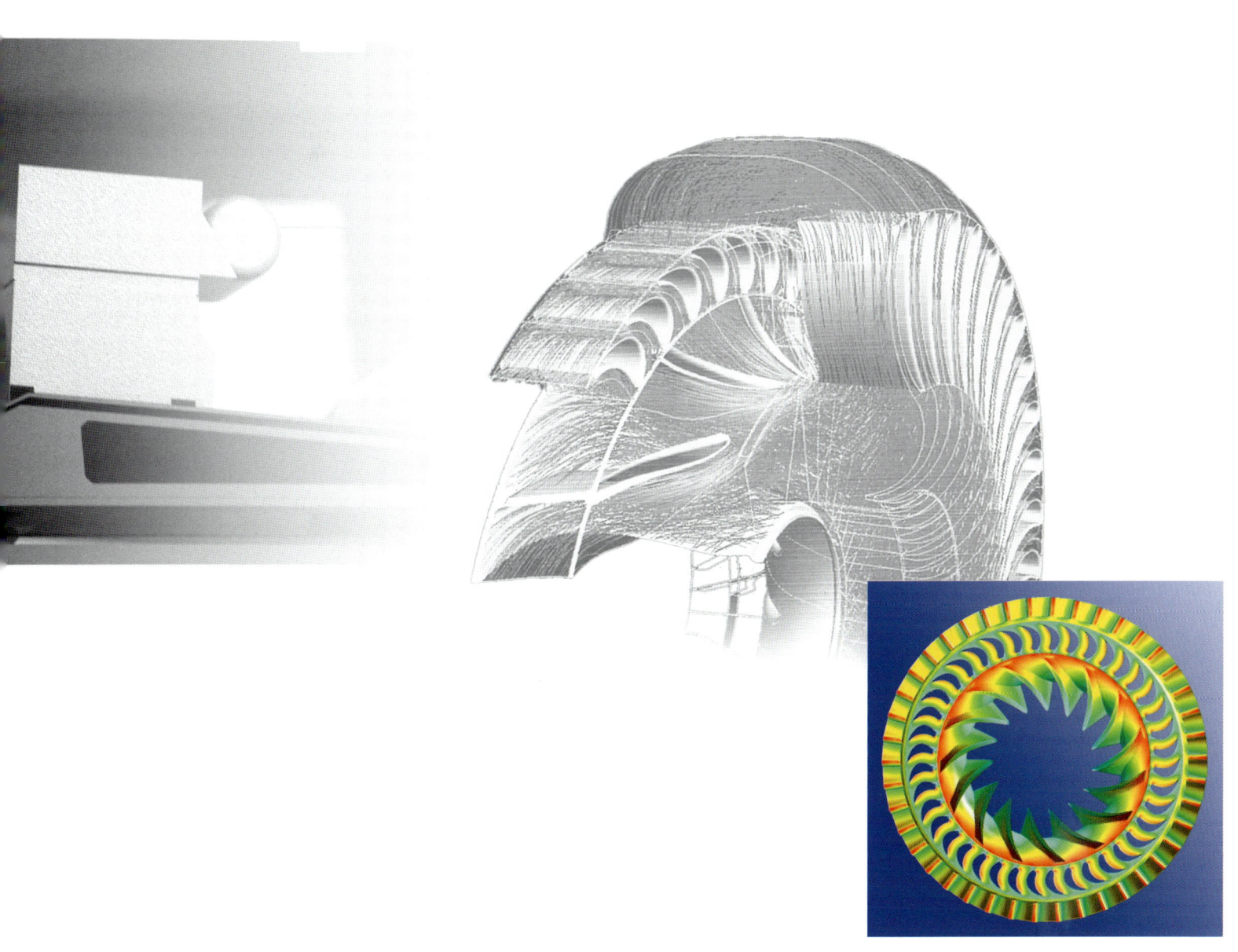

Technological Outlook

The development of hydrodynamic power transmission at Voith as outlined in the previous chapters, has always been characterized by a creative tension between market and technology. Be it new or changing applications – Voith is utilizing its own technological competence to develop suitable products with hydrodynamic core components that stand out from the competition by their high customer benefit. Even one hundred years after the registration of the Föttinger patent, the saying still applies: "There are no obsolete technologies, there is only the sin of omitting further development".

This sin will never be committed by Voith. Voith is prepared for new tasks. The following chapters describe some future-oriented core competences that have been established within the company, together with the necessary tools and resources. The market will then decide, which products are actually being developed.

Further Improved Torsional Vibration Models for Couplings and Converters
The excellent vibrational isolation of the input side from the output side by the intermediary hydrodynamic transmission element converter or coupling had already been recognized during the development phase and was proven in tests with a coupling in 1923. In the past, this proof and experiences from practical operation led to the assumption that hydrodynamics provide a complete vibrational isolation of the two machine systems. Today, engineers know that this assumption, especially when it comes to drivelines with low damping, is correct only up to a certain degree.

Early attempts of describing torsional vibration behaviour analytically have been documented by F. Söchting in 1938. He anticipated that the interrelation between the transmittable torque and the input speed follows the same model laws, both at constant and variable speeds. Fritz Kugel, on the other hand, recognized in 1959, that Söchting's assumption of identical transmission behaviour both in stationary and instationary situations was incorrect due to the mass inertia of the circulating fluid. He formulated a pragmatic approach towards calculating this inertia in dependence on the excitation frequency of the input side.

Scientific pioneering work regarding behavioral patterns of hydrodynamic couplings in connection with vibration-prone drives was carried out by J. Frömder

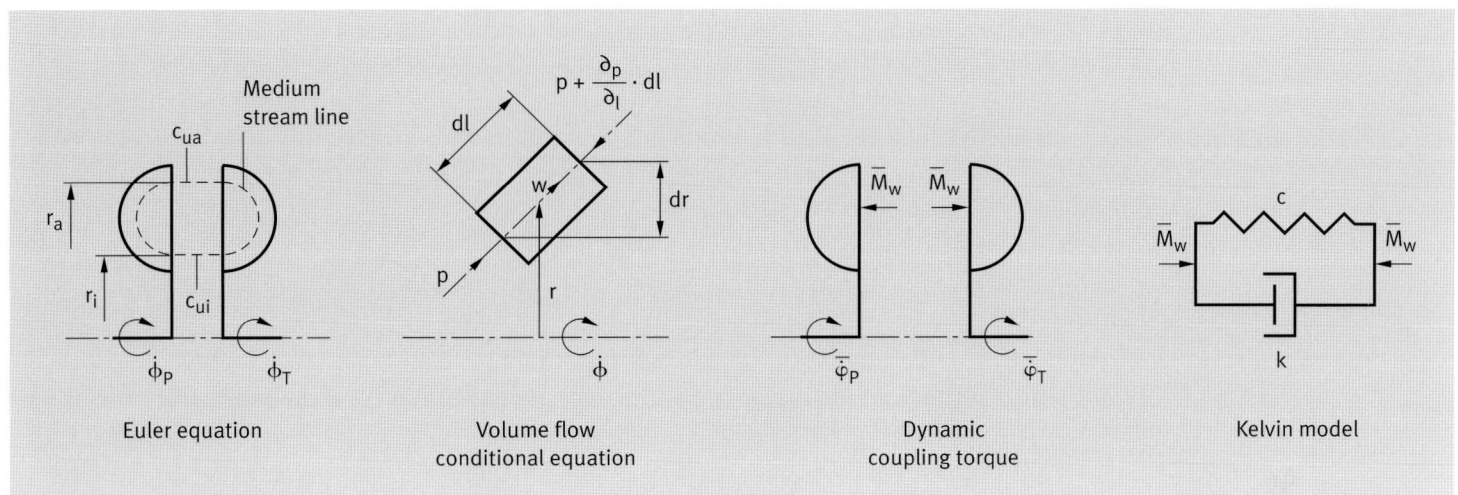

| Euler equation | Volume flow conditional equation | Dynamic coupling torque | Kelvin model |

and H. Hasselgruber after 1961 within the course of their research activities at Technical University Hanover. Both derive from Euler's turbine equation, as well as an instationary frictional laminar flow and arrive at frequency-dependent stiffness and damping values that are equivalent to fluid couplings.

Deduction of the numeric calculation model from the volume flow conditional equation via the dynamic coupling torque to the Kelvin model.

Basing upon the work of Frömder and Hasselgruber, the Voith engineer Helmut Worsch carried out an analytical comparison of the Kelvin model for flexible couplings (parallel connection of spring and damper) and deduced analytical relations for the equivalent stiffness and damping characteristics of hydrodynamic couplings. The applicability of his models was verified i. a. in experiments by Achim Menne.

If one describes the temporarily variable quantities contained in the volume flow target model by a stationary and an instationary share, the alternating torque M_K transmitted by the coupling is proportional to the instationary angular velocity $\varphi_{P, T}$ and a complex transmission form h_f, if slip values are low.

The dynamic transmission behaviour can thus be compared with the behaviour of flexible couplings, although, with hydrodynamic couplings, the parameters stiffness and damping depend on the exciter frequency. The complex transmission function can be determined in experiments by applying the torque balances. This allows the calculation of the equivalent stiffness and damping values.

267

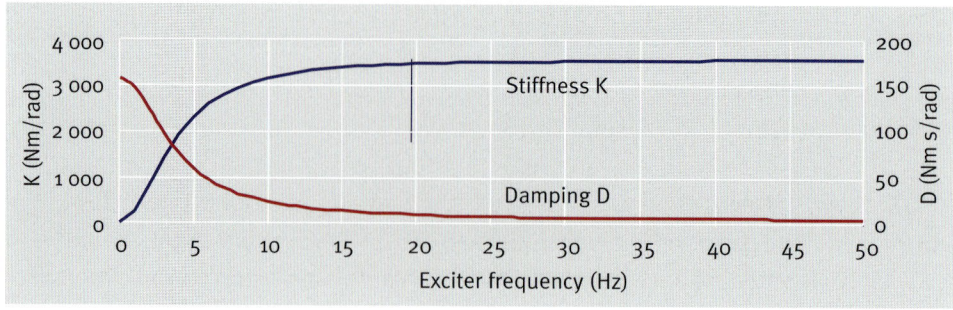

Equivalent stiffness K and damping D as function of the exciter frequency examined at a VTC 487 T. The maximum stiffness value of 3 500 Nm/rad for the coupling is very low and corresponds to that of a steel shaft with a diameter of 30 mm and a length of 1 863 mm.

Test series and field trials proved that the stiffness of conventional drive components is higher by at least a decade than that of hydrodynamic couplings. From this it follows that the first natural frequency of a driveline is determined by the hydrodynamic coupling and usually does not exceed 20 Hz; often it is even below 10 Hz.

The characteristic dependence of the stiffness and the damping of the exciter frequency results in a low-pass behaviour of the hydrodynamic coupling. With low exciter frequencies up to the first natural frequency, torque fluctuations that are introduced into the hydrodynamic coupling are transmitted, yet strongly damped. Above the first natural frequency, the instationary share of the coupling torque declines rapidly.

Over a wide range, hydrodynamic converters behave identically in stationary and instationary operation. This behaviour, which deviates from that of a hydro-dynamic coupling, can be explained by the circulating fluid flow that changes only insignificantly. From this it follows that hydrodynamic converters with relatively constant λ curves have no or only very low stiffness. Damping values can be established from the slopes of the stationary performance graphs.

Numeric Simulation of Flow Fields by Computational Fluid Dynamics
Numeric simulation of flow fields is the result of the combined application of physical, mathematical and computational methods.

The basics for this were established at the end of the 19th century by the differential equation of Louis Marie Henri Navier and Georg Gabriel Stokes, which allows the mathematical description of an instationary, frictional, three-dimensional flow field.

An analytic solution of this differential equation can only be provided in simple cases. The direct numerical solution of complicated flow fields is not suitable for industrial applications due to the enormous requirements on computer output and storage capacities.

It was only after engineers succeeded in describing the instationary flow spectrum by suitable statistic turbulence models that IT requirements could be reduced to practicable and commercially acceptable dimensions. From the nineties, three-dimensional and frictional flow fields could, for the first time, be calculated with the CFD method.

Independent of the individual task, this calculation process includes several fixed stages. These are illustrated by using a three-rim converter as an example.

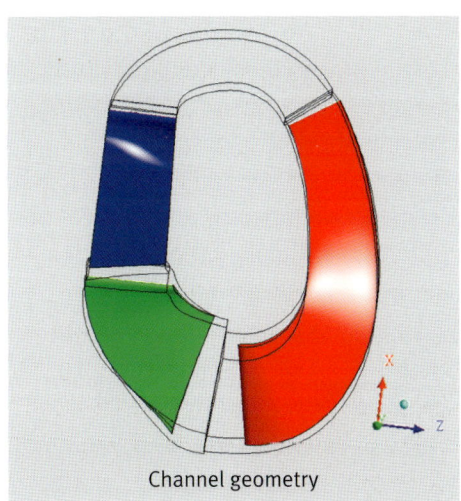
Channel geometry

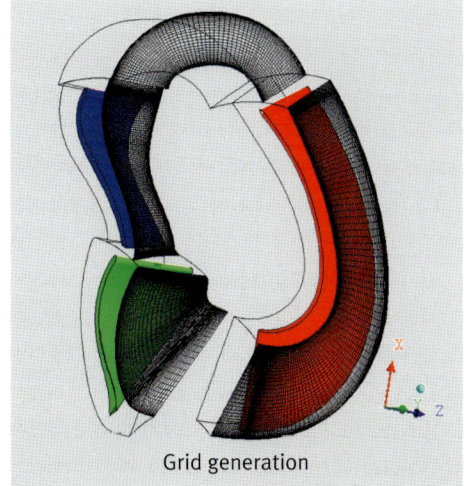
Grid generation

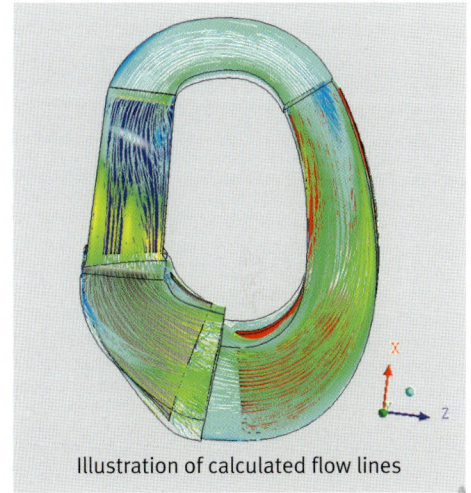
Illustration of calculated flow lines

The calculations are carried out on high-performance computers or a cluster of PCs. Currently available statistic turbulence models can usually be applied only for a certain class of flows. The continuous optimization of these turbulence models is and will be a focal point of further developments in CFD technology.

Due to the geometrically defined flow channels and the fact that these circuits are always operated when completely filled, torque converters are the most accessible Föttinger unit for numeric flow simulation at the current state of CFD technology. The deviations between numerically and experimentally established values are very low, especially at the nominal operating point, and show adequate comparative results across the further development of the curve.

CDF stages:
- *Determination of channel geometry (flow field)*
- *Grid generation (computing volume cells)*
- *Preprocessing (operating media, operating data, marginal conditions)*
- *Numerical calculation*
- *Postprocessing (evaluation and illustration of operands, pressures, torques, speeds, etc.)*

CFD technology has therefore become an integral part of the development process of hydrodynamic torque converters, offering the following advantages:

– CFD contributes to the understanding of relevant flow phenomena and design parameters and opens up new possibilities of utilizing flow characteristics.
– CFD allows comparing and analyzing various design variations, even if absolute qualitative statements might deviate
– CFD contributes to shorter development times: developments can be realized faster and more cost-effectively and require less experimentation
– CFD allows the supplementation of design regulations with new, effective variations by "copying" positive flow characteristics of well running machines.

Unlike with torque converters, the flow paths of couplings and brakes are not rigidly determined by guiding channels. Additionally, for the purpose of characteristic curve adaptation, these units are operated partly filled, resulting in two-phase flows (oil-air or water-air) with free surfaces. Attempts at simulating this condition numerically have so far not shown satisfactory results. For this reason, supplementary experimental examinations are carried out, in order to allow the application of CFD technology also for these Föttinger units and illustrate the separate fluid phases.

Gamma-Tomographic Evidence of Fluid Distribution in Partly Filled Couplings

After it was discovered in the thirties that couplings and brakes show stable characteristic curves with free flow surfaces in the blade grid even if they are only partly filled, design engineers have always wanted to observe this flow development and describe its relevant effects. Apart from optical methods of flow monitoring in the form of sight glasses and evaluations by video recordings, engineers today use quantitative speed measurements after the Laser Doppler method.

Gamma tomography is a modern qualitative process that can visualize the highly disrupted fluid flow during low partial fillings and high slip. This method has been developed by the Institut of Safety Research at the Rossendorf research center in Dresden.

With a razor-sharp (Greek; tomos) bundle from a source of gamma rays, the test object is penetrated vertically to the rotational axis. In a detector arc, the hitting gamma quantums are converted into electric impulses and evaluated statistically on the basis of a defined measuring cycle. An image of the metal contours has been provided by previous measurements without operating fluid. The qualitative fluid distribution is created by extractions from the operative measurement and the pre-measurement.

Work is underway to refine this method and achieve higher picture resolutions. Through these activities Voith expects to gain an additional tool for qualifying the CFD method for partly filled couplings and brakes.

Test coupling 422 SVW.
Operating medium water.
The image shows the fluid distribution on level 1 (impeller) and 2 (turbine wheel).

Qualitative evaluation:
Red = Water (high density)
Yellow = Water/air mixture
Light blue = Air with water
Blue = Air

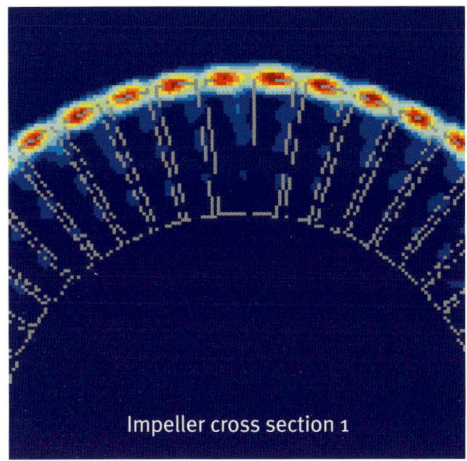

Impeller cross section 1

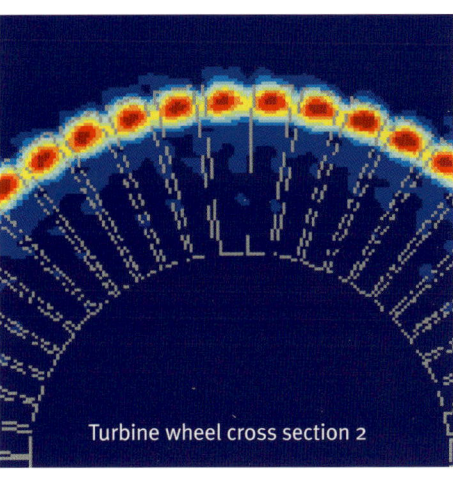

Turbine wheel cross section 2

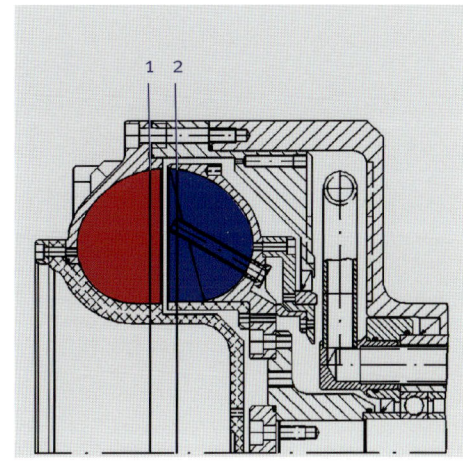

Multi-Body Simulation of Hydrodynamic Drive Units

Multi-body simulation (MBS) allows making statements about the dynamic behaviour of complete drivelines and systems already during the development stage. Depending on the depth of integration of Voith units into the overall system, the application of MBS varies from Product Group to Product Group.

Following requests by vehicle manufacturers, the Market Division "Rail" has evolved into a systems supplier in recent years. Voith now designs and supplies complete drive systems for rail vehicles with mechanical and hydrodynamic transmissions including engine, cooling system, carrier frame and electronic control. The configuration, adaptation and optimization of such systems required the application of appropriate developing and simulation methods and building up necessary systems-related know-how.

Multi-body simulation is a key element in this context, allowing statements about dynamic loads, as well as the vibrational and acoustic behaviour of a complete system already at an early stage. In a subsequent test phase, this dynamic behaviour can then be examined further by technical measurements and, where necessary, improved.

As an extension of pure torsional vibration calculations that have been carried out by Voith for many years, multi-body simulation does not just calculate the rotational variances of the driveline. It also enables the integration of all other variances, especially those of a translatory nature, and their influence on the overall systems behaviour.

Apart from isolated observations, multi-body simulation also allows incorporating the complete driveline into the vehicle and examining the interaction between vehicle and drive system. A complete MBS model is composed in stages from the components driveline, bogie and vehicle as a whole.

The example on the next page shows the simulation of the secondary driveline of a railcar with turbo transmission, cardan shafts and axle drives, integrated into the vehicle including bogie and chassis. On the basis of such a model, the kinematics of the cardan shafts can be illustrated, and loads in the driveline during different driving situations interacting with the dynamics of the chassis can be established. Additionally, special load conditions, for example wheel spin or sliding, can be simulated.

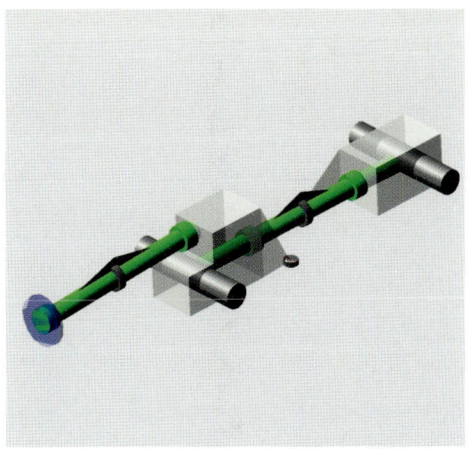
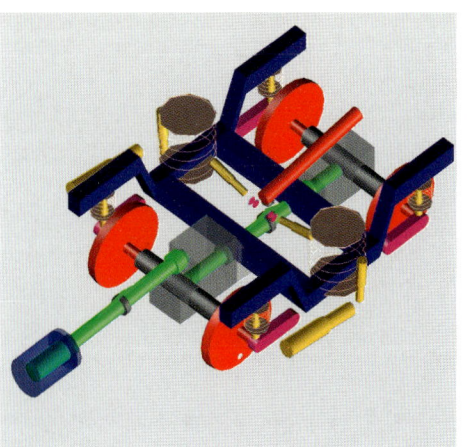

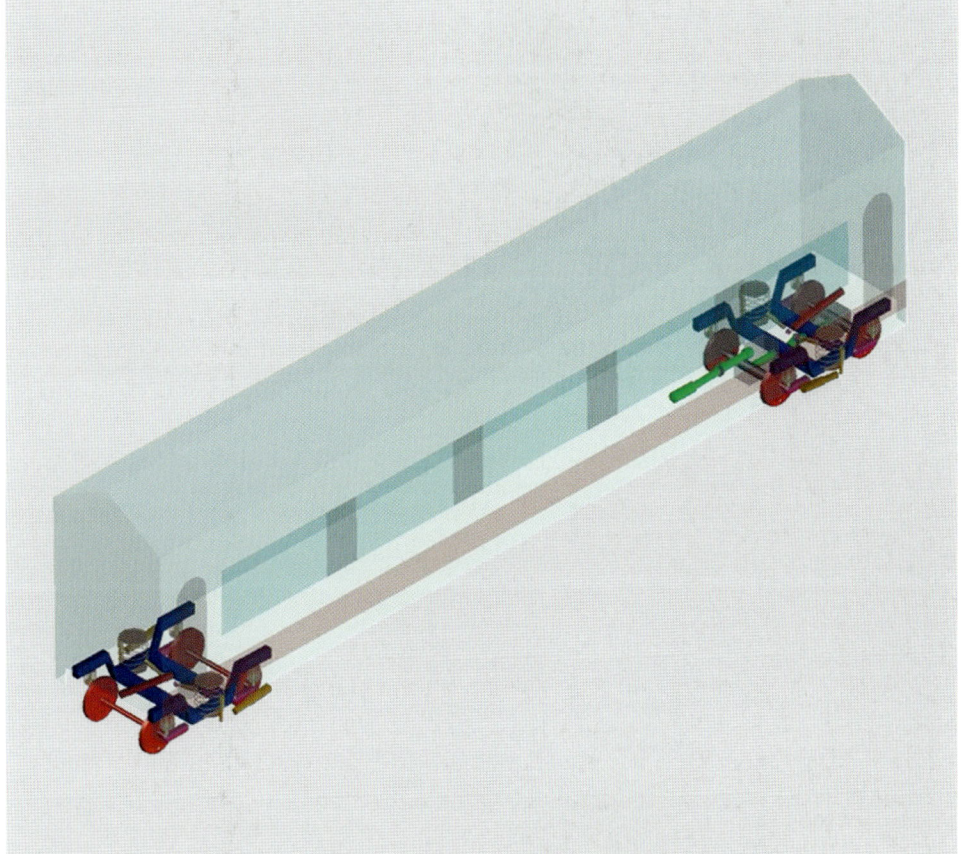

Simulation of cardan shaft kinematics when cornering.

During the examination and optimization of the dynamic behaviour of a diesel-hydraulic drive system for railcars on a test stand, development times could be significantly shortened and results could be greatly improved as a result of multi-body simulation combined with observations of structure-born sound. Owing to the extensive modeling of the support structure, optimizations can be simulated and their effects on the systems behaviour can be tested. In practical operation, these tests would have been highly cost and labor-intensive. On the prototype, only the most effective optimizations have to be realized.

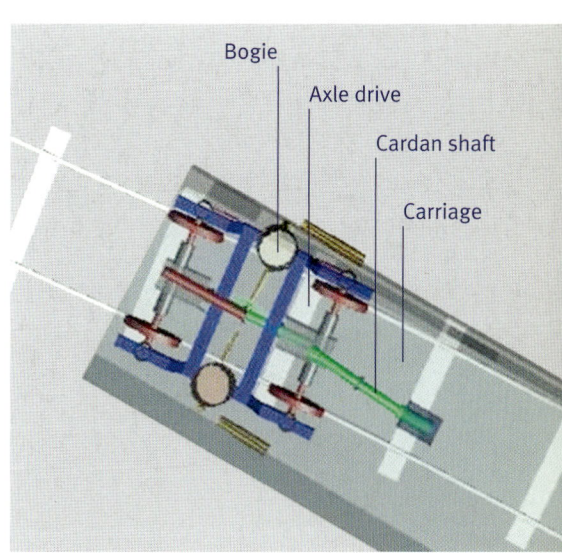

Bogie
Axle drive
Cardan shaft
Carriage

273

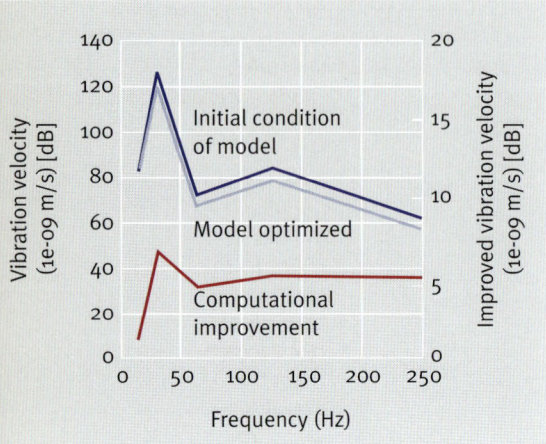

Octave band spectrum of vibration velocity. Modeled initial situation, compared with test stand measurements. Optimized model and computational improvement. The improvement was examined at the test stand by comparative measurements of airborne sound.

Right: Hybrid MBS model (modeled construction of test stand).

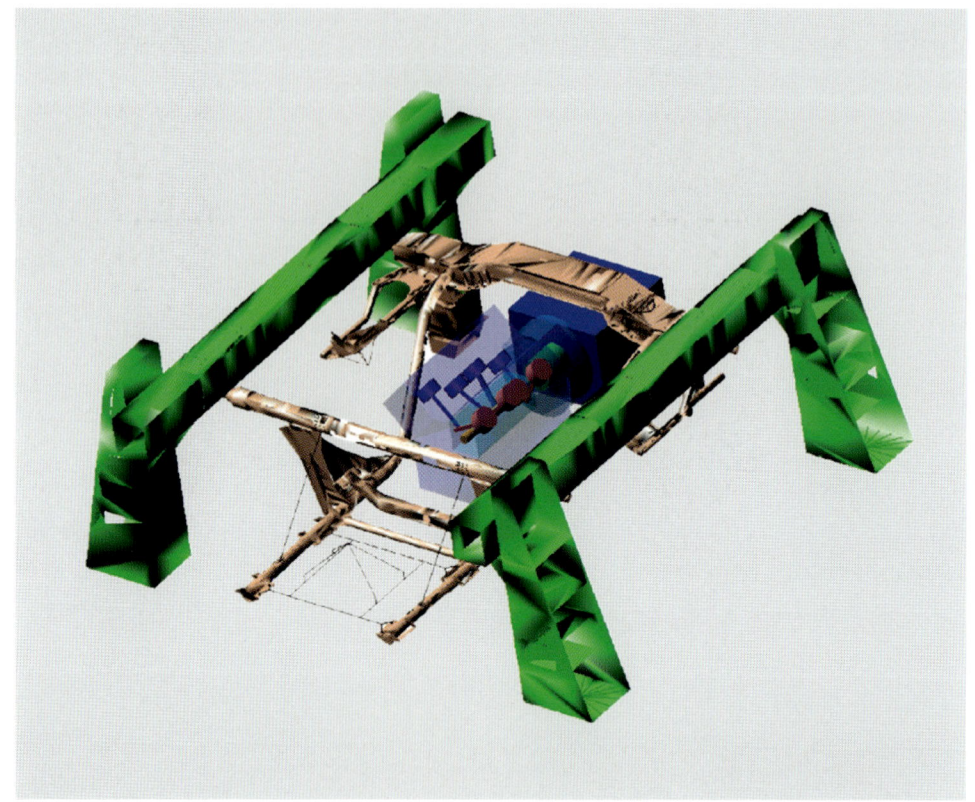

Extended Development Competence by Cutting-Edge Test Stand Technology

Friedrich Voith recognized at an early stage that the development of large turbines with high outputs was impossible without appropriate testing equipment. Small beginnings eventually resulted in the test laboratory for water turbines, the "Brunnenmühle". This was also the place where the first Föttinger units were developed, to which Voith's practical observation equally applied. Soon after, independent test fields were built for the products of the Power Transmission Division in the Heidenheim and later also the Crailsheim works.

Today, the tasks are far more complex. Apart from pure product and function tests, complete systems concepts and structures have to be tried out prior to regular operation. Against this background, one of the world's most modern, multi-functional tests stands with an output of 6.3 MW was commissioned in Crailsheim during the Föttinger anniversary year 2005. The planning concept for the stand envisaged a universal system for a wide variety of products and applications.

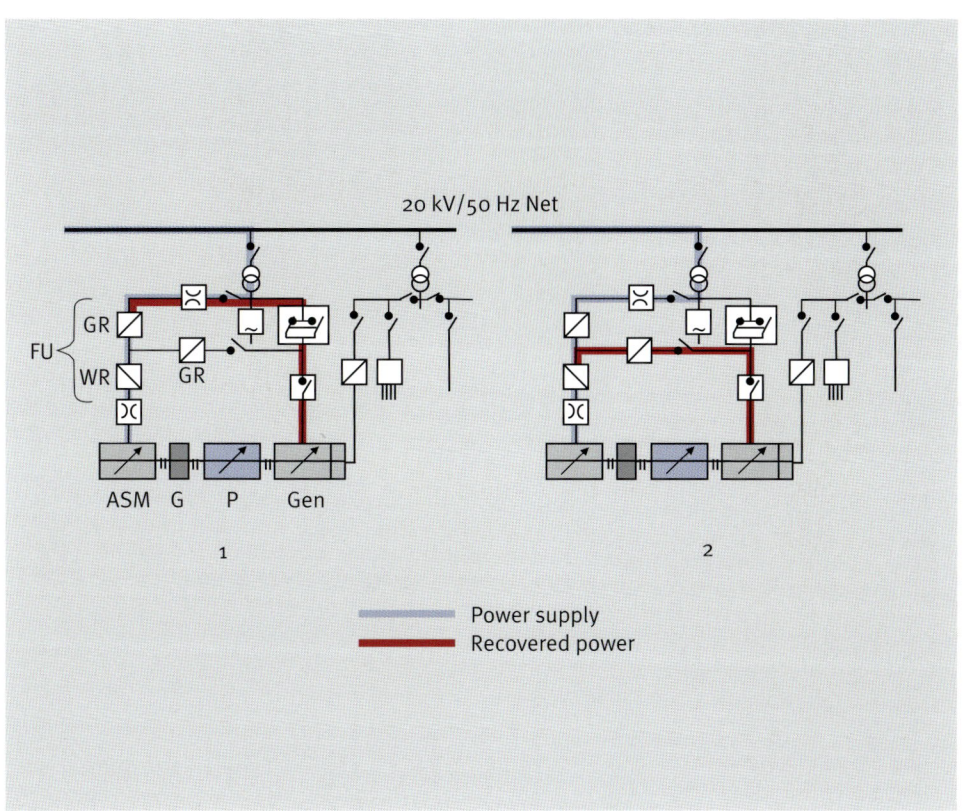

20 kV/50 Hz Net

GR
FU
WR   GR

ASM   G   P   Gen

1   2

━━━ Power supply
━━━ Recovered power

At the foreground was the task of simulating the operation of a wind power station with the variable-speed, super-imposed transmission WinDrive under quasi-realistic conditions. For this purpose, the drive of the test stand has to provide a dynamic wind profile with load peaks of up to 5.0 MW and be capable of generating speed variations of at least 300 min^{-1} per second. As in subsequent real operation, the synchronous generator as energy converter should be coupled solidly to the net frequency during these function and development tests.

With this design, the variable-speed characteristics of the WinDrive with their load-reducing effect on the driveline, as well as the control links from synchronization up to blade adjustment (pitch control of the propeller), can be simulated in overload operation and subsequently optimized.

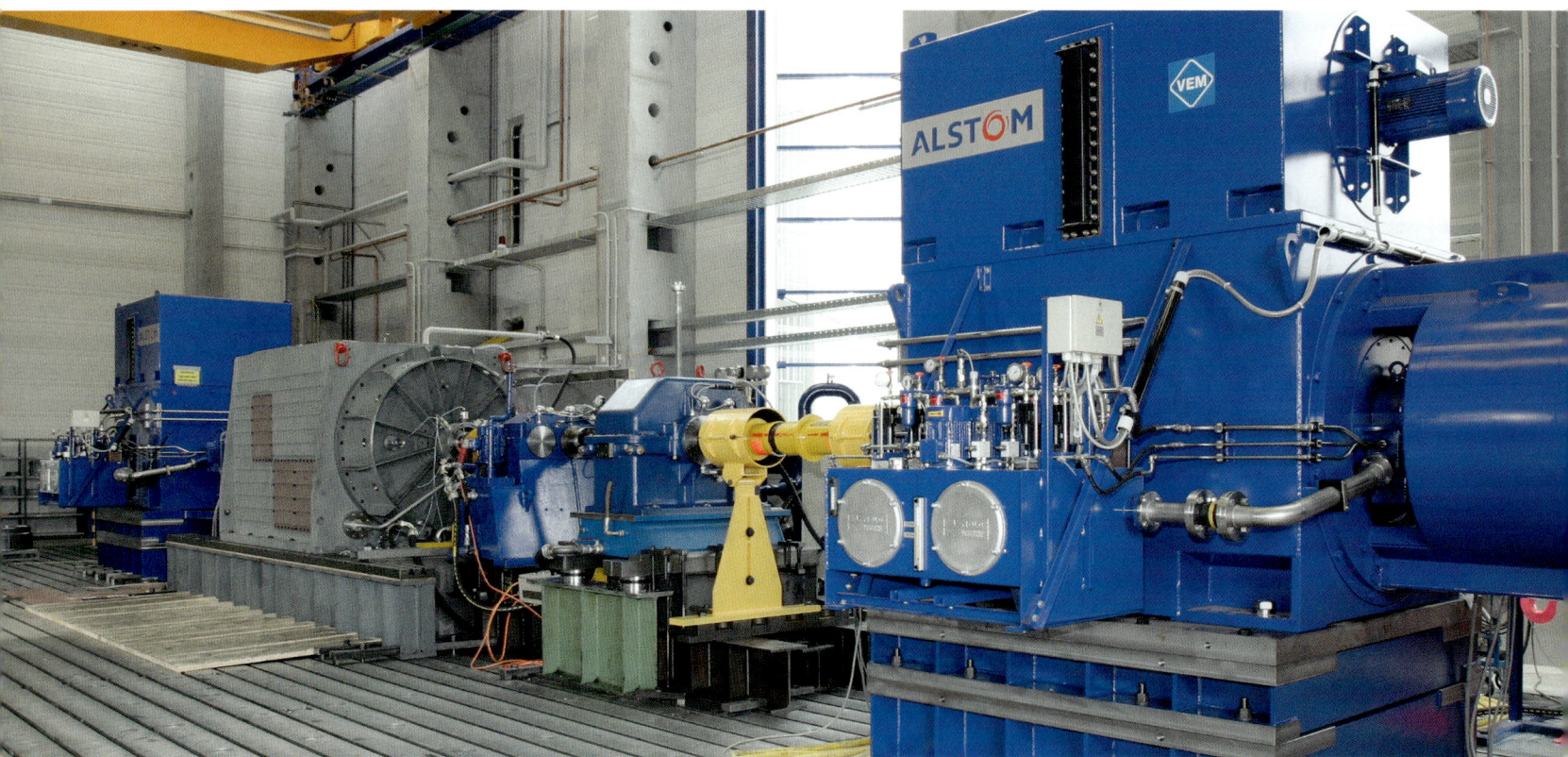

VORECON test run using the generator as a load unit. Multi-functional 6.3 MW test stand for drive components.

Further planning requests for higher transmission outputs were made by other products in anticipation of future developments and innovations. For such tests, the motor can be run primarily at constant speed, while the generator acts as a variable load unit. The electricity produced by the generator with variable frequency is converted in the rectifier module and fed back to the direct current level of the frequency inverter on the motor side.

After the analysis of future tasks, the test stand had to fulfill the following demands:

- Generator output must be 6.0 MW
- Dynamic variations ($n \geq 300$ min^{-1}/s) must be possible
- Characteristic control in the four quadrant area required
- Plant and test stand control must be integrated (linkage of control structure)
- Energy recovering load unit with two operating modes.

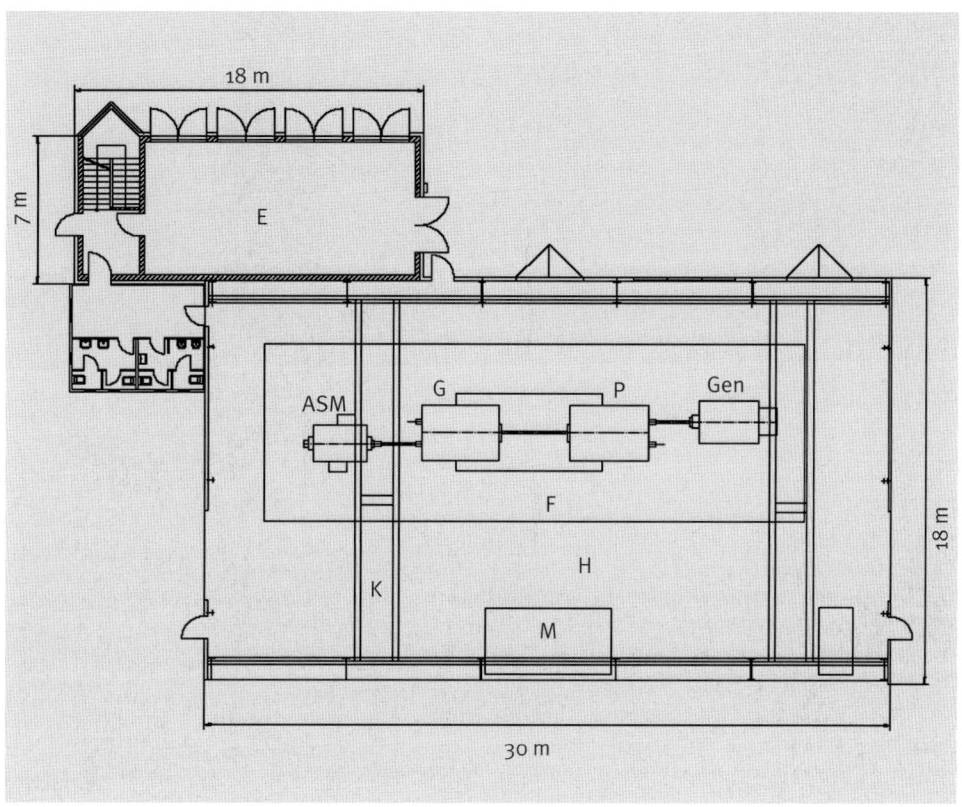

18 m

7 m

E

ASM

G   P   Gen

F

H

K

M

18 m

30 m

Building outline with arrange-
ment sketch.
E = Energy building (4 storeys)
H = Test hall 545 m²
M = Measuring cabin 6 x 3 m
F = Plate area 192 m²
       Concrete thickness 3 m
       Point load ± 3.2 MN
K = Bridge crane 100 t
       Lifting height 10 m
P = Test object
G = Adaptation transmission

The wide-ranging planning directives resulted in a highly flexible overall test stand concept. Both electric machines can be fixed on the 24 m x 8 m slotted plate area in any position; stepped sub-modules allow variations in height. Adopting a method that has proven itself in mining environments, the main machines are linked to the 3 kV supply cables by plug-in connections, so that the assembly groups are not impeded by cable harnesses. The measurements and load values stated in the hall plan signify the high flexibility of the test hall for different test projects.

The measuring signals are set, converted and digitalized on site in a measuring cabinet close to the test object and then forwarded to the measuring cabin, where they are processed and displayed. The machine control and the measuring unit of the test object operate as separate units, coupled only on the basis of safety-related aspects.

Rolf Besserer

Voith Turbo

The Development into a Group Division

With Technical Progress Toward Success

The Group Division Voith Turbo evolved from the turbine construction business of the Heidenheim plant. Today, in fiscal year 2004/2005, it is the second largest Division within the Voith Group after Paper Technology, with sales of EUR 818 million and 3900 employees. With a share of some EUR 550 million, hydrodynamic components, including turbo transmissions, DIWA transmissions, all kinds of turbo couplings and retarders, represent the core business in the Power Transmission Division of the Voith Group.

The last decade of the 20th century at Voith was characterized by joint ventures, acquisitions, and corporate reorganization. After the corporate assets were split between the families of Hermann and Hanns Voith in 1992, the joint venture of Voith and Sulzer in the Paper Technology Division in 1994 was the first step toward the corporate restructuring from a centrally managed organization into a group of companies with Voith AG in Heidenheim as their holding company. The Business Divisions of the then J. M. Voith GmbH in Heidenheim were separated into independent enterprises. This resulted in the Group Divisions Voith Paper Technology, Voith Hydro (later Voith Siemens Hydro Power Generation), Voith Turbo and Voith Industrial Services.

Voith Turbo GmbH & Co. KG comprises the plants at Heidenheim, Crailsheim and Garching. This company, based in Heidenheim since 1996, operates as the head organization for Voith Turbo and as holding company for the individual companies of the power transmission division. The designation "Turbo" was first used by Voith for hydrodynamic transmissions and couplings in 1932 – long before power transmission acquired its current status and importance within the Voith Group.

This new group organization, based on the knowledge and the expertise of the key technologies represented within Voith Turbo, went hand in hand with the restructuring into Market Divisions: in order to ensure optimum internal and external organization, competencies and responsibilities were allocated in line with these Market Divisions. The Market Divisions "Rail" and "Road", as well as "Marine", were assigned to the location Heidenheim with the plant in Garching. The Market Division "Industry" is based in Crailsheim.

As a result of the technical progress and the commercial successes of Voith hydrodynamics, new production sites and plants were added over the years. To illustrate sales volumes, the following table shows the total number of individual products delivered to date:

Market Division	Product	Application	Number of Units
Rail	Turbo transmissions	Rail vehicles	40 000
Road	DIWAbus, DIWA transmissions	Buses	200 000
	DIWAmatic transmissions	Fork lift trucks	80 000
	Retarders	Coaches, Trucks	350 000
Industry	Turbo couplings	Engineering,	1 100 000
	Variable-speed couplings and Torque converters	Energy technology, Transport industry, etc.	30 000

Today, Voith power transmission technology is represented by over 30 Marketing and Service Companies.

The Beginnings of Hydrodynamics in Hydro Turbine Construction in Heidenheim

The establishment of the department for turbo transmissions in 1932 marked the introduction of the prefix "Turbo" as a product name, which was later also used for couplings. During the first few years, high development costs and innumerable tests were unavoidable, and the commercial success of hydrodynamics was doubted. However, the company owners, Walther, Hermann and Hanns Voith continued to pursue the new path. When demand began to rise after 1937, they saw their entrepreneurial vision confirmed.

From 1932, the design and development department for turbo transmissions and Voith-Sinclair TG/VSK couplings was located in the "Barracks" behind the old administration building, while they were produced in two manufacturing buildings next to the large turbine halls. In 1957, these halls were replaced by sheds 330 and 331. Tests were carried out in the "Kriegerstadel", a small outbuilding in the adjacent former "Gärtnerei Krieger" in Ulmer Straße. This temporary arrangement lasted until the transmission plant was built in 1953.

Test stand in the "Kriegerstadel".

New Transmission Plant in Alexanderstraße in Heidenheim

After 1945, the demand for turbo transmission increased rapidly. "When the old steam locomotives in Germany were gradually withdrawn after the Second World War to make way for modern diesel locomotives, Voith experienced a real boom", writes Hugo Rupf, the future Chairman of the Board.

The new "Transmission plant" in Heidenheim.

Soon the production capacities could no longer handle the demand from locomotive manufacturers, and late deliveries were the order of the day.

To eliminate this unsatisfactory situation, Voith decided in 1953 to build a new plant for transmissions and couplings away from Alexanderstraße. A new modern factory with a production area of 12 000 m² that was later extended to 18 000 m² was built. Standard components were manufactured in series and in advance, which resulted in drastically reduced delivery times.

Partial view of the Voith plant in Heidenheim with the new transmission factory on the left.

Three years later, another plant was built, due to the indication from the mining industry that there would be a massive demand for turbo couplings.

A lack of available space and the exhausted labor market in Heidenheim did not permit another extension at the company's headquarters. At the time, Voith employed 7,000 people in Heidenheim. The search for a suitable location ended with the decision to build a plant for turbo couplings in Crailsheim, some 60 kilometers north of Heidenheim. Production at the new plant started as early as 1957.

With the establishment of Voith Getriebe KG in 1961, the transmission plant in Heidenheim was legally and organizationally separated from the company's hydro power and paper machine engineering activities. Unlike plant engineering that is characterized by individual production, transmissions could be mass-produced and required a different organization, in order to survive in the market.

This requirement also applied to the next manufacturing plant in Garching near Munich. A "Special Factory", as Hanns Voith called it, for automatic transmissions for the vehicle industry was needed. The transmission plant thus had a separate branch for this product line. Production in Garching started in early 1963.

This means that within ten years, three new manufacturing plants were built for hydrodynamics. The production area increased from 12 000 m² in 1953 (transmission plant) to 40 000 m² (transmission plant 18 000 m², Crailsheim 10 000 m², Garching 12 000 m²).

In 1982, Voith Getriebe KG was re-integrated into J. M. Voith GmbH, "in order to emphasize the special importance of power transmission as one of the company mainstays", as the company magazine "Voith-Mitteilungen" described it at the time.

The two factories in Heidenheim and Garching remained "Independent Business Divisions" of J. M. Voith GmbH until 1996.

Today, Voith Turbo GmbH & Co. KG employs 1 050 people in Heidenheim.

1953 New "Transmission plant" at Alexanderstraße in Heidenheim for turbo transmissions and turbo couplings
1957 Production of turbo couplings transferred to Voith Turbo KG, Crailsheim
1961 The "Transmission plant" in Heidenheim now operates under Voith Getriebe KG, Heidenheim
1963 Voith Getriebe KG opens a new plant in Garching near Munich
1982 Re-integration of Voith Getriebe KG into J. M. Voith GmbH in Heidenheim

Hydrodynamics in St. Pölten

The impetus of entering the market for hydrodynamic power transmission for rail vehicles was provided by the works in St. Pölten in 1932. Only one year later, the first delivery of 24 turbo transmissions for diesel locomotives of Austrian State Railways (ÖBB) took place. By 1945, 350 turbo transmissions had been built in St. Pölten.

In 1946, the plant was occupied by the Soviet Army (USIA operation). Contact with the head quarters in Heidenheim was stopped, the existing equipment appeared to be insufficient for an immediate restart of the production activities. It took until 1954 before the situation changed, when locomotive manufacturers in Austria and the Eastern Block countries needed turbo transmissions for mod-

The plant of J. M. Voith AG in St. Pölten, Austria.

ernization projects. Transmission production eventually started again in Trais-mauer, another plant in St. Pölten under the administration of the Red Army.

In 1968, when the modernization wave tapered off, the plant in Traismauer was closed. Between 1954 and 1968, some 1 700 turbo transmissions had been built there. The production equipment was transferred to the Voith plant in St. Pölten.

After the state contract and the withdrawal of the occupying forces, the plant in St. Pölten was handed back to the Austrian Republic in 1955.

In 1958, J. M. Voith AG, St. Pölten was founded by Österreichische Landes-bank with the support of the Voith family. This created a new foundation for cooperation with the headquarters in Heidenheim. As it had done before the war for paper machines and turbines, J. M. Voith AG now assumed responsibility for the "Eastern markets", including power transmission products. In 1962, a new agreement with Voith Turbo KG Crailsheim for variable-speed turbo couplings was formed. In the 2003/2004 fiscal year, production of these couplings in St. Pölten stopped and reincorporated into the program of the Crailsheim facto-ry. By 2003, St. Pölten had delivered 400 variable-speed turbo couplings for the drives of boiler feed pumps in thermal power stations in China, paving the way for today's highly successful business in this country. Until May 2005, the plants in St. Pölten and Traismauer had produced some 4 000 turbo transmissions.

As a result of the political changes in the Eastern Block and the fall of the Iron Curtain, Voith was confronted by a completely new situation in November 1989. Markets had to be reorganized – an undertaking that could now finally be real-ized. Voith in Heidenheim had begun at an early stage to acquire the shares in J. M. Voith AG, St. Pölten from Bank Austria (formerly Österreichische Landes-bank) and also purchased free-floating shares in the stock market.

The restructuring of the Voith Group also meant that the business divisions at St. Pölten had to be represented by new companies. The power transmission division of J. M. Voith AG (later "Voith Austria Holding AG") in St. Pölten was integrated in 1998 into the newly founded "Voith Turbo GmbH & Co. KG, St. Pölten" and allocated to the Group Division "Voith Turbo".

1903 Establishment of J. M. Voith Maschinen-fabrik und Gießerei, St. Pölten
1933 Production of the first turbo transmissions for railbuses
1946 Plant is put under Soviet administration (USIA)
1954 Turbo transmission production starts at Traismauer (near St. Pölten)
1955 Plant under Austrian control after the state contract comes into force
1958 Establishment of J. M. Voith AG in St. Pölten
1968 Plant in Traismauer is closed
1998 Power transmission converted into an independent business division and estab-lishment of Voith Turbo GmbH & Co. KG, St. Pölten

Crailsheim plant in 1957.

Voith Turbo KG, Crailsheim

In 1957, after experiencing severe war damage in April 1945, the town of Crailsheim was in the process of reconstruction. Voith had chosen this location due to its proximity to Heidenheim, the availability of sufficient land alongside a former air base and particularly because of the high number of qualified workers. For the then relatively undeveloped region, the establishment of Voith was a godsend.

On a site in the western part of the town, a factory with a production area of 4 000 m² was built during the first construction phase. It was initially used for manufacturing turbo couplings for mining applications and road vehicles ("Daimler Coupling"). As early as 1962, the plant was extended by another 6 000 m². Since then the location has been continuously expanded by new buildings and the acquisition of additional land. The reasons for this ongoing growth were the good business development of turbo couplings, the transfer of new hydrodynamic products from Heidenheim to Crailsheim (1962 variable-speed turbo couplings, 1963 geared variable-speed turbo couplings, 1973 retarders for coaches and trucks, 1986 starting converters for gas turbines), as well as the expansion of stationary applications with other power transmission components. Each of the products has its own success story.

In 1967, ten years after its foundation, Crailsheim employed a workforce of 550.

The production of variable-speed and gear variable-speed couplings for higher outputs and speeds, as well as the search for new applications and variations of turbo couplings, instigated by the mining crisis in the second half of the sixties, provided Crailsheim with the hydrodynamic competence that proved so beneficial in 1973, when it came to designing new retarder concepts, and from 1986 for the technical revision of starting converters for gas turbines.

Since the seventies, business was increasingly dominated by export. This development went hand in hand with the establishment of marketing companies in foreign countries. Alongside the marketing companies that already existed throughout Europe, the first such subsidiary in the USA in 1974 represented an important milestone.

Increasing energy requirements and the construction of thermal power stations all require hydrodynamic products made by Voith Turbo – from turbo couplings for underground mining to multi-stage variable-speed drives for off-shore technology. In 1986, a production site for variable-speed couplings for applications in thermal power stations was established in Hyderabad, India, which currently employs 200 people.

Today, the plant in Crailsheim is the head of the Market Division "Industry" of Voith Turbo GmbH & Co. KG. When founded in 1957, it employed 229 people; this number has now increased to approximately 900. During this time, sales rose from DM 13 million to EUR 250 million.

1956 Establishment of Voith Turbo GmbH & Co. KG. Building work begins.
1957 Production of turbo couplings for coal mining applications starts
1962 Production of variable-speed turbo coupling starts
1963 Production of geared variable-speed turbo coupling starts
1973 Retarders for coaches and new retarder developments for trucks
1974 Land acquisition at Hardtstraße for the Product Group "Retarders" and construction of the retarder workshop
1981 Staff moves into new administration building at Hardtstraße
1985 Development of the Vorecon multi-stage drive
1986 Establishment of Voith India Private Ltd, Hyderabad, India, for local production of variable-speed couplings
1992 Relocation of retarder production to Garching
2004 Developments for utilization of wind energy start
2005 Official opening of the test stand for wind power transmissions with an output of up to 6 300 kW

The Crailsheim plant today.

The Plant in Garching

The decision to build a new plant for automatic transmissions was made following Voith's entry in the market as a supplier to the automotive industry. "In this particular case we had to anticipate that our capacities would be determined by the automobile industry", Hugo Rupf wrote later. A site close to a large city was chosen, "Where a few thousand people could be employed if needed." This size of the newly acquired land corresponded to this anticipation.

Yet staff recruitment turned out to be problematic after all. Voith built company apartments, organized commuter services for nearby villages and eventually carried out promotional campaigns in Italy, Yugoslavia and Greece, in order to cover its staff requirements. The newly recruited employees underwent special training programs.

At the beginning of 1963, Garching saw the first completed transmission. The opening ceremony for the new plant took place in the presence of Hanns Voith and local dignitaries on 30 September 1963. Today, 140 people work on a production area of 12 000 m².

Initially, the plant concentrated exclusively on the production of DIWAbus transmissions types 200S/501. In 1965, this was followed by fork lift truck transmissions and in 1969 by transmissions for construction machines. In 1970, internal gear pumps were added, followed by the new Voith DIWA automatic in 1975 and the Certomatic C 845 transmission in 1983. With the "Hydrodamp" torsional vibration damper for cars and the Voith Retarder for coaches and trucks, further series products from the power transmission sector were moved to Garching.

Garching plant in 1968.

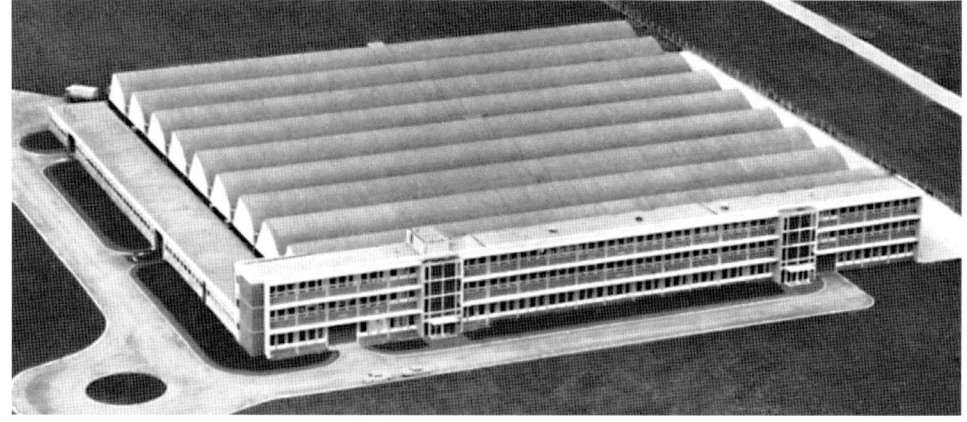

Simultaneously to the expansion of the production program, several extensions and extra buildings were necessary and manufacturing area was increased to just under 35,000 m². Today, the Garching plant employs 400 people.

The Future of Hydrodynamic Power Transmission Within the Voith Group

Machine building at Voith has developed across more than 150 years, based on essential human requirements and in line with technical progress. It is no coincidence that Voith uses its technical expertise ranging from papermaking over hydropower to power transmission equally successfully. Paper, energy and drives create the foundation for this success in a market, where demands from users and operators alike on technical efficiency are constantly growing.

Hydrodynamics are an excellent example of this. From Föttinger's first theoretical considerations of a marine drive up to today's wide variety of applications in mobile and stationary areas, we have come a long way, a way that will continue to lead Voith Turbo into a successful future of hydrodynamics.

Wilhelm Hahn (1882 – 1939)
Born in Stuttgart-Bad Cannstatt.
Engineering studies and doctorate in engineering at Stuttgart Technical University.
1921 Joined Voith in Heidenheim as Manager of Department for Water Turbine Development ("Impeller Office").
1926 Member of the management team of the Turbine Construction Department. This team, consisting of several people with equal rights, succeeded the previous individual directorship of Albert Ungerer.
1928 Additional responsibility as head of the Design Department for water turbines, pumps, hydrodynamic couplings (from 1932 also hydrodynamic transmissions) and Voith Schneider propulsions.
1930 Technical Director of the Turbine Construction Department.
1937 Hahn leaves Voith.

Wilhelm Hahn was a highly creative engineer not just in his special field – water turbine construction. The first start-up and synchronizing coupling for Herdecke pumped storage power station was built under his auspices in 1929. From then on he focused specifically on the development of turbo transmissions and couplings. In 1932, it was down to his initiative that the Department TG was founded, and he selected Hans Faic Canaan as its manager. The co-operation with Harold Sinclair and the expansion of TG by the division VSK in 1934 are also a result of his initiative. Yet the ensuing development of transmissions for special vehicles turned out to be highly expensive. Hahn's research ambitions, robustly and assertively put into practice in several areas, did not lead to the commercial results expected by the Board of Voith and eventually resulted in him leaving the company.

There are numerous publications by Wilhelm Hahn. From 1925 to 1927, he was Chairman of the "Brenz Group of the Baden-Württemberg Association of Engineers" – a sub-organization of the Association of German Engineers (VDI).

Heinz Höller (born in 1941)
Born in Hagen, Westphalia.
After an apprenticeship as a locksmith from 1961 to 1964 engineering studies at the State Engineering College in Hagen.
1964 Joined Voith Turbo KG in Crailsheim as design engineer of the S-Coupling Department. Contributed to the development of the new SV series of self- and externally supported fill-controlled couplings and to the new construction of geared variable-speed couplings.

1979 Manager of the Test Department: Intensive development of constant-fill couplings for belt conveyors by optimizing the mixed profile. Development and introduction of the annular chamber, patent holder.
1993 Manager of the Central Engineering Department, coordinator of hydrodynamic core competences for the Group Division Voith Turbo.

Heinz Höller has been highly instrumental in the successful further development of all three hydrodynamic drive components – couplings/variable-speed couplings, converters/variable-speed converters, hydrodynamic brakes/retarders. He consistently pursued the vision of hydrodynamic applications that had been started by Willibald Meyer. In 1997 he became the central contact for all questions related to this technology. The establishment of the "Hydrodynamic core competence" is closely associated with his name. His work has been accompanied by countless lectures and publications.

Rolf Keller (1913 – 2005)
Born in Stuttgart.
Engineering studies at Stuttgart Technical University.
1939 Joined Voith in Heidenheim to work in the TG Department under Wilhelm Gsching.
1944 Head of Hydraulic Developments, in charge of commissioning turbo transmissions in the "Kriegerstadel".
1950 Construction and responsibility of a new test field in hall 53.
1953 Construction and responsibility of a new, larger test field in the new transmission plant.
1965 – 1978 Manager of the Division Turbo Transmissions and Final Drives.

From 1945, Rolf Keller has firmly put his stamp on the development of converters for turbo transmissions. New starting and cruising converters were designed. During the twelve years of his management, transmission construction experienced a massive upswing, in which he had a major share due to his prudent and visionary decisions. His language skills and his diplomatic manner gained him the trust of German and overseas business partners. He wrote numerous technical essays and held lectures at home and abroad, taking on and continuing the hydrodynamic legacy of Fritz Kugel.

From 1958 to 1967, Rolf Keller was Chairman of the "Brenz Group of the Baden-Württemberg Association of Engineers" – a sub-organization of the Association of German Engineers (VDI).

Fritz Kugel (1899 – 1962)
Born in Esslingen, Neckar.
Engineering studies at Stuttgart Technical University.
1927 Joined Voith in Heidenheim. Worked in the Development Department for Water Turbines ("Blade Office"): Development of turbines and pumps, contributions to the "Herdecke" coupling.
1932 Development of turbo couplings for turbo transmissions and industrial applications.
1934 In charge of Voith Sinclair couplings (VSK). Manager of the Department Turbo Transmissions/VSK from 1937, succeeding Ernst Seibold. From 1935 to 1938 Chairman of the "Brenz Group of the Baden-Württemberg Association of Engineers" – a sub-organization of the Association of German Engineers (VDI).
1954 Appointment as Director of the transmission plant built one year earlier.
1961 – 1962 Technical Director of the newly established, legally independent company Voith Getriebe KG, completing his move away from the water turbine division. In 1961 Honorary doctorate (Dr. Ing. E. h.) from Berlin Technical University.

Fritz Kugel was a charismatic yet impatient and not always easy personality. Quite early he pushed the further development of the products that derived from the license product "Voith Sinclair Coupling", i. e. Voith turbo couplings. After taking over the department in 1937, he was instrumental in the continuous development of turbo transmissions and turbo couplings and their application in all-terrain and track vehicles. From approximately 1948 this was followed by comprehensive research on coupling type SV, resulting in many successful designs. The combination of a transmission and the coupling type SV into the R.K. geared variable-speed coupling was one of Kugel's decisions. It was thanks to his personal efforts that DB German Railways decided in favor of diesel-hydraulics and against diesel-electrics developed in the USA, when steam locomotives at DB were phased out.

His crowning achievement are undoubtedly the large 2 000 HP three-converter transmissions with hydrodynamic brakes for the 4 000 HP locomotives of Southern Pacific in the USA. In addition to his passion for rail vehicle transmissions, Kugel also supported the further development of Voith turbo couplings for applications in the mining and power generation industry, after the responsibility for these products had been transferred to the new plant in Crailsheim. He personally and regularly looked after "his" couplings. In Heidenheim he was always appreciated as a restless provider of ideas, and his colleagues in Crailsheim held him in high esteem for exactly the same reason.

In his acceptance speech for his honorary doctorate, he looked thoughtfully and perhaps knowingly into the future: "Even today, the Föttinger principle of power transmission holds many a secret whose unveiling might be attractive for an engineer. There are enough young people who will push Föttinger's legacy further, in order to apply it as a useful tool for the benefit of mankind." On 4 November 1962, Fritz Kugel died in a traffic accident just outside Heidenheim.

Fritz Kugel was a prolific writer of technical essays and an inexhaustible lecturer at technical events. He had committed himself to the task of passing on scientific and technical findings quite early as a young engineer when he was Chairman of the "Brenz Group of the Baden-Württemberg Association of Engineers" – a sub-organization of the Association of German Engineers (VDI) – from 1935 to 1938.

Willibald Meyer (1921 – 1993)
Born in Poznan.
Engineering studies at Berlin Technical University.
1951 Joined Voith in Heidenheim as design engineer in the department "Hydrodynamic Automatic Transmission" under Wilhelm Gsching. Worked in the test field for turbo transmissions. Contributed to the starting converter for Lünersee pumped storage power plant.
1958 Change to the newly established company Voith Turbo KG in Crailsheim.
1962 Appointment as Technical Director of Voith Turbo KG to succeed Fritz Kugel. Gradual relocation of the Product Group "Hydrodynamic Couplings" from Heidenheim to Crailsheim.
Until 1966 Establishment of an independent development, testing and construction department for hydrodynamic couplings and brakes. State-of-the-art measuring and evaluation equipment.
1971 Taking over the Product Group "Hydrodynamic Brakes" from the Heidenheim plant.
Until 1986 Executive Vice President of Voith Turbo KG in Crailsheim.

Willibald Meyer, one of the generation recruited by Hans Faic Canaan, was a classic, internally effective manager, carrying out daily factory tours and regular weekly meetings. Soon after his appointment as Director he recognized that Crailsheim could be commercially successful in the long term only if it was granted unlimited product responsibility. His high technical competence and his good personal relationship with Fritz Kugel helped him to put his ideas into practice. By assuming responsibility and development competence for two Product

Groups, he created the technical basis for building up good business, especially for couplings. Therefore, the mining market with the quickly developed TVF coupling and the power generation industry with drives for boiler feed pumps, fans and coal mills could be successfully handled. Today, the retarder business is one of the mainstays of the company. His introduction of a quality assurance system in construction further underlined the excellent reputation of the products.

The work of Willibald Meyer had far-reaching consequences. He had created sound prerequisites for new applications that became possible only long after his departure. With his unshakable belief in hydrodynamics as a universally applicable technology on the one hand and his very individual leadership style on the other hand, he has definitely put his stamp on the activities in Crailsheim.

Helmut Müller (born in 1918)
Born in Netzschkau, Vogtland.
Engineering studies at the Engineering College in Zwickau.
In August 1939 he joined Junkerswerke in Dessau. Worked on turbo charger drives for aeronautical engines. First contact with Fritz Kugel.
1950 Joined Voith in Heidenheim after two years at PIV in Bad Homburg.
Manager of the Turbo Transmission Division from 1978 until 1983 as successor of Rolf Keller.

When Helmut Müller, at the time working at Junkers, examined various possibilities of speed regulation for turbo chargers at great heights, he also came across the Voith Sinclair coupling. In his report he nevertheless described it as "fairly unsuitable". At high slip, the "Jet outlets" would only allow a low oil flow. He therefore suggested using an adjustable scoop tube and hence became a co-inventor of the Voith coupling type SV. Fritz Kugel liked the imaginative young engineer. In 1950 he invited him to join the turbo transmission department of Voith.

His activities in this department widened quickly. His first task concentrated on improving an unsatisfactory transmission control system, but subsequently he worked alongside Fritz Kugel at the further development of SVL couplings until 1962, in addition to his regular duties. The results of this co-operation provided the basis for the future development of variable-speed couplings type SC and the geared variable-speed couplings type R.K, for which Helmut Müller was responsible in the early sixties. Another challenge were the torsional vibrations created by diesel engines and the correct design of bogie locomotives with four cardanshaft driven final drives. In the transmission plant, Müller soon became a universally recognized expert on vibration issues. The development of the hydro-dynamic brake for heavy locomotives can also be traced back to him, paving the way for the first retarder for road applications.

Helmut Müller's wealth of ideas and his pragmatic approach have resulted in many technical innovations in transmission and coupling design that are universally applied.

Johannes Peltner (1913 – 2002)
Born in Wittkowitz, Moravia.
Engineering studies, concentrating on power transmission technology, at Berlin Technical University.
1939 Post as Design Engineer in the aeronautical construction department of BMW in Berlin until 1945.
1945 – 1952 Design Engineer at the state enterprise SNECMA in France. Contributions to the jet engines of the Mystère and Mirage planes.
1952 Position at ILO in Pinneberg reporting to Ernst Biefang. Encouraged the company to resume the development of Helmut Weinrich's design of a hydro-dynamic transmission. It should become his life task.
1960 Co-inventor of Weinrich's "Transmission, specifically for vehicles".
1961 After ILO had been acquired by Voith, post as Senior Engineer and head of the department DIWAmatic Transmissions.
1971 – 1978 Manager of the department DIWA Transmissions.

From 1961, Johannes Peltner established the department DIWAmatic at Voith. The principle of the D 851 transmission has been developed by him. Unlike the DIWAbus transmission it had two mechanical gears. This contra-rotating converter that had been considerably improved compared to the original ILO version eventually replaced the conventional Voith synchronized converter. Its efficiency was significantly higher and resulted in drastically improved starting conversion. One of Peltner's special strengths was the strict application of value analysis. The success of the product was largely based on meticulous cost overview and cost control methods introduced by Peltner.

Elmar Rohne (born in 1924)

Born in Riga, Latvia. High school graduation, then war service in Africa followed by captivity.

1946 – 1950 Engineering studies at Munich Technical University. Worked for Siemens until 1953.

1953 – 1958 Scientific assistant at the Institute for Fluid Mechanics at Munich TU, doctorate in 1957.

1958 Joined Voith as development engineer in the transmission plant under Fritz Kugel.

1965 – 1989 Manager of the test field and the converter development department as successor of Rolf Keller. During this period also several years in charge of the DIWA test field at Erchenstraße and the aerodynamic research team at the Brunnenmühle test laboratory, concentrating on tunnel ventilation.

With countless tests, publications and lectures on hydrodynamic power transmission and the conversion of theoretical findings into practical use, Elmar Rohne has made a substantial impact. His conversion formula, the "Rohne formula" on the influence of the size, speed and viscosity of operating fluids on the efficiency of torque converters, is widely known. As a classic work in this field, his book "Hydrodynamics in Power Transmission" published in 1987 is still regarded as an essential reference book. Rohne enjoys high scientific respect among his colleagues. Employees and business partners have always appreciated his calm and well-balanced co-operation.

In 1996, the Association of German Engineers (VDI) awarded him with the Fritz Kesselring Medal for special services in the field of hydrodynamic power transmission.

Hubert Schmölz (born in 1940)

Born in Meckenbeuren near Friedrichshafen.

Engineering studies at Constance Technical School.

1964 Joined Voith in Heidenheim as a design engineer in the transmission construction department.

1966 Design of a large two-converter transmission for 3 000 HP.

1972 Design of a reversing section with cardan joint of the actuating shaft and a new rotary actuator. Hubert Schmölz had this excellent and innovative solution patented. Design of a 720 HP wheel loader transmission with specially mounted bearings for the adjustable guide blades, stiff housing and high performance gears and converters.

1988 Appointed as Chief Design Engineer. Additional responsibility for final drives.

1994/95 Railcar drives for the 460 and 600 kW high power range.
Hubert Schmölz was soon regarded as one of the most gifted junior design engineers in the transmission department. His extraordinary creative abilities went hand in hand with an outstanding talent for realistic market assessment. His high-performance transmissions were commercially successful.

Ernst Seibold (1900 – 1977)
Born in Stuttgart-Bad Cannstatt.
Engineering studies at Stuttgart Technical University.
1923 Joined Voith in Heidenheim as design engineer for Kaplan turbines and large pumps.
1929 New concept of the Herdecke starting and synchronization coupling. Meeting with Hermann Föttinger in Heidenheim.
1932 Head of the new TG group. Development of a hydrodynamic transmission for the railcars of Austro-Daimler-Werke.
1934 Manager of the TG/VSK Department, extended by Voith Sinclair couplings after the licensing agreement with Harold Sinclair.
1937 Left Voith for family reasons to join Ploucquet in Heidenheim.
1953 Rejoined Voith as Chief Engineer and Manager of the department "Turbo Transmissions for Rail Vehicles", responsible for contacts to the central office of DB German Railways in Munich.
1960 Appointment as Divisional Director.
1962 – 1965 Successor of Fritz Kugel as Technical Director of Voith Getriebe KG.

Ernst Seibold was one of the first engineers at Voith who came in contact with the Föttinger principle. He was still a young man when he developed a functioning hydrodynamic start-up coupling and clutch together with Wilhelm Hahn as a replacement of the non-functioning project of such a unit for Herdecke pump storage power station. Föttinger units fascinated him for the rest of his working life. With his calm nature that was so different from that of his superior Fritz Kugel, Ernst Seibold enjoyed the respect of his employees and business partners. He had a special and trustful relationship with the key customer DB German Railways.

Klaus Vogelsang (born in 1939).
Born in Anklam, Pomerania.
Apprenticeship as a locksmith, 1962 – 1965 Engineering studies at Bergische Universität Wuppertal.
1965 Joined Voith Getriebe KG. Development of rail components.
First contact to the development team for hydrodynamic brakes.

Basic development of a new retarder system for commercial vehicles.
1971 Group Manager for retarder development at Voith Turbo KG in Crailsheim for the application areas industrial brakes, measuring brakes and special designs for shipbuilding and railcars.
1985 Manager of the retarder development department, concentrating on road retarders and their series production.
2001 – 2004 Manager for retarder technology with overall responsibility for development, series production, component and driving tests.

Klaus Vogelsang holds a wide range of patents that are decisive for the success of the Voith Retarder. His ingenuity has turned the retarder for road vehicles into a product that is applied and recognized all over the world. His development methods that were based on new analyses right from the beginning, deviated more and more from obsolete procedures. New approaches to problem solving and consistent, user-oriented strategies that were greatly supported by his openness, paved the way to success. He adopted the same tactics when the market revealed a trend from inline to offline systems presenting many new challenges. The new development of the Aquatarder using the cooling water of the vehicle engine as operating fluid is highly acclaimed in expert circles.

The commercial success of the retarder generations is not least due to the focused discipline with which Klaus Vogelsang realized his ideas.

Georg Wahl (born in 1938).
Born in Ulm.
Engineering studies at Rudolf-Diesel Polytech in Augsburg.
1965 Joined Voith Turbo KG in Crailsheim after working for MAN in Nuremberg and Daimler-Benz in Untertürkheim.
1993 Manager for variable-speed drive technology.
1996 – 2003 Additional sales responsibility for applications in the oil and gas industry.

By improving and extending the functions, increasing outputs and speeds, reducing vibrations and optimizing combinations of variable-speed couplings, transmissions and oil supplies, Georg Wahl succeeded in extending the existing applications of these units and opening up new fields. The fact that he was put in charge of quality assurance in design which led to a significant improvement of the operating safety of these drives, has certainly contributed to this success. From 1982, he was highly instrumental in the development of the VORECON that was afterwards patented in his name and the name of another engineer. In 1986, the Crailsheim plant took over the design and the production of hydrodynamic synchronous converters from the Heidenheim works. After targeted development work, Georg Wahl expanded the application range of this product for sub-sea use ("Subsea") and utilization as starting device for gas turbines ("VOSYCON").

As a senior design engineer, Georg Wahl has played a substantial role in the further development and the success of hydrodynamic variable-speed drives from Voith.

Helmut Weinrich (1909 – 1988)
Born in Greiz, Thuringia.
1927 – 1931 Electric engineering and fluid technology studies at Chemnitz Technical University.
1931 Establishment of his own machine plant with a wind channel in Chemnitz. Research projects for marine and aviation.
1945 Arrested by the Soviet occupying forces.
From 1950 freelance employee at Gutehoffnungshütte (GHH) in Sterkrade. Development of steam turbines and axial compressors.
From 1954 Co-operation with ILO in Pinneberg to prepare patents for contra-rotating converters. Measurements at a trial converter with Professor Karl Kollmann, Karlsruhe TU. First prototypes for motor scooter transmissions.
1956 Move to Pinneberg, continuing as freelance employee for GHH.
1958 Patent application, which was granted in 1960 for "Transmissions, specifically for vehicles" (among the co-inventors is also Johannes Peltner). At ILO development of transmission prototypes for the HS-30 tank and for Ford cars. The developments were later discontinued. The S 833 fork lift truck transmission enters series production.
1961 ILO is taken over by Rockwell, USA. Voith Getriebe KG in Heidenheim acquires the transmission sector. Helmut Weinrich moves to Heidenheim, working as freelance employee in the DIWAmatic Department for Johannes Peltner. In his house in Zang near Heidenheim he also produces very small expansion turbines with high speeds. Helmut Weinrich was a passionate engineer who went everywhere with his slide rule and solved existing problems ad hoc with numerous approximation formulas developed by himself. After his arrest by the Soviets he even built his own slide rule from cardboard strips whilst in captivity. He knew the required logarithms by heart. During this time he had already started thinking about the design of the contra-rotating converter.

The Authors

Wolfgang von Berg
Born in Gerstetten near Heidenheim in 1941.
Apprenticeship as a locksmith at Ansbach Engineering College, followed by engineering studies.
1962 Joined Voith Turbo KG in Crailsheim.
1964 – 1968 Engineering studies at Würzburg-Schweinfurt Technical University.
1968 Rejoined Voith as design engineer in the department "Variable-Speed Drives".
1982 Group Manager in this department.
1991 Manager for customer-specified designs.
1997 – 2003 Engineering manager for variable-speed drives.

Rolf Besserer
Born in Nussdorf near Ludwigsburg in 1939.
Apprenticeship as an industrial clerk, 1963 – 1966 economic studies at Pforzheim College, finishing as graduate business administration (FH).
1971 Joined Voith Turbo KG, Crailsheim, to head the Department Finance and Accounting, Organization, EDP and General Administration.
1986 Manager of the Central Commercial Division.
1994 – 2004 Vice President of the Group Central Division Balances and Taxes (until 1998), Investments and Asset Administration at J. M. Voith GmbH, Heidenheim (since May 1997 J. M. Voith AG, today Voith AG).

Helmut Fleuchaus
Born in Osterburken, Neckar-Odenwald, in 1955.
1976 – 1980 Engineering studies at Ulm Technical University.
1982 Joined Voith Turbo KG in Crailsheim, project engineer and area manager for constant-fill couplings in the Product Group "Start-up Components".
2002 Marketing manager for the Product Group "Start-up Components".

Heinz Höller
Born in Hagen, Westphalia, in 1941.
Apprenticeship as a locksmith, 1961 – 1964 engineering studies at the
State Engineering College in Hagen.
1964 Joined Voith Turbo KG in Crailsheim as a design engineer in the department
"S-Couplings", assistant of the plant management and the technical board.
1979 Manager of the test department.
1993 Manager of central engineering.
1997 Coordinator of the hydrodynamic core competences for the Group Division
Voith Turbo.

Klaus Nolz
Born in Bruck at the River Leitha, Austria, in 1932.
Engineering studies at the State Academy for Technology, Vienna.
1951 Joined Voith in St. Pölten as design engineer in the water turbine
department.
1955 Change to Heidenheim to Voith Getriebe KG, afterwards to Voith Turbo KG
in Crailsheim, to work as a design engineer.
1959 Sales and design manager "Turbo Couplings for Combustion Engines".
1966 Design manager "Industrial Power Transmission Technology".
1986 – 1993 Manager of the Central Engineering Division.

Wolfgang Paetzold
Born in Chemnitz in 1930.
Advanced level in 1949.
Qualification as skilled worker 1951.
Studies of Mechanical Engineering and applied fluid technology 1951 – 1956
at the Technical University of Dresden.
1956 Joined Voith Turbo KG in Crailsheim as sales and design engineer,
responsible for TD couplings.
1959 Change to J. M. Voith GmbH in Heidenheim, turbo transmissions.
1969 Head of project department.
1983 Design manager for turbo transmissions.
1988 – 1995 Marketing and sales manager for turbo transmissions and
final drives.

Hermann Schweickert

Born in Alzey, Rheinhessen, in 1935.

1954 – 1959 Engineering studies at Karlsruhe Technical University.

1959 Joined J. M. Voith GmbH in Heidenheim as design engineer in the hydraulic test laboratory of the water turbine department.

1964 – 1969 Scientific assistant at the chair for fluid machines at Karlsruhe University.

1968 Doctorate in engineering.

1969 – 1998 Rejoined Voith in Heidenheim to take up senior posts in the design and sales departments of the Business Division "Water Turbines".

1998 – 2001 Post-graduate studies at the Historic Institute of Stuttgart University, Department for Science and Engineering History.

2002 Doctorate in philosophy.

Klaus Vogelsang

Born in Anklam, Pomerania, in 1939.

Apprenticeship as a locksmith, 1962 – 1965 Engineering studies at Bergische Universität Wuppertal.

1965 Joined Voith Getriebe KG. Development of rail components. First contacts to the development team for hydrodynamic brakes.

1971 Group Manager for retarder development at Voith Turbo KG in Crailsheim for various application areas (industrial brakes, measuring brakes and special designs for shipbuilding and railcars).

1985 Manager of the retarder development department, concentrating on road retarders and their series production.

2001 – 2004 Manager for retarder technology with overall responsibility for development, series production, components and driving tests.

Bernhard Wüst

Born in Pforzheim in 1949.

1969 – 1976 Engineering studies at Karlsruhe University.

1982 Doctorate in engineering at the Institute for Construction Machines at Karlsruhe University.

1986 Joined Voith Getriebe KG in Heidenheim as design engineer for commercial vehicle transmissions.

1990 Engineering manager for commercial vehicle transmissions.

1999 Development manager for commercial vehicle transmissions.

2001 Manager for strategic projects/industrialization.

Definition of Technical Terms

The definition of the special hydrodynamic terms follows the stipulations per VDI-Regulation 2153 (Edition April 1994) and the structures applied in this publication. Some of the numerous terms that gradually developed, can be found in literature or are in common use at Voith, have been added as general designations, without sub-divisions and design variations, as "Other Names" for the information of the reader. Terms that are protected by Voith are marked (®,™).

Basic Terms of Hydrodynamic Power Transmission

Per VDI 2153, including other names *(in italics)*:

Föttinger Unit
Generic term for hydrodynamic converters, couplings and brakes.

Föttinger Circuit, Föttinger Transformer, Turbo Transformer

Hydrodynamic Converter
A minimum of three bladed wheels that redirect flow in a closed housing
– Single and multi-phase converters
– Variable-speed converters
– Adjustable torque converters
– Clutch converters

Converter (stepless transmission), Torque Converter, Föttinger Transformer, Föttinger Converter, Hydraulic Converter, Hydrodynamic Transmission, Transformer, Turbo Transmission, Turbo Converter).

Hydrodynamic Coupling
Transmission of power by two bladed wheels redirecting flow (pump and turbine). Torque output figures are equal.

Inline Retarder
Hydrodynamic brake in the driveline of a vehicle with coaxial flow of the traction torques, arranged either in front of or behind the transmission.

Input Power
The amount of power available at the input of a hydrodynamic power transmission system required by the driven machine, including losses within the power flow of the transmission process.

Lock-up Coupling
See "Start-up Coupling with Lock-up Device".

Offline Retarder
Hydrodynamic brake in the driveline of a vehicle, installed at a suitable point, similar to a PTO.

Operating Fluid
Energy-transmitting fluid that also fulfills auxiliary functions, depending on the application: heat dissipation, lubrication of mechanical components, control tasks, protection against corrosion. Examples: mineral oils, water, special fluids.

Power to Weight Ratio
Figure composed of the relation of mass (weight) to the output that can be a minimum-value criterion for commercial considerations.

Power-Shift Transmission
Transmission for high operating requirements with output stages that can be shifted under load without splitting the power flow.

Primary Retarder
Hydrodynamic brake in the driveline of a vehicle, arranged in front of the transmission.

Retarder
See "Hydrodynamic Brake".

Safety Coupling

Hydrodynamic coupling that limits the transmitted torque by influencing the flow circuit, or switches off by draining in the event of an overload of the driven machine.

Scoop Tube

Hydrodynamic couplings with external fluid circuits, especially fill-controlled couplings, are almost exclusively fitted with this static head pump acting in a circumferential side chamber.

S-Coupling

The "S" (scoop tube) has been adopted from Sinclair; afterwards used by Voith as the first letter of scoop-tube activated coupling (fill-controlled couplings) designations that can vary, depending on the type. Example: SVL.

Secondary Retarder

Hydrodynamic brake in the driveline of a vehicle, arranged behind the transmission.

Self-Inducing Speed

Speed to which a single-shaft gas turbine must be run up, after which it can accelerate by itself.

Slide Valve

Adjustable device for narrowing the flow channels of fluid converters for the periodic reduction of the transmission rate.

Slip

Parameter (%) of the speed difference (pump – turbine) of hydrodynamic couplings, related to the pump speed.

Starting Converter

Hydrodynamic converter that reaches optimum efficiency at low speeds.

Start-up Coupling

Coupling that fulfills a required function during the start of the driving machine and/or the run-up of the driven machine.

Start-up Coupling with Lock-up Device
Hydrodynamic coupling combined with a mechanical coupling that couples the driveline after the run-up phase of the driven machine, for nonslip operation.

Static Head Pump
Displacement unit at a rotor for the external circuit of the operating fluid arranged in a side chamber, utilizing the static pressure of th circulating fluid ring (see "Scoop Tube").

Step-Up Gear
Speed-increasing gear for adapting the engine speed to the Föttinger circuits.

Storage Chamber
Space in the hub area of the impeller of constant-fill couplings, frequently used for operating fluid, that can be utilized for manipulating the transmission rate, and also during overloads.

Synchronous Converter
Hydrodynamic converter with pump and turbine rotating in the same direction.

T-Coupling
The "T" (traction) has been adopted from Sinclair; afterwards used by Voith as the first letter for constant-fill coupling designations (start-up, safety and adjustable couplings), which can vary depending on the type. Example: TV, TVV, or DT for double circuits.

Torque Converter
See "Hydrodynamic Converter"

Transmission
Device for transmitting and converting mechanical energy, powers or movement (mechanical, hydrostatic, hydrodynamic, pneumatic transmissions).

Turbo Converter
Term coined and frequently used by Föttinger, see "Hydrodynamic Converter".

Turbo Coupling
Term coined and frequently used by Föttinger, see "Hydrodynamic Coupling".

Turbo Reversing Transmission

Turbo transmission with hydrodynamic torque converters for each direction of travel for quick, wear-free directional changes and braking by draining and filling the circuits.

Turbo-Split Transmission

Turbo transmission with hydrodynamic power-split for separate driving of the two bogies of a diesel locomotive.

Turbo Transmission

Hydrodynamic multi-circuit transmission for rail vehicles with several Föttinger circuits that are switched on or off by filling and draining. Depending on requirements, they contain gears (input and output gears, mechanical reversing section).

Turboflexx™

Modular turbo transmission concept for diesel locomotives up to 1 500 kW.

TurboSyn™

Hydrodynamic coupling as start-up coupling with lock-up function, whose turbine wheel couples (synchronizes) automatically to the pump rotor after the run up of the driven machine is completed.

Variable-Speed Coupling

Hydrodynamic coupling for stepless variation of the transmission behavior by fluid adaptation with an adjustable scoop tube.

Variable-Speed Turbo Coupling

Voith-specific term, see "Variable-Speed Coupling".

Voith Sinclair Coupling

Constant-fill or fill-controlled coupling ("T" and "S") for industrial applications, with its function being taken over from the licensing agreement with HCP, Isleworth/UK (Harold Sinclair's company), and supplied by Voith, supplemented with Voith technology, between 1934 to 1949 (to some degree even until 1965).

VORECON®

Hydrodynamic super-imposed gear as variable-speed industrial drive, mostly as multi-circuit transmission by a combination of Föttinger circuits and power-split via planetary gears. This allows optimized efficiencies in the main operating area.

VOSYCON®

Hydrodynamic converter as starting device for large industrial plants (especially compressors) in variable-speed design and with lock-up coupling (synchronized converter).

Water Retarder

Hydrodynamic brake, suitable for using water (or an anti-freeze mix) as operating fluid.

WinDrive®

Hydrodynamic super-imposed gear for wind power stations based on the VORECON® principle with power-split via planetary gears and adjustable converters for the economical adaptation of the operating speed of the propeller to the prevailing wind speed at constant generator speed.

Bibliography

This bibliography provides a list of books and publications used for writing this book. Further titles refer to suppementary literature.

Bauer, Gustav: "Entstehung und Entwicklung des Turbowandlers "Föttinger Transformator", der Turbokupplung "Vulcankupplung" und des "Vulcangetriebes" (Origins and development of the turbo converter "Föttinger Transformer", the turbo coupling "Vulcan Coupling" and the "Vulcan Transmission").
In: Schiff und Hafen, year 4, H. 9, 1952, p. 361 – 367.

Bek, M.: "Elektronisches Steuerungssystem für automatische Getriebe" (Electronic control system for automatic transmissions). In: Voith publication G 1225.

Brandstetter, Ernst: "Von der Traisen zum Huang Pu Jiang – 100 Jahre Voith St. Pölten" (From the Traisen to the Huang Pu Jiang – 100 years of Voith St. Pölten). St. Pölten/Vienna 2003.

Depping, H., Körner, T., Wüst, B.: "Kraftstoffsparen durch optimierten Antriebsstrang" (Saving fuel by driveline optimization). In: Der Nahverkehr 3/1997.

Deschimag Publication: "Vulcangetriebe und hydraulische Kupplungen" (Vulcan transmissions and hydraulic couplings). Publisher: Deutsche Schiff- und Maschinenbau AG, Bremen 1932/33.

Dick, Heinrich: "Ein anpassungsfähiges Steuerungssystem zur Automatisierung von Differential-Wandler-Omnibusgetrieben" (An adaptable control system for the automation of differential converters).
In: ATZ 79 (1977) 9, p. 351 – 355.

Dillmann, A., Nowacki, H., Siekmann, H.: "Hermann Föttinger und die Strömungstechnik" (Hermann Föttinger and fluid technology). In: "1799 – 1999 – Von der Bergakademie zur Technischen Universität Berlin" (From the mining academy to Berlin Technical University – History and Future), p. 339 – 343. Berlin 2000.

Fechner, Gerd: "Hermann Föttinger – eine Erfindung hat 100. Geburtstag. Weg einer deutschen Erfindung im Maschinenbau von Stettin über Isleworth nach Heidenheim und Crailsheim" (Hermann Föttinger – an invention celebrates its 100th birthday – journey of a German engineering invention from Stettin via Isleworth to Heidenheim and Crailsheim). 77 pages, Crailsheim 2005 (unpublished).

Georgi, B.: "40 Jahre Koepchen-Werk Pump-Speicherwerk Herdecke" (40 years of Koepchen-Werk pumped storage power station Herdecke).

Gößler, Dieter: "Hilfsmittel zum Kuppeln von Speicherpumpen" (Auxiliary systems for coupling storage pumps). In: Die Wasserwirtschaft, XLIII. Year 1952/53, p. 181 – 189.

Gsching, Wilhelm: "Das Voith-DIWA-Busgetriebe" (The Voith DIWA bus transmission). In: ATZ 55 (1953) 3, p. 53 – 60.

Gsching, Wilhelm: "Über die Entwicklung der automatischen Voith Strömungsgetriebe für Kraftfahrzeuge" (About the development of automatic Voith hydrodynamic transmissions for vehicles). In: ATZ 69 (1967) 5.

Häckert, Hans: "Wohin mit dem Nachtstrom?" (What to do with off-peak electricity?).
In: Energie und Technik, Nov. 1958, p. 345 – 352.

Hahn, Wilhelm: "Der Entwicklungsstand der Speicherpumpen unter besonderer Berücksichtigung der Maschinen des Großkraftwerkes Herdecke" (The state of development of storage pumps under special consideration of the machines in Herdecke power station).
In: ZVDI, Volume 74 (1930), No. 25, p. 881.

Helfer, F.: "Voith Strömungsgetriebe für Straßen- und Bau-fahrzeuge" (Voith hydrodynamic transmissions for road and construction vehicles). In: ATZ 68 (1966), H. 4.

Jahrbuch der Schiffbautechn. Gesellschaft:
Volume 11/1910, Volume 31/1930, Volume 46/1952, Berlin.

Mann, Martina: "Meine Beziehung zur Familie und Firma Voith – Ein Kapitel meiner Lebenserinnerungen" (My relationship to Voith as a family and a company – a chapter from my memoirs), October 2000 (unpublished).

Paetzold, Wolfgang: "Voith Turbogetriebe 1930 – 1985" (Voith turbo transmissions 1930 – 1985), Volume 1: Locomotive transmissions, Volume 2: Railcar transmissions.
Voith Turbo GmbH & Co. KG, Heidenheim 2002/2004.

Peltner, Johannes: "Aufbau und Merkmale der Getriebe D 851/D 854" (Design and characteristics of D 851/D 854 transmissions).
In: Schriftenreihe für Verkehr und Technik, H. 65, p. 23 – 31.

Peltner, Johannes: "Das neue automatische Voith Nutz-fahrzeuggetriebe D 851" (The new automatic Voith D 851 commercial vehicle transmission).
In: ATZ 75 (1973), p. 442 – 447.

Rupf, Hugo: "Vom Glück verwöhnt – ein Leben für Voith" (Spoilt by fortune – a life for Voith). Stuttgart/Leipzig 2001.

Schweickert, Hermann: "Der Wasserturbinenbau bei Voith zwischen 1913 und 1939 und die Geschichte der Eingliederung neuer Strömungsmaschinen" (Water turbine construction at Voith between 1913 and 1939 and the history of the integration of new fluid machines).
PhD Thesis, Heidenheim 2002.

Sinclair, Harold: "The Romance of Fluidrive".
Company publication Hydraulic Coupling Patents Ltd.
(approx. 1952/53).

Spannhake, Wilhelm: "Die neueste Ausführung des Föt-tinger Transformators" (The latest design of the Föttinger transformer). ZVDI No. 19, Volume 57, May 1913.

VDI-Berichte 1592: "Hydraulische Leistungsübertragung – hydrodynamische und hydrostatische Systeme im Wettbe-werb" (Hydraulic power transmission – hydrodynamic versus hydrostatic systems).
Düsseldorf 2001, ISBN 3-18-091592-7.

VDI-Richtlinie 2153: "Hydrodynamische Leistungsübertra-gung (Begriffe – Bauformen – Wirkungsweise)" (Hydrodynamic power transmission (Terms – Designs – Operation).
Publisher: Verein Deutscher Ingenieure (Association of German Engineers), April 1994.

Voith Forschung und Konstruktion: "Entwicklungen in der Antriebstechnik" (Developments in power transmission), H. 33, March 1989

Voith, Hanns et al.: "Erfahrung aus der Vergangenheit – Leistung in der Gegenwart – Aufgabe für die Zukunft" (Experience from the past – achievements in the present – challenge of the future).
Company brochure to mark 100th anniversary, 1967.

Voith, J. M.: "Geschichte – Produkte – Sozialwesen" (History – Products – Social Services).
Publisher: J. M. Voith GmbH, Heidenheim, approx. 1956).

Voith, J. M.: "Hydrodynamik in der Antriebstechnik" (Hydrodynamics in power transmission). Paperback DIN A5.
Publisher: J. M. Voith GmbH, Heidenheim. Mainz 1987.

Voith, Walther: "Kurzer Rückblick und Rechtfertigung über meine Tätigkeit als Seniorchef der Firma J. M. Voith in Heidenheim a. d. Brenz, Württemberg und St. Pölten, Öster-reich" (Short retrospective and justification of my activities as Chairman of J. M. Voith in Heidenheim/Brenz, Württemberg and St. Pölten, Austria).
Pruggern, February 1947 (not published).

Supplementary Literature

Fister, Werner: "Fluidenergiemaschinen" (Fluid energy machines), Vol. 1/Vol. 2, Berlin 1984/1986.

Förster, Hans Joachim: "Stufenlose Fahrzeuggetriebe" (Stepless vehicle transmissions).
Cologne 1996, ISBN 3-8249-0268-0.

Pfleiderer, Carl/Petermann, Hartwig: "Strömungs-maschinen" (Fluid machinery).
6th Edition, Berlin 1991, ISBN 3-540-53037-1.

Siegloch, Herbert: "Strömungsmaschinen. Grundlagen und Anwendungen" (Fluid machinery. Fundamentals and applications), Vienna 1984, ISBN 3–446–14049–2.

Siekmann, Helmut: "Strömungslehre. Grundlagen" (Fluid mechanics. Fundamentals),
Berlin 2000, ISBN 3–540–66851–9.

Siekmann, Helmut: "Strömungslehre für den Maschinen-bau. Technik und Beispiele" (Fluid mechanics for engineering. Technology and examples).
Berlin 2001, ISBN 3–540–42041–X.

Name Index

People

Companies